ENVIRONMENTAL POLICY

Eleventh Edition

To Sandy and Dana,
For their love and support, and
To Sean and Nico
For their future

ENVIRONMENTAL POLICY

New Directions for the Twenty-First Century

Eleventh Edition

Edited by

Norman J. Vig
Carleton College

Michael E. Kraft
University of Wisconsin–Green Bay

Barry G. Rabe
University of Michigan

Los Angeles | London | New Delhi
Singapore | Washington DC | Melbourne

FOR INFORMATION:

SAGE Publications, Inc.
2455 Teller Road
Thousand Oaks, California 91320
E-mail: order@sagepub.com

SAGE Publications Ltd.
1 Oliver's Yard
55 City Road
London, EC1Y 1SP
United Kingdom

SAGE Publications India Pvt. Ltd.
B 1/I 1 Mohan Cooperative Industrial Area
Mathura Road, New Delhi 110 044
India

SAGE Publications Asia-Pacific Pte. Ltd.
18 Cross Street #10-10/11/12
China Square Central
Singapore 048423

Printed in the United States of America

ISBN 978-1-5443-7801-5

This book is printed on acid-free paper.

Acquisitions Editor: Scott Greenan
Editorial Assistant: Lauren Younker
Production Editor: Olivia Weber-Stenis
Copy Editor: TNQ
Typesetter: TNQ
Indexer: TNQ
Cover Designer: Karine Hovsepian
Marketing Manager: Jennifer Jones

20 21 22 23 24 10 9 8 7 6 5 4 3 2 1

CONTENTS

PREFACE

At the beginning of the third decade of the twenty-first century, environmental policy faces significant challenges both at home and around the world. New demands for dealing with the risks of climate change and threats to biological diversity, and for meeting the rising aspirations of the planet's nearly eight billion people, will force governments everywhere to rethink policy strategies and tools. They need to find effective and acceptable ways to reconcile environmental and economic goals and values through new approaches to sustainable development.

In the United States, the early part of the 2010s saw a stagnant economy and persistently high unemployment. Over the following decade the economy improved greatly, albeit with highly unequal results; some benefited while others did not. Despite these economic gains, new criticism emerged in the United States over the perceived adverse economic impacts of environmental and energy policies and regulations. As a result, the Donald Trump presidency pursued the most aggressive anti-environmental agenda of any administration since Ronald Reagan's in the 1980s. By fall 2020, the global coronavirus pandemic created enormous economic damage and renewed concerns over a severe and possibly prolonged recession. As economies around the world struggle to regain these losses over the next several years, environmental policies may well face continued skepticism both in the United States and in other nations if they are seen as adding to economic burdens.

Many of these criticisms in the United States divide members of the two major parties deeply, as Republicans have continued to call, as they long have, for repealing, reducing, and reining in environmental policies and regulations in the face of strong Democratic defense of the same policies and actions. The result has been intense and relentless partisan debate on Capitol Hill and at the state and local levels where many of the same conflicts arose. Some environmentalists have blamed Democrats as well for what they see as an often timid defense of environmental policy or for their reluctance to embrace new and far-reaching proposals for a Green New Deal favored by some within the party. Political debate over the next few years may continue to be framed in these terms even as leading businesses, the scientific community, and increasing numbers of public officials recognize that the real challenge today is to find ways to meet economic and other human needs while also protecting the environment on which we depend, that is, on how to foster sustainable development.

The election of President Donald Trump in November 2016 brought a dramatic change in policy positions and priorities after eight years of the Barack Obama administration. Particularly in his second four-year term, Obama sought through executive action to strengthen protection of public health and the environment, foster the development of clean energy resources, and establish a viable and broadly supported path toward global action on climate change through the Paris Agreement of 2015. In contrast, decisions in the Trump administration from 2017 through January 2021 aimed to reverse many of Obama's major policy initiatives, especially on energy use and climate change. The differences between the two presidencies on societal values and policy priorities, key appointments to administrative agencies, regulatory decision-making, budgetary support for established programs, and the use of science in decision-making could hardly have been greater. These differences are explored at length in the chapters that follow.

When the first environmental decade was launched in the early 1970s, protecting our air, water, and other natural resources seemed a relatively simple proposition. The polluters and exploiters of nature would be brought to heel by tough laws requiring them to clean up or get out of business within five or ten years. But preserving the life support systems of the planet now appears a far more daunting task than anyone imagined back then. Not only are problems such as global climate change much more complex than recognized by early efforts to control air and water pollution, but now more than ever, the success of US policies is tied to the actions of other nations.

This book seeks to explain the most important developments in environmental policy and politics since the 1960s and to analyze the central issues that face us today. Like the previous editions, it focuses on the underlying trends, institutional strengths and shortcomings, and policy dilemmas that all policy actors face in attempting to resolve environmental controversies. All chapters have been thoroughly revised and updated, and five of them are new to this edition. We have also attempted to compare the positions and actions of the Trump administration to those of the George W. Bush and Barack Obama administrations and to put these differing approaches to environmental policy in the context of ongoing debates over the cost and effectiveness of past policies, as well as the search for ways to reconcile and integrate economic, environmental, and social goals through sustainable development. As such, the book has broad relevance for the environmental community and for all concerned with the difficulties and complexities of finding solutions to environmental problems at the beginning of the third decade of the twenty-first century.

Part I provides a retrospective view of policy development as well as a framework for analyzing policy change in the United States. Chapter 1 serves as an introduction to the book by outlining the basic issues in US environmental policy since the late 1960s, the development of institutional capabilities for addressing them, and the successes and failures in implementing policies and achieving results. In Chapter 2, Barry G. Rabe considers the evolving role of the states in environmental policy at a time when the devolution of responsibilities has faced sharp scrutiny from new federal leaders. He focuses on innovative policy approaches used by the states and the promise of—as well as the constraints on—state action on the environment. Part I ends with a chapter by Christopher Borick and Erick Lachapelle that analyzes public opinion on the environment, drawing from a diversity of recent surveys to explain the nature of the public's views and the support or opposition they offer for policy proposals.

Part II analyzes the role of federal institutions in environmental policymaking. Chapter 4, by Norman J. Vig, discusses the role of recent presidents as environmental actors, evaluating their leadership on the basis of several common criteria. In Chapter 5, Michael E. Kraft examines the role of Congress in environmental policy, giving special attention to partisan conflicts over the environment and continued policy gridlock. In Chapter 6, Kimberly Smith describes how federal courts shape environmental policy and often play a critical role in limiting the actions of the other branches of government. In Chapter 7, Richard N. L. Andrews examines the EPA and the way it uses the policy tools granted to it by Congress, especially its regulatory authority, to address environmental challenges.

Some of the broader dilemmas in environmental policy formulation and implementation are examined in Part III. Chapter 8, by Sanya Carley, examines national energy policy, exploring technical, economic, and political controversies surrounding the major energy sources used today as well as the challenges and opportunities that their use presents for climate change. In Chapter 9, William Lowry and John Freemuth describe the major policies and agencies affecting the use of the nation's natural resources, their historical development, and controversies over their implementation today.

In Chapter 10, Sheila M. Olmstead introduces economic perspectives on environmental policy, including the use of benefit–cost analysis, and she assesses the potential of market forces as an alternative or supplement to conventional regulation. Chapter 11, by Kent E. Portney and Bryce Hannibal, examines the intriguing efforts by communities throughout the nation to integrate environmental sustainability into policy decisions in areas as diverse as energy use, housing, transportation, land use, and urban social life.

Part IV shifts attention to selected global issues and controversies. In Chapter 12, Henrik Selin and Stacy D. VanDeveer survey the key scientific evidence and major disputes over climate change, as well as the evolution of the issue since the late 1980s. They also assess government responses to the problem of climate change and the outlook for public policy actions. Chapter 13 examines the plight of developing nations that are struggling with a formidable array of threats brought about by rapid population growth and resource exploitation. Richard J. Tobin surveys the pertinent evidence, recounts cases of policy success and failure, and outlines the remaining barriers (including insufficient commitment by rich countries) to achieving sustainable development in these nations. Finally, Chapter 14, by Daniel Fiorino, explores the political economy of green growth, recounting the history of the key ideas and the broad debates taking place today over how to promote sustainable development that meets people's needs and yet is compatible with environmental constraints.

In the final chapter of the book we review the many environmental challenges that continue to face the nation and the world at an unusually critical time, and discuss innovative policy instruments that might help us to better address these issues in the future. We offer a postscript at the end of this volume that addresses results from the November 2020 presidential and congressional elections.

We thank the contributing authors for their generosity, cooperative spirit, and patience in response to our editorial requests. It is a pleasure to work with such a conscientious and punctual group of scholars. Tragically, two of our authors passed away unexpectedly shortly after completing their chapters for this book. We consider ourselves most fortunate to have worked with Kent Portney and John Freemuth over the years, and we will miss them both.

Special thanks are also due to the staff of CQ Press/SAGE, including Charisse Kiino, Scott Greenen, Anna Villarruel, Olivia Weber-Stenis, Lauren Younker, Tiara

Beatty, Kenzie Offley, Jennifer Jones, Erica DeLuca, and Agnes Preethi. We also thank Hannah Smith of the University of Michigan for her assistance on several chapters. We owe a great debt to our colleagues who graciously agreed to review previous editions of the book and offered many useful suggestions: William G. Holt, Birmingham-Southern College; Daniel Fiorino, American University; Jeff W. Justice, Tarleton State University; Jack Rasmus, St. Mary's College; Ninian Stein, Smith College; Gerald A. Emison, Mississippi State University; Rebecca Bromley-Trujillo, University of Kentucky; Sarah Anderson, University of California, Santa Barbara; Irasema Coronado, University of Texas at El Paso; Robert Duffy, Colorado State University; Erich Frankland, University of Wyoming; Raymond Lodato, University of Chicago; Melissa K. Merry, University of Louisville; and John W. Sutherlin, University of Louisiana at Monroe; Meghann N. Smith; Montclair State University; Deborah M. Steketee, Aquinas College. We also gratefully acknowledge support from the Department of Public and Environmental Affairs at the University of Wisconsin–Green Bay. Finally, we thank our students for forcing us to rethink our assumptions about what really matters. As always, any remaining errors and omissions are our own responsibility.

Norman J. Vig
Michael E. Kraft
Barry G. Rabe

ABOUT THE EDITORS

Norman J. Vig is the Winifred and Atherton Bean Professor of Science, Technology, and Society emeritus at Carleton College. He has written extensively on environmental policy, science and technology policy, and comparative politics and is coeditor with Michael G. Faure of *Green Giants? Environmental Policies of the United States and the European Union* (2004) and with Regina S. Axelrod and David Leonard Downie of *The Global Environment: Institutions, Law, and Policy*, 2nd ed. (2005).

Michael E. Kraft is a professor of political science and the Herbert Fisk Johnson Professor of Environmental Studies emeritus at the University of Wisconsin–Green Bay. He is the author of *Environmental Policy and Politics*, 8th ed. (2022), and coauthor of *Coming Clean: Information Disclosure and Environmental Performance* (2011, winner of the Lynton K. Caldwell award for best book on environmental politics and policy that year) and of *Public Policy: Politics, Analysis, and Alternatives*, 7th ed. (2021). In addition, he is coeditor of both the *Oxford Handbook of Environmental Policy* (2013) and *Business and Environmental Policy* (2007) with Sheldon Kamieniecki and of *Toward Sustainable Communities: Transition and Transformations in Environmental Policy*, 2nd ed. (2009), with Daniel A. Mazmanian.

Barry G. Rabe is the J. Ira and Nicki Harris Family Professor of Public Policy and the Arthur F. Thurnau Professor of Environmental Policy at the Gerald R. Ford School of Public Policy at the University of Michigan. He also serves as a nonresident senior fellow at the Brookings Institution and chaired the US Environmental Protection Agency Assumable Waters Committee from 2015 to 2017. He is the author of numerous books and articles, including *Statehouse and Greenhouse: The Emerging Politics of American Climate Change Policy*, which received the 2017 Martha Derthick Book Award from the American Political Science Association for making a lasting contribution to the study of federalism. His latest books are *Can We Price Carbon?* (MIT Press, 2018) and *Trump, the Administrative Presidency, and Federalism* (Brookings, 2020), coauthored with Frank J. Thompson and Kenneth K. Wong, and he is currently working on a book examining the politics of methane emissions.

ABOUT THE CONTRIBUTORS

Richard N. L. Andrews is a professor emeritus of public policy, environmental studies, environmental sciences and engineering, and city and regional planning at the University of North Carolina at Chapel Hill. A primary focus of his research and writing is the history of US environmental policy. He is the author of *Managing the Environment, Managing Ourselves: A History of American Environmental Policy*, 3rd ed. (2020), "The EPA at 40: An Historical Perspective" (*Duke Environmental Law and Policy Forum*, 2011), "Reform or Reaction: EPA at a Crossroads" (*Environmental Sciences & Technology*, 1995), and many other articles on related topics.

Christopher Borick is a professor of political science and public health, and Director of the Institute of Public Opinion at Muhlenberg College in Allentown, Pennsylvania. His primary areas of research include public opinion, environmental policy, and state and local politics. He is the author of four books, and over 40 articles and book chapters, and he has directed over 350 large scale surveys on an array of policy matters. Since 2008 he has been a Director of the National Surveys on Energy and Environment (NSEE), which includes multiple surveys of Americans each year on matters including their perceptions of climate change and views regarding energy and climate policy.

Sanya Carley is a professor and director of the Master of Public Affairs programs at the Paul H. O'Neill School of Public and Environmental Affairs at Indiana University. Her research focuses on energy policy, the justice and equity implications of the energy transition, and public perceptions of energy infrastructure. She is the author of *Energy-based Economic Development: How Clean Energy Can Drive Development and Stimulate Economic Growth* (2014), "Empirical Evaluation of the Stringency and Design of Renewable Portfolio Standards" (*Nature Energy*, 2018), "A Framework for Evaluating Geographic Disparities in Energy Transition Vulnerability" (*Nature Energy*, 2018), and "An Analysis of the Macroeconomic Effects of 2017–2025 Federal Fuel Economy and Greenhouse Gas Emissions Standards" (*Journal of Policy Analysis and Management*, 2019), among other articles.

Daniel J. Fiorino is the founding Director of the Center for Environmental Policy and Distinguished Executive in Residence in the School of Public Affairs at American University, and a faculty member in the Department of Public Administration and Policy. He served for many years in a variety of senior positions at the US Environmental Protection Agency. He is the author of *A Good Life on a Finite Earth: The Political Economy of Green Growth* (2018); *Can Democracy Handle Climate Change?* (2018), and *The New Environmental Regulation* (2006), and co-editor of *Conceptual Innovation in Environmental Policy* (2017) and *Environmental Governance Reconsidered* (2nd ed., 2017).

John Freemuth was Distinguished Professor of Public Policy and Cecil Andrus Endowed Chair of Environment and Public Lands at Boise State University. His primary interest was with the public lands of the United States. He wrote *Islands Under Siege: National Parks and the Politics of External Threats* (1991), contributed chapters to two editions of *Environmental Politics and Policy in the West* (2007 and 2016), and wrote "A Happy Combination? Great Interests, Particular Interests, and State-Federal Conflicts over Public Lands" (*Publius*, 2018) as well as many other articles.

Bryce Hannibal is a research scientist in the Institute for Science, Technology, and Public Policy at the Bush School of Government and Public Service, Texas A&M University. He has published in a number of scholarly journals, including *Public Management Review*, *Frontiers of Environmental Science*, *Social Science Quarterly*, *Environmental Sociology*, *Sociological Spectrum*, *International Journal of Social Economics*, *Environmental Science and Policy*, and *Environmental Behavior*.

Erick Lachapelle is an associate professor in the Department of Political Science at the University of Montreal. He is the principal investigator for the Canadian Surveys on Energy and the Environment (CSEE). He has written widely on climate change public opinion in Canada and the United States, and has published articles in *Global Environmental Politics*, *PLoS ONE*, *Environmental Politics*, *Energy Policy*, *Policy Studies Journal*, *Climate Policy*, *Perspectives on Global Development and Technology*, and *Canadian Foreign Policy*, among other journals.

William R. Lowry is a professor of political science at Washington University in St. Louis. He studies environmental issues and natural resource policies, particularly those involving public lands and waters. He is the author of five books, including *Repairing Paradise* (Brookings, 2009) and numerous articles. He is also a US Navy veteran and a former employee of the National Park Service.

Sheila M. Olmstead is a professor of public affairs at the Lyndon B. Johnson School of Public Affairs at the University of Texas at Austin, a university fellow at Resources for the Future (RFF) in Washington, DC, and a senior fellow at the Property and Environment Research Center in Bozeman, Montana. She was previously a fellow and senior fellow at RFF (2010–2013) and an associate professor (2007–2010) and assistant professor (2002–2007) of environmental economics at the Yale School of Forestry and Environmental Studies. Her research has been published in leading journals such as the *Journal of Economic Perspectives*, *Proceedings of the National Academy of Sciences*, *Journal of Environmental Economics and Management*, and *Journal of Urban Economics*. With Nathaniel Keohane, she is the author of the book *Markets and the Environment*. From June 2016 to June 2017, she served on the President's Council of Economic Advisers.

Kent E. Portney was a professor and the director of the Institute for Science, Technology, and Public Policy at the Bush School of Government and Public Service at Texas A&M University. Previously, he taught for many years at Tufts University. He was the author of *Sustainability* (2015), *Taking Sustainable Cities Seriously: Economic Development, the Environment, and Quality of Life in American Cities*, 2nd ed. (2013),

Approaching Public Policy Analysis (1986), *Siting Hazardous Waste Treatment Facilities: The NIMBY Syndrome* (1991), and *Controversial Issues in Environmental Policy* (1992). He also was the coauthor of *Acting Civically* (2007) and *The Rebirth of Urban Democracy* (1993).

Henrik Selin is an associate professor in the Frederick S. Pardee School of Global Studies at Boston University. His authored, coauthored, and coedited books include *Mercury Stories: Understanding Sustainability through a Volatile Element* (MIT Press, 2020), *The European Union and Environmental Governance* (Routledge, 2015), *Global Governance of Hazardous Chemicals: Challenges of Multilevel Management* (MIT Press, 2010), *Changing Climates in North American Politics: Institutions, Policymaking, and Multilevel Governance* (MIT Press, 2009), and *Transatlantic Environment and Energy Politics: Comparative and International Perspectives* (Ashgate, 2009). In addition, he has authored and coauthored more than four dozen peer-reviewed journal articles and book chapters, as well as numerous reports, reviews, and commentaries on issues relating to the environment and sustainability.

Kimberly Smith is a professor of environmental studies and political science at Carleton College. Her research and teaching interests include American environmental policy, American constitutional law, American environmental thought, and environmental ethics. Among her publications are *Wendell Berry and the Agrarian Tradition: A Common Grace* (University Press of Kansas, 2003); *African American Environmental Thought* (University Press of Kansas, 2007); *Governing Animals: Animal Welfare and the Liberal State* (Oxford University Press, 2012); and *The Conservation Constitution: The Conservation Movement and Constitutional Change, 1870–1930* (University Press of Kansas, 2019), awarded the Forest History Society's 2020 Charles A. Weyerhaeuser Book Award. In 2020 she was awarded the William R. Freudenburg Lifetime Achievement Award by the Association for Environmental Studies and Sciences (AESS).

Richard J. Tobin has spent much of his career working on international development. After retiring from the World Bank, he has served as a consultant to UNICEF, the United Nations Development Programme, the United Nations Population Fund, the African Development Bank, the Asian Development Bank, the Arab Administrative Development Organization, and the Organization for Security and Co-operation in Europe. He continues to serve as consultant to the World Bank and has also worked on development projects funded by the US Agency for International Development, the United Kingdom's Department for International Development, Australia's Department of Foreign Affairs and Trade, and the Bill and Melinda Gates Foundation.

Stacy D. VanDeveer is a professor in the Department of Conflict Resolution, Human Security, and Global Governance at the University of Massachusetts Boston. His research and teaching interests include the global politics of resource overconsumption, international environmental policymaking and institutions, connections between environmental and security issues, and comparative and EU environmental politics. In addition to authoring and coauthoring over 100 articles, book chapters, working papers, and reports, he is the coeditor or coauthor of nine books, including *EU Enlargement and the Environment* (2005), *Changing Climates in North American Politics* (2009), *Transatlantic Environment and Energy Politics* (2009), *Comparative Environmental Politics* (2012), *Transnational Climate Change Governance* (2014), *The European Union and Environmental Governance* (2015), *Want, Waste or War?* (2015), and *The Global Environment*, 5th ed. (2020).

ENVIRONMENTAL POLICY AND POLITICS IN TRANSITION

1

US ENVIRONMENTAL POLICY

A Half-Century Assessment

Michael E. Kraft and Norman J. Vig

The United States and the world have come a long way since the beginning of the modern environmental movement, around 1970. Progress was particularly evident for the first half of this period, from 1970 to 1995, when environmental policy enjoyed considerable bipartisan support despite occasional political reaction to prevailing policies and programs, most evident during Ronald Reagan's presidency (1981–1989). As we recount in this introductory chapter, during that 25-year span, the US Congress approved a broad array of new and expansive public policies and guided and funded the development of governmental institutions capable of putting them into effect. The result was striking improvements in environmental quality and public health throughout the nation. We review these achievements at the end of the chapter as we do the policies' collective limitations.

Since 1995, however, the bipartisan consensus on environmental policy has broken down, replaced by an increasingly acrimonious debate between the two major parties. This new partisan polarization emerged in the mid- to late-1990s in Congress and intensified during the George W. Bush administration from 2001 through January 2009. It has been central to environmental policy debate since then, as evident in the stark differences in policies, regulatory actions, budgetary priorities, and personnel appointments during the two most recent presidencies: those of Barack Obama and Donald Trump.

Whether the issue is clean air, clean water, energy use, or climate change, the two parties typically have found little common ground. What one administration struggles to achieve under the sharply critical eye of its partisan adversaries is reversed by the next. Even agreement on the core scientific facts can no longer be assured, adding to public confusion and dismay, and hindering the development of much needed new policies, as evident in debates over how to deal with climate change.

Despite the substantial progress the nation has made in the first half century of environmental policy, there is a real question of whether it can continue if deep partisan differences remain and block the adoption of new and innovative solutions. The timing could hardly be worse because the nation and world desperately need to develop effective policies to combat climate change while also modernizing the environmental, natural resources, and energy policies that we have developed over the past half century. These policies could be made more effective, efficient, and equitable, and more appropriate for the twenty-first century as the chapters in this volume make clear.

How can we best learn from our collective experience of the past 50 years? And how should we respond to the demands of scientists and environmental leaders who call for a new generation of public policies to address the momentous challenges we face today?

National and global actions taken in response to the deadly coronavirus pandemic in 2020 suggest both what is possible as well as the obstacles we must overcome.

This chapter provides a historical and institutional analysis that seeks to explain how policymakers have addressed environmental problems previously and the policy choices they have made. We review the activities of government in addressing environmental problems, including the structure of US government that can facilitate or hinder decisions, the processes of agenda setting and policymaking, major policy decisions made over the past five decades, and what those policies have achieved since their adoption. In the concluding chapter in this volume (Chapter 15), we return to the many remaining challenges of the twenty-first century, and we explore the need for a fresh examination of environmental governance and the possible new directions in policies that better match the problems that the nation and world now face as well as the public's willingness to support them.

THE CHALLENGES OF CONTEMPORARY ENVIRONMENTAL PROBLEMS

In the late 1960s and early 1970s, environmental issues soared to a prominent place on the political agenda in the United States and other industrial nations. The new visibility was accompanied by abundant evidence, domestically and internationally, of heightened public concern over environmental threats and broad support for governmental action.[1] By the 1990s, policymakers around the world had pledged to deal with a range of important environmental problems, from protection of biological diversity to air and water pollution control. Such commitments were particularly manifest at the 1992 United Nations Conference on Environment and Development (the Earth Summit) held in Rio de Janeiro, Brazil, where an ambitious agenda for redirecting the world's economies toward sustainable development was approved, and at the December 1997 Conference of the Parties in Kyoto, Japan, where delegates agreed to a landmark treaty on global climate change. Although it received far less media coverage, the World Summit on Sustainable Development, held in September 2002 in Johannesburg, South Africa, reaffirmed the commitments made a decade earlier at the Earth Summit, with particular attention to the challenge of alleviating global poverty. The far-reaching goals of the Earth Summit and the 2002 Johannesburg meeting were revisited at the 2012 Rio+20 United Nations Conference on Sustainable Development held once again in Brazil, where global commitments were reaffirmed once more.

Despite the positions taken at these and many other comparable meetings in recent decades, rising criticism of environmental programs also was evident throughout the 1990s and in the first two decades of the twenty-first century, both domestically and internationally. So too were a multiplicity of efforts to chart new policy directions. For example, intense opposition to environmental and natural resource policies arose in the 104th Congress (1995–1997), when the Republican Party took control of both the House and Senate for the first time in 40 years. Ultimately, much like the earlier efforts in Ronald Reagan's administration, that antiregulatory campaign on Capitol Hill failed to gain much public support.[2] Nonetheless, pitched battles over environmental and energy policy continued in every Congress since then (see Chapter 5).

Both antiregulatory actions and fights over them were equally evident in the executive branch, particularly during the George W. Bush administration, as it sought to rewrite environmental rules and regulations to favor industry and to increase development of US oil and natural gas supplies on public lands, and even more directly in the Donald Trump administration, which shared many of the same priorities (see Chapter 4). Yet growing dissatisfaction with the effectiveness, efficiency, and fairness of environmental policies was by no means confined to congressional conservatives and the Bush and Trump administrations. It could be found as well among a broad array of interests, including the business community, environmental policy analysts, environmental justice groups, and state and local government officials, although not always with the ideological agenda that was so evident in the Bush and Trump administrations.[3]

Since 1992, governments at all levels have struggled to redesign environmental policy for the twenty-first century. Under Presidents Bill Clinton and George W. Bush, the Environmental Protection Agency (EPA) tried to "reinvent" environmental regulation through the use of collaborative decision-making involving multiple stakeholders, public–private partnerships, market-based incentives, information disclosure policies, and enhanced flexibility in rulemaking and enforcement (see Chapters 7, 10, and 14).[4] Many state and local governments have pursued similar goals with the adoption of innovative policies that promise to address some of the most important criticisms directed at contemporary environmental policy (see Chapters 2 and 11).[5] The election of President Barack Obama in 2008 brought additional attention to new policy ideas, especially in his second term of office when he pursued strong policies on clean energy and climate change (see Chapter 4).

The precise way in which Congress, the White House, the states, and local governments—and other nations—will change environmental policies in the years to come remains uncertain. The prevailing partisan polarization and policy gridlock of recent years may give way to greater consensus on the need to act; yet policy change rarely comes easily in the US political system. Its success likely depends on the conditions that affect all policymaking: the saliency of the issues, public support for action, media coverage, the relative influence of opposing interests, and the state of the economy. Political leadership, as always, will play a critical role, especially in articulating the problems and potential solutions, mobilizing the public and policy actors, and trying to reconcile the deep partisan and ideological divisions that exist today on environmental protection and natural resource issues. Political conflict over the environment is not likely to vanish anytime soon. Indeed, it may well increase as the United States and other nations struggle to define how they will respond to the latest generation of environmental challenges, particularly climate change.

THE ROLE OF GOVERNMENT AND POLITICS

The high level of political conflict over environmental protection efforts in the past several decades, particularly evident during the Trump administration, underscores the important role a government plays in devising solutions to the nation's and the world's mounting environmental ills. Global climate change, the spread of toxic and hazardous chemicals, loss of biological diversity, air and water pollution, and the continued growth of the world's population and its economic needs require diverse and often demanding actions by individuals and institutions at all levels of society and in both the public and private sectors. These range from scientific research and technological innovation to strong public policy initiatives and significant changes in both individual and corporate behavior. As political scientists, we believe the government has an indispensable role to play in environmental protection and improvement. Because of this conviction, we have commissioned chapters for this volume that focus on environmental policies and the governmental institutions and political processes that affect them. Our goal is to illuminate that role as well as to suggest needed changes and strategies for making them.

Government plays a preeminent role in this policy arena primarily because environmental threats, such as urban air pollution and climate change, pose risks to the public's health and well-being that cannot be resolved satisfactorily through private actions alone. That said, there is no question that individuals and nongovernmental organizations, such as environmental groups and scientific research institutes, can do much to protect environmental quality and promote public health. There is also no doubt that business and industry can do much to promote environmental quality and foster the pursuit of national energy goals, such as improved energy efficiency and increased reliance on renewable energy sources. We see evidence of extensive and often creative individual, nonprofit, and corporate actions of this kind regularly, for example, in sustainable community efforts and sustainable business practices, as discussed in Chapters 11 and 14.

Yet such actions often fall short of national needs without the backing of public policy, without, for example, laws mandating control of toxic chemicals that are supported by the authority of government or standards for drinking water quality and urban air quality that are developed and enforced by the EPA, the states, and local governments. The justification for government intervention lies partly in the inherent limitations of the free market system and the nature of human behavior. Self-interested individuals and a relatively unfettered economic marketplace guided mainly by a concern for short-term gains or profits tend to create spillover effects, or externalities; pollution and other kinds of environmental degradation are examples. As economists have long recognized, collective action is needed to correct such market failures (see Chapter 10). In addition, the scope and urgency of environmental problems typically exceed the capacity of private markets and individual efforts to deal with them quickly and effectively. For these reasons, among others, the United States and other nations have relied on government policies—at local, state, national, and international levels—to address environmental and resource challenges.

Adopting public policies does not imply, of course, that the voluntary and cooperative actions by citizens in their communities or the many environmental initiatives undertaken by corporations cannot be the primary vehicle of change in many instances. Nor does it suggest that governments should not consider a full range of policy approaches—including market-based incentives, new forms of collaborative decision-making, and information provision strategies—to supplement conventional regulatory policies where needed. Public policy intervention should be guided by the simple idea that we ought to use those policy approaches that offer the greatest promise of working to resolve the problem at hand. Sometimes that will mean governments setting and enforcing public health or environmental standards (regulation), and sometimes it will mean relying on market incentives, such as carbon taxes, or information disclosure, such as data provided to the public on toxic chemicals and drinking water quality. Typically, governments employ a combination of policy tools to reach agreed-upon objectives: improving environmental quality, minimizing health and ecological risks, and helping to integrate and balance environmental and economic goals.

Political Institutions and Public Policy

Public policy is a course of government action or inaction in response to social problems. It is expressed in goals articulated by political leaders; in formal statutes, rules, and regulations; and in the practices of administrative agencies and courts charged with implementing or overseeing programs. Policy states the intent to achieve certain goals and objectives through a conscious choice of means, usually within a specified period of time. In a constitutional democracy like the United States, policymaking is distinctive in several respects: It must take place through constitutional processes, it requires the sanction of law, and it is binding on all members of society.

The constitutional requirements for policymaking were established well over 200 years ago, and they remain much the same today. The US political system is based on a division of authority among three branches of government and between the federal government and the states. Originally intended to limit government power and to protect individual liberty, this division of power translates today into a requirement that one build an often elusive political consensus among members of Congress, the president, and key interest groups for any significant national policymaking to take place. Such fragmented authority may impede the ability of the government to adopt timely and coherent environmental policy, as has been evident for some of the most challenging of modern environmental problems. Weak national climate change policy is something of a poster child for such governmental gridlock, which is an inability to act on problems because of divided authority and prevailing political conflict (see Chapter 5).

Dedication to principles of federalism means that environmental policy responsibilities are distributed among the federal government, the fifty states, and tens of thousands of local governments. Here, too, strong adherence to those principles may result in no agreement on national policy action. Yet a federal structure also means that states often are free to adopt environmental and energy policies as they see fit, as has been the case for natural gas "fracking" where no major national policies have been in force. Some of the states have a track record of favoring environmental policies that go well beyond what is possible politically in Washington, DC. California's adoption of a strong climate change policy and Minnesota's successful encouragement of renewable energy sources are two notable illustrations of the considerable power that states have in the US political system (see Chapter 2).[6] The flip side of that coin is that some states will choose to do far less than others in the absence of national requirements.

Responsibility for the environment is divided within the branches of the federal government as well, most notably in the US Congress, with power shared between the House and Senate and jurisdiction over environmental policies scattered among dozens of committees (see Table 1.1). For example, approximately twenty Senate and

TABLE 1.1 ■ Major Congressional Committees With Environmental Responsibilities[a]

Committee	Environmental Policy Jurisdiction
HOUSE	
Agriculture	Agriculture generally; forestry in general and private forest reserves; agricultural and industrial chemistry; pesticides; soil conservation; food safety and human nutrition; rural development; water conservation related to activities of the Department of Agriculture
Appropriations[b]	Appropriations for all programs
Energy and Commerce	Measures related to the exploration, production, storage, marketing, pricing, and regulation of energy sources, including all fossil fuels, solar, and renewable energy; energy conservation and information; measures related to general management of the Department of Energy and the Federal Energy Regulatory Commission; regulation of the domestic nuclear energy industry; research and development of nuclear power and nuclear waste; air pollution; safe drinking water; pesticide control; Superfund and hazardous waste disposal; toxic substances control; health and the environment
Natural Resources	Public lands and natural resources in general; irrigation and reclamation; water and power; mineral resources on public lands and mining; grazing; national parks, forests, and wilderness areas; fisheries and wildlife, including research, restoration, refuges, and conservation; marine affairs and oceanography, international fishing agreements, and coastal zone management; US Geological Survey
Science, Space, and Technology	Environmental research and development; marine research; energy research and development in all federally owned nonmilitary energy laboratories; research in national laboratories; NASA, National Weather Service, and National Science Foundation
Transportation and Infrastructure	Transportation, including civil aviation, railroads, water transportation, and transportation infrastructure; Coast Guard and marine transportation; federal management of emergencies and natural disasters; flood control and improvement of waterways; water resources and the environment; pollution of navigable waters; bridges and dams
SENATE	
Agriculture, Nutrition, and Forestry	Agriculture in general; food from fresh waters; soil conservation and groundwater; forestry in general; human nutrition; rural development and watersheds; pests and pesticides; food inspection and safety

TABLE 1.1 ■ Major Congressional Committees With Environmental Responsibilities[a]

Committee	Environmental Policy Jurisdiction
Appropriations[b]	Appropriations for all programs
Commerce, Science, and Transportation	Interstate commerce and transportation generally; coastal zone management; inland waterways; marine fisheries; oceans, weather, and atmospheric activities; transportation and commerce aspects of outer continental shelf lands; science, engineering, and technology research and development; surface transportation
Energy and Natural Resources	Energy policy, regulation, conservation, research and development; coal; oil and gas production and distribution; civilian nuclear energy; solar energy systems; mines, mining, and minerals; irrigation and reclamation; water and power; national parks and recreation areas; wilderness areas; wild and scenic rivers; public lands and forests; historic sites
Environment and Public Works	Environmental policy, research, and development; air, water, and noise pollution; climate change; construction and maintenance of highways; safe drinking water; environmental aspects of outer continental shelf lands and ocean dumping; environmental effects of toxic substances other than pesticides; fisheries and wildlife; Superfund and hazardous wastes; solid waste disposal and recycling; nonmilitary environmental regulation and control of nuclear energy; water resources, flood control, and improvements of rivers and harbors; public works, bridges, and dams

[a]In addition to the standing committees listed here, select or special committees may be created for a limited time. Each committee also operates with subcommittees (generally five or six) to permit further specialization. Committee webpages offer extensive information about jurisdiction, issues, membership, and pending actions, and include both majority and minority views on the issues. See www.house.gov/committees/ and www.senate.gov/committees.

[b]Both the House and Senate appropriations committees have interior and environment subcommittees that handle all Interior Department agencies as well as the Forest Service and the EPA. The Energy Department, Army Corps of Engineers, and Nuclear Regulatory Commission fall under the jurisdiction of the subcommittees on energy and water development. Tax policy affects many environmental, energy, and natural resource policies and is governed by the Senate Finance Committee and the House Ways and Means Committee.

Sources: Compiled from descriptions of committee jurisdictions reported in Rebecca Kimitch, "CQ Guide to the Committees: Democrats Opt to Spread the Power," *CQ Weekly Online* (April 16, 2007): 1080–83, http://library.cqpress.com/cqweekly/weeklyreport110-000002489956, and from current House and Senate committee websites.

twenty-eight House committees and subcommittees have some jurisdiction over EPA activities.[7] The executive branch is also institutionally fragmented, with at least some responsibility for the environment and natural resources located in twelve cabinet departments and in the EPA, the Nuclear Regulatory Commission, and other agencies (see Figure 1.1). Most environmental policies are concentrated in the EPA and in the

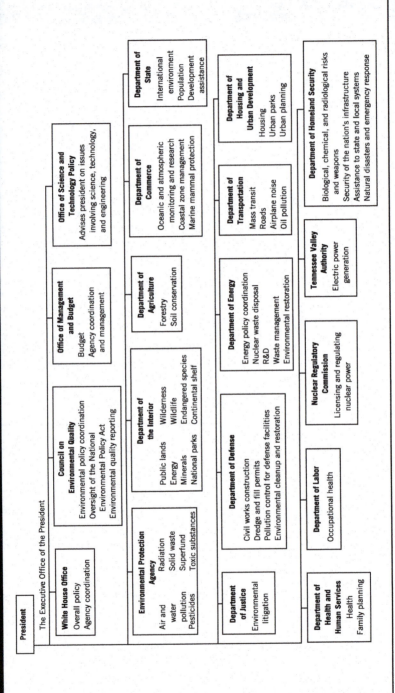

FIGURE 1.1 ■ Executive Branch Agencies With Environmental Responsibilities

President

The Executive Office of the President

White House Office
Overall policy
Agency coordination

Council on Environmental Quality
Environmental policy coordination
Oversight of the National Environmental Policy Act
Environmental quality reporting

Office of Management and Budget
Budget
Agency coordination and management

Office of Science and Technology Policy
Advises president on issues involving science, technology, and engineering

Environmental Protection Agency
Air and water pollution   Radiation
Pesticides   Solid waste
   Superfund
   Toxic substances

Department of the Interior
Public lands   Wilderness
Energy   Wildlife
Minerals   Endangered species
National parks   Continental shelf

Department of Agriculture
Forestry
Soil conservation

Department of Commerce
Oceanic and atmospheric monitoring and research
Coastal zone management
Marine mammal protection

Department of State
International environment
Population
Development assistance

Department of Justice
Environmental litigation

Department of Defense
Civil works construction
Dredge and fill permits
Pollution control for defense facilities
Environmental cleanup and restoration

Department of Energy
Energy policy coordination
Nuclear waste disposal
R&D
Waste management
Environmental restoration

Department of Transportation
Mass transit
Roads
Airplane noise
Oil pollution

Department of Housing and Urban Development
Housing
Urban parks
Urban planning

Department of Health and Human Services
Health
Family planning

Department of Labor
Occupational health

Nuclear Regulatory Commission
Licensing and regulating nuclear power

Tennessee Valley Authority
Electric power generation

Department of Homeland Security
Biological, chemical, and radiological risks and weapons
Security of the nation's infrastructure
Assistance to state and local systems
Natural disasters and emergency response

Sources: Council on Environmental Quality, *Environmental Quality: Sixteenth Annual Report of the Council on Environmental Quality,* (Washington, DC: Government Printing Office, 1987); *United States Government Manual 2020,* available at www.usgovernmentmanual.gov/.

Interior and Agriculture Departments; yet the Departments of Energy, Defense, Transportation, and State are increasingly important actors as well. Finally, the more than one hundred federal trial and appellate courts play key roles in interpreting environmental legislation and adjudicating disputes over administrative and regulatory actions (see Chapter 6).

The implications of this constitutional arrangement for policymaking were evident in the early 1980s as Congress and the courts checked and balanced the Reagan administration's efforts to reverse environmental policies of the previous decade. They were just as evident in Barack Obama's presidency when the Republican House of Representatives frequently took strong exception to the president's budget recommendations and proposals for new rules and regulations in the agencies, especially the EPA's efforts to reduce toxic pollution from coal-fired power plants and to restrict release of greenhouse gases linked to climate change. Similarly, during the Trump administration, members of the president's own party often voiced objections to the severity of his proposed budget cuts for environmental programs and scientific research, and routinely rejected them, often increasing science budgets over White House objections (see Chapter 5).[8]

During the last two decades, the conflict between the two major parties on environmental issues had one striking effect. It shifted attention to the role of the states in environmental policy. As Barry G. Rabe discusses in Chapter 2, the states often have been at the center of the most innovative actions on environmental and energy policy, including climate change, when the federal government remained mired in partisan disputes. By 2020, for example, well over half of the states had adopted some form of climate change policy, particularly to favor use of renewable energy sources, whereas Congress and the White House could reach no agreement on what to do.[9]

Generally, after broad consultation and agreement among diverse interests, both within and outside of the government, divided authority typically produces slow and incremental alterations in public policy. Such political interaction and accommodation of interests enhance the overall legitimacy of the resulting public policies. Over time, however, the cumulative effect often results in disjointed policies that fall short of the ecological or holistic principles of policy design so often touted by environmental scientists, planners, and activists.

Nonetheless, when issues are highly visible or salient, the public is supportive, and political leaders act cohesively, the US political system has proved flexible enough to permit substantial and fairly rapid policy advancement.[10] Quick and bipartisan congressional actions in response to the coronavirus pandemic in spring 2020 is a recent example. As we shall see, this also was the case in the early to mid-1970s, when Congress enacted major changes in US environmental policy, and in the mid-1980s, when Congress overrode objections of the Reagan administration and greatly strengthened policies on hazardous waste and water quality, among others. Passage of the monumental Clean Air Act Amendments of 1990 is an example of the same alignment of forces. With bipartisan support, Congress adopted the act by a margin of 401 to 25 in the House and 89 to 10 in the Senate. Comparable bipartisanship during the mid-1990s produced major changes in the Safe Drinking Water Act and in regulation of pesticide residues in food. In 2005 and 2007, the same kind of bipartisan cooperation allowed Congress to approve new national energy policies and significantly expand protection of wilderness areas. In 2016, it also led to approval of major changes to the Toxic Substances Control Act, and in 2019 and 2020 to approval of sweeping land conservation measures (see Chapter 5).

Policy Processes: Agendas, Streams, and Cycles

Students of public policy have proposed several models for analyzing how issues get on the political agenda, how they are defined or framed, and how they move through the policy processes of government. These theoretical frameworks help us to understand both long-term policy trends and short-term cycles of progressive action and political reaction. One set of essential questions concerns *agenda setting*: How do new problems emerge as political issues that demand the government's attention, if they do achieve such recognition, and how are they defined or framed in the public mind?

For example, why was it so difficult for climate change to gain the attention of policymakers over the years, and why did various policy actors frame the issue so differently and interpret climate science in such disparate ways? Climate change's rise on the political agenda was quite slow, and then it became a significant issue by the 2008 presidential election campaign, only to fade again in prominence as the nation's attention was fixed on the economy and persistently high unemployment. In the 2016 elections, it returned as a prominent issue and was given a central role in the Trump administration as it sought to overturn the core elements in Barack Obama's climate policy (see Chapters 3, 4, 5, and 12).

As the case of climate change illustrates, hurdles almost always must be overcome for an issue to rise to prominence. The issue must first gain societal recognition as a problem, often in response to economic, technological, or social changes. It must be defined or framed as a particular kind of problem, which in turn affects the way possible solutions are developed and whether they are seen as acceptable.[11] Organized interest groups strongly affect this process, as do the media. Finally, governmental policymakers must consider the issue to be salient enough to warrant action. An issue is not likely to reach this latter stage unless conditions are ripe—for example, a triggering event that focuses public opinion sharply, as occurred with the Deepwater Horizon oil spill in the Gulf of Mexico in 2010 and with wildfires in the West in recent years, which reinforced public concern about the effects of climate change.[12]

John Kingdon describes this kind of agenda setting as the convergence of three streams of information and activity that flow through the political system at any time: (1) evidence of the existence of problems, (2) available policies to deal with them, and (3) the political climate or willingness to act. Although largely independent of one another, these problem, policy, and political streams can be brought together at critical times when policy entrepreneurs (key activists and policymakers) are able to take advantage of the moment and make the case for policy action.[13]

Once an issue is on the agenda, it must pass through several more stages in the policy process. These stages are often referred to as the *policy cycle*. Although terminology varies, most students of public policy suggest five stages of policy development beyond agenda setting itself. These are (1) *policy formulation* (designing and drafting policy goals and strategies for achieving them, which may involve extensive use of environmental science, economics, and policy analysis), (2) *policy legitimation* (mobilizing political support and formal enactment by law or other means), (3) *policy implementation* (putting programs into effect through provision of institutional resources such as agency budgets and staffs and making key administrative decisions, such as regulatory advances or retreats, as well as judicial rulings on them), (4) *policy evaluation* (how well policies are working in terms of meeting their goals at a reasonable cost), and (5) *policy change* (modifying program goals or the means used to achieve them, or ending programs altogether).[14]

The policy cycle model is useful because it emphasizes all phases of policymaking. For example, how well a law is implemented by agencies such as the EPA or the Interior Department and how the courts rule on it is as important as the goals and motivations of those who designed and enacted the legislation. The model also suggests the continuous nature of the policy process. No policy decision or solution is final because changing conditions, new information, and shifting opinions will require policy reevaluation and change. Other short-term forces and events, such as presidential or congressional elections or environmental accidents, can profoundly affect the course of policy over its life cycle. Thus, policy at any given time is shaped by the interaction of long-term social, economic, technological, and political forces as well as short-term fluctuations in the political climate. These factors are manifest in the development of environmental policy.

THE DEVELOPMENT OF ENVIRONMENTAL POLICY FROM THE 1970S TO THE TWENTY-FIRST CENTURY

As implied in the policy cycle model, the history of environmental policy in the United States is not one of steady improvement in human relations with the natural environment. Rather, it has been highly uneven, with significant discontinuities, particularly since the late 1960s. The pace and nature of policy change, as is true for most areas of public policy, reflect the dominant social values at any given time, the saliency of the issues, and the prevailing economic and political conditions.

Sometimes, as was the case in the 1970s, the combination facilitates major advances in environmental policy, and at other times, such as during the early 1980s, early 2000s, and late 2010s, we have periods of reaction and retrenchment. A third possibility, evident in the 2010s during President Obama's second term, is that no political consensus exists on what to do and consequently no major legislative actions take place. Yet, even in times like this, we see governments responding to changing environmental challenges through executive authority, rulemaking in administrative agencies, state-level actions, and court decisions. These responses were evident both in the Obama administration and in the Trump administration. As noted earlier in the chapter, Trump sought to reverse many Obama initiatives through executive orders and through deep cuts in agency budgets even as Congress resisted those cuts (see Chapters 4 and 5). That is, policy change need not come only through the adoption of new legislation; it can be accomplished through administrative actions as well, and that route to policy change may be preferred when a presidential administration seeks rapid change with minimal public visibility.[15]

Despite these variations in political conditions and policy responses, it is fair to say that since the late 1960s, generally we have seen substantial public support for environmental protection and expanding government authority to act (see Chapter 3).[16] We focus here on the major changes from that time through the middle of the second decade of the twenty-first century, and we discuss the future challenges for environmental politics and policy in the concluding chapter of the book.

Policy Actions Prior to 1970

Until about 1970, the federal government played a sharply limited role in environmental policymaking—public land management being a major exception to this pattern. For nearly a century, Congress had set aside portions of the public domain for preservation as national parks, forests, grazing lands, recreation areas, and wildlife refuges. The multiple use and sustained yield doctrines that grew out of the conservation movement at the beginning of the twentieth century, strongly supported by President Theodore Roosevelt, ensured that this national trust would contribute to economic growth under the stewardship of the Interior and Agriculture Departments.

Steady progress was also made, however, in managing the lands in the public interest and protecting them from development.[17] After several years of debate, Congress passed the Wilderness Act of 1964 to preserve some of the remaining forestlands in pristine condition, "untrammeled by man's presence." At the same time, it approved the Land and Water Conservation Fund Act of 1965 to fund federal purchases of land for conservation purposes and the Wild and Scenic Rivers Act of 1968 to protect selected rivers with "outstandingly remarkable features," including biological, scenic, and cultural value.[18]

During the mid-1960s, the United States also began a major effort to reduce world population growth in developing nations through financial aid for foreign population programs, chiefly voluntary family planning and population research. President Lyndon B. Johnson and congressional sponsors of the programs tied them explicitly to a concern for "growing scarcity in world resources."[19]

Despite this longtime concern for resource conservation and land management, and the new interest in population and development issues, federal environmental policy was only slowly extended to the control of industrial pollution and human waste. Air and water pollution were long considered to be strictly local or state matters, and they were not high on the national agenda until around 1970. In a very early federal action, the Refuse Act of 1899 required individuals who wanted to dump refuse into navigable waters to obtain a permit from the Army Corps of Engineers; however, the agency largely ignored the pollution aspects of the act.[20] After World War II, policies to control the most obvious forms of pollution were gradually developed at the local, state, and federal levels, although some of the earliest local actions to control urban air pollution date back to the 1880s and the first limited state actions to the 1890s.

By the late 1940s and 1950s, we see the forerunners of contemporary air and water pollution laws. For example, the federal government began assisting local authorities in building sewage treatment plants and initiated a limited program for air pollution research. Following the Clean Air Act of 1963 and amendments to the Water Pollution

Control Act of 1948, Washington began prodding the states to set pollution abatement standards and to formulate implementation plans based on federal guidelines.[21]

Agenda Setting for the 1970s

The first Earth Day was April 22, 1970. Nationwide "teach-ins" about environmental problems demonstrated the environment's new place on the nation's social and political agendas. With an increasingly affluent and well-educated society placing new emphasis on the quality of life, concern for environmental protection grew apace and was evident across the population, if not necessarily to the same degree among all groups.[22] The effect was a broadly based public demand for more vigorous and comprehensive federal action to prevent environmental degradation. In an almost unprecedented fashion, a new environmental policy agenda rapidly emerged. Policymakers viewed the newly salient environmental issues as politically attractive, and they eagerly supported tough new measures, even when the full impacts and costs were unknown. As a result, laws were quickly enacted and implemented throughout the 1970s but with a growing concern over their costs and effects on the economy and an increasing realization that administrative agencies at all levels of government often lacked the capacity to assume their new responsibilities.

Congress set the stage for the spurt in policy innovation at the end of 1969 when it passed the National Environmental Policy Act (NEPA). The act declared that

> it is the continuing policy of the Federal Government, in cooperation with State and local governments, and other concerned public and private organizations, to use all practicable means and measures, including financial and technical assistance, in a manner calculated to foster and promote the general welfare, to create and maintain conditions under which man and nature can exist in productive harmony, and fulfill the social, economic, and other requirements of present and future generations of Americans.[23]

The law required detailed environmental impact statements for nearly all major federal actions and established the Council on Environmental Quality to advise the president and Congress on environmental issues. President Richard Nixon then seized the initiative by signing NEPA as his first official act of 1970 and proclaiming the 1970s as the "environmental decade." In February 1970, he sent a special message to Congress calling for a new law to control air pollution. The race was on as the White House and congressional leaders vied for environmentalists' support.

Policy Escalation in the 1970s

By the spring of 1970, rising public concern about the environment galvanized the 91st Congress (1969–1971) to action. Sen. Edmund Muskie, D-Maine, the then leading Democratic hopeful for the presidential nomination in 1972, emerged as the dominant policy entrepreneur for environmental protection issues. As chair of what is now called the Senate Environment and Public Works Committee, he formulated proposals that went well beyond those favored by the president. Following a process of policy escalation, both houses of Congress approved the stronger measures and set the tone of environmental policymaking for much of the 1970s. Congress had frequently played a more dominant role than the president in initiating environmental policies, and that pattern continued in the 1970s. This was particularly so when the Democratic Party controlled Congress during the Nixon and Ford presidencies. Although support for environmental protection was bipartisan during this era, Democrats provided more leadership on the issue in Congress and were more likely to vote for strong environmental policy provisions than were Republicans.[24]

The increase in new federal legislation in the next decade was truly remarkable, especially since, as we noted earlier, policymaking in US politics usually takes place through incremental change. Appendix 1 lists the major environmental protection and natural resource policies enacted from 1969 to 2020. They are arranged by presidential administration primarily to show a pattern of significant policy development throughout the period, not to attribute chief responsibility for the various laws to the president.

These landmark measures covered air and water pollution control (the latter enacted in 1972 over a presidential veto), pesticide regulation, endangered species protection, control of hazardous and toxic chemicals, ocean and coastline protection, improved stewardship of public lands, requirements for the restoration of strip-mined lands, the setting aside of more than one hundred million acres of Alaskan wilderness for varying degrees of protection, and the creation of a "Superfund" (in the Comprehensive Environmental Response, Compensation, and Liability Act, or CERCLA) for cleaning up toxic waste sites. Nearly all of these policies reflected a conviction that the federal government must have enough authority to compel polluters and resource users to adhere to demanding national pollution control standards and new decision-making procedures that ensure responsible use of natural resources. There were other signs of commitment to environmental policy goals as Congress and a succession of presidential administrations (through Jimmy Carter's term) cooperated on land conservation issues, such as wilderness protection, national parks, and wildlife refuges. Throughout the 1970s, the Land and Water Conservation Fund, financed primarily through royalties from offshore oil and gas leasing, was used to purchase additional private land for park development, wildlife refuges, and national forests.

Congress maintained its strong commitment to environmental policy throughout the 1970s, even as the salience of these issues for the public seemed to wane. For example, it revised the Clean Air Act of 1970 and the Clean Water Act of 1972 through amendments approved in 1977. Yet, by the end of the Carter administration, concerns over the impact of environmental regulation on the economy and specific objections to implementation of the new laws, particularly the Clean Air Act, began creating a backlash.

Political Reaction in the 1980s

The Reagan presidency brought to the federal government a markedly different environmental policy agenda (see Chapter 4). Virtually all environmental protection and resource policies enacted during the 1970s were reevaluated in light of the president's desire to reduce the scope of government regulation, shift responsibilities to the states, and depend more on the private sector. Whatever the merits of Reagan's new policy agenda, it was put into effect through a risky strategy that relied on ideologically committed presidential appointees to the EPA and the Agriculture, Interior, and Energy Departments and on sharp cutbacks in budgets for environmental programs.[25]

Congress initially cooperated with Reagan, particularly in approving budget cuts, but it soon reverted to its accustomed defense of existing environmental policy, frequently criticizing the president's management of the EPA and the Interior Department under Anne Gorsuch (later Burford) and James Watt, respectively; both Burford and Watt were forced to resign by the end of 1983. Among Congress's most notable achievements of the 1980s were its strengthening of the Resource Conservation and Recovery Act (Hazardous and Solid Waste Amendments, 1984); the enactment of the Superfund Amendments and Reauthorization Act (1986), which toughened the act and also established the federal Toxics Release Inventory (TRI); and amendments to the Safe Drinking Water Act (1986) and the Clean Water Act (1987; see Appendix 1 for a list of major federal environmental laws from 1969 to 2020).

As we discuss later in this chapter, budget cuts and the loss of capacity in environmental institutions took a serious toll during the 1980s. Yet even the determined efforts of a popular president could not halt the advance of environmental policy. Public support for environmental improvement, the driving force for policy development in the 1970s, increased markedly during Reagan's presidency and represented the public's stunning rejection of the president's agenda.[26] Paradoxically, Reagan strengthened environmental forces in the nation. Through his lax enforcement of pollution laws and prodevelopment resource policies, he created political issues around which national and grassroots environmental groups could organize. These groups appealed successfully to a public that was increasingly disturbed by the health and environmental risks of industrial society and by threats to ecological stability. As a result, membership in national environmental groups soared, and new grassroots organizations developed, creating further political incentives for environmental activism at all levels of government.[27]

By the fall of 1989, there was little mistaking congressional receptivity to continuing the advance of environmental policy into the 1990s. Especially in his first 2 years as president, George H. W. Bush was eager to adopt a more positive environmental policy agenda than his predecessor; this eagerness was particularly evident in his support for the demanding Clean Air Act Amendments of 1990. Bush's White House, however, was deeply divided on environmental issues for both ideological and economic reasons.

SEEKING NEW POLICY DIRECTIONS: FROM THE 1990S TO THE TWENTY-FIRST CENTURY

Environmental issues received considerable attention during the 1992 presidential election campaign. Bush, running for reelection, criticized environmentalists as extremists who were putting Americans out of work. The Democratic candidate, Bill Clinton, took a far more supportive stance on the environment, symbolized by his selection of Sen. Al Gore, D-Tennessee, as his running mate. Gore was the author of a best-selling book, *Earth in the Balance*, and had one of the strongest environmental records in Congress.

Much to the disappointment of environmentalists, Clinton exerted only sporadic leadership on the environment throughout his two terms in office. However, he and Gore quietly pushed an extensive agenda of environmental policy reform as part of their broader effort to "reinvent government," making it more efficient and responsive to public concerns. Clinton was also generally praised for his environmental appointments and for his administration's support for initiatives such as the restoration of the Florida Everglades and other actions based on new approaches to ecosystem management. Clinton reversed many of the Reagan- and Bush-era executive actions that were widely criticized by environmentalists, and he favored increased spending on environmental programs, alternative energy and conservation research, and international population policy.

Clinton also earned praise from environmental groups when he began speaking out forcefully against the antienvironmental policy decisions of Republican Congresses (see Chapters 4 and 5), for his efforts through the President's Council on Sustainable Development to encourage new ways to reconcile environmental protection and economic development, and for his "lands legacy" initiatives.[28] Still, Clinton displeased environmentalists as often as he gratified them.

The environmental policy agenda of George W. Bush's presidency is addressed in Chapter 4 and at points throughout the rest of the book, as are actions taken during Barack Obama's presidency from January 2009 through January 2017. As widely expected from statements Bush made on the campaign trail and from his record as governor of Texas, he and his cabinet departed significantly from the positions of the Clinton administration. The economic impact of environmental policy emerged as a major concern, and the president gave far more emphasis to economic development than he did to environmental protection or resource conservation.

Like his father, Bush recognized the political reality of popular support for environmental protection and resource conservation. Yet as a conservative Republican, he was also inclined to represent the views of the party's core constituencies, particularly industrial corporations and timber, mining, agriculture, and oil interests. He drew heavily from those constituencies, as well as from conservative ideological groups, to staff the EPA and the Interior, Agriculture, and Energy Departments, filling positions with what the press termed industry insiders, a practice that reappeared by 2017 in the Trump administration.[29] In addition, Bush sought to further reduce the burden of environmental protection through the use of voluntary, flexible, and cooperative programs and to transfer to the states more responsibility for the enforcement of federal laws. Bush also withdrew the United States from the Kyoto Protocol on global climate change, significantly weakening US leadership on global environmental issues.

The administration's tendency to minimize environmental concerns was equally clear in its 2001 proposal for a national energy policy (which concentrated on the increased production of fossil fuels) and, throughout Bush's two terms, in many decisions on clean air rules, water quality standards, mining regulations, and the protection of national forests and parks, decisions that were widely denounced by environmentalists.[30] Many of these decisions received considerably less media coverage than might have been expected. In part, this neglect appeared to reflect the administration's strategy of keeping a low profile on potentially unpopular environmental policy actions, a pattern that also was evident in the Trump administration from 2017 through 2020. But President Bush benefited further from the sharply altered political agenda after the terrorist attacks of September 11, 2001, as well as from the decision in 2003 to invade Iraq.[31]

Barack Obama's environmental policy priorities and actions are described in some detail in Chapter 4 and in many of the chapters that follow. Hence, we leave much of that appraisal until later in the volume. However, we address budgetary and administrative changes during the Obama presidency in the next section as well as comparable changes in budgets, institutions, and staffing in the Trump administration.

BUDGETS AND POLICY IMPLEMENTATION

In this review of environmental policy development since 1970, we have highlighted the adoption of landmark policies and the political conflicts that shaped them. Another part of this story is the changes over time in budgetary support for the agencies responsible for implementing the policies.

Agency budgets are an important part of institutional capacity, which in turn affects the degree to which public policies might help to improve environmental quality. Although spending more money hardly guarantees policy success, substantial budget cuts can significantly undermine established programs and hinder the achievement of policy goals. For example, the massive reductions in environmental funding during the 1980s had long-term adverse effects on the government's ability to implement environmental policies. Equally sharp budget cuts proposed by Congress in the mid- to late 1990s, by the Bush administration in the 2000s, and by the Trump administration from 2017 through 2020.

Changes since the 1980s in budgetary support for environmental protection merit brief comment here. The appendices offer more detail. In constant 2020 dollars (that is, adjusting for inflation), the total spending authorized by the federal government for natural resource and environmental programs was about 27 percent higher in 2020 than it was in 1980 (see Appendix 4). However, in some program areas reflecting the core functions of the EPA, such as pollution control and abatement, spending declined substantially (about 32 percent) from 1980 to 2020, in constant dollars. In contrast, spending on conservation and land management rose appreciably between 1980 and 2020, more than tripling, again in constant dollars. For most budget categories, spending decreased during the 1980s before recovering under the administrations of George H. W. Bush and Bill Clinton, and to some extent under George W. Bush and Barack Obama. A notable exception, other than the case of pollution control, is spending on water resources, where the phaseout of federal grant programs resulted in a significant decline in expenditures between 1980 and 2010 before recovering somewhat by 2020; overall, spending dropped by about 17 percent between 1980 and 2020.

Even when the budget picture was improving, most agencies faced important fiscal challenges. Agencies' legal responsibilities rose substantially under environmental policies approved between the 1970s and 2020, and the agency staffs often found themselves with insufficient resources to implement new policies fully and to achieve the environmental quality goals they embodied.

These constraints can be seen in the budgets and staffs of selected environmental and natural resource agencies. For example, in constant dollars, the EPA's operating budget as we calculate it (the EPA determines it somewhat differently) was only a little higher in 2020 than it was in 1980, despite the many new duties Congress gave the agency during this period (see Appendix 2). The agency's budget authority rose from 2000 to 2010, enjoying a big boost in Obama's first year in office. It then declined in 2011, rose modestly to $10.8 billion in 2012, but declined again in his last few years in office, ending at $8.3 billion in proposed spending in fiscal year 2017.

In early 2017, the Trump administration proposed a 31 percent cut from this level of funding, including a 26 percent cut in the environmental program and management budget. However, Congress approved a budget agreement that year that included only a 1 percent cut in EPA's budget and no further reduction in its staff. The administration proposed similar agency budget cuts in succeeding years, with much the same result. By fiscal year 2019, EPAs reported that its budget stood at $8.85 billion, and for fiscal year 2020 at an estimated $9.1 billion. The Trump administration proposed for fiscal year 2021 that it be reduced by 26 percent to $6.7 billion, a level that Congress once again is likely to reject.[32]

The EPA's staff grew by a greater percentage than its budget, rising from slightly fewer than 13,000 in 1980, the last year of the Carter administration, to around 17,360 by 2011; however, the agency saw its staff decline substantially after 2011; by 2016, the fiscal year 2017 budget, it stood at about 15,400.[33] Most other agencies saw a decrease in staff from 1980 through 2010, some remained at about the same level, and a few enjoyed an increase in staff size (see Appendix 3).

The Trump White House proposed a large reduction in the EPA's staff of some 2,500 people below its modern low point in 2016, as the administration sought to diminish the agency's role in environmental protection significantly and turn over many of its responsibilities to the states. As noted, however, Congress did not go along fully. Even so, by 2019, EPA staff levels had declined to about 14,000, a level not seen since the late 1980s. The administration's fiscal year 2021 budget proposal put the desired staffing much lower yet, at 12,610.[34] For the near term, it seems likely that both budgets and staffing levels will remain at their present levels, which almost certainly will continue to adversely affect the capacity of the agency to achieve the objectives set out for it by Congress.

IMPROVEMENTS IN ENVIRONMENTAL QUALITY

It is difficult, both conceptually and empirically, to measure the success or failure of environmental policies.[35] Yet one of the most important tests of any public policy is whether it achieves its stated objectives. For environmental policies, we should ask if air and water quality are improving, hazardous waste sites are being cleaned up, and biological diversity is protected adequately. Almost always, we also want to know what these improvements cost, not just to government, but for society as a whole. There is no simple way to answer those questions, and it is important to understand why that is, even if some limited responses are possible.[36]

Measuring Environmental Conditions and Trends

Environmental policies entail long-term commitments to broad social values and goals that are not easily quantified. Short-term and highly visible costs are easier to measure than long-term, diffuse, and intangible benefits, and these differences often lead to intense debates over the value of environmental programs. For example, should the EPA toughen air quality standards to reduce adverse health effects or hold off out of concern for the economic impacts of such a move (see Chapter 7)? The answer often seems to depend on which president sits in the White House and how sensitive the EPA is to public concerns over the relative benefits and costs.

Variable and often unreliable monitoring of environmental conditions and inconsistent collection of data over time also make it difficult to assess environmental trends. The time period selected for a given analysis can affect the results, and many scholars discount some data collected prior to the mid-1970s as unreliable. One thing is certain, however. Evaluation of environmental policies depends on significant improvements in monitoring and data collection at both state and federal levels. With better and more appropriate data, we should be able to speak more confidently in the future about policy successes and failures. Of course, any such judgments require that policymakers examine the data objectively and evaluate programs on the evidence as opposed to acting on ideological leanings. This assumption is difficult to make today as scientific data and professional expertise are not necessarily valued as consistently by policymakers and other political actors as both were previously.

In the meantime, scientists and pundits continue to debate whether environmental conditions are deteriorating or improving, and for what reasons. Many state-of-the-environment reports that address such conditions and trends are issued by government agencies and environmental research institutes. For the United States, EPA and other agency reports, discussed below, are available online and offer authoritative data. Not surprisingly, interpretations of the data may differ. For instance, critics of environmental policy tend to cite statistics that show rather benign conditions and trends (and therefore little reason to favor public policies directed at them), whereas most environmentalists focus on what they believe to be indicators of serious environmental decline and thus a justification for government intervention. The differences sometimes become the object of extensive media coverage and public debate.

Despite the many limitations on measuring environmental conditions and trends accurately, it is nevertheless useful to examine selected indicators of environmental quality. They tell us at least something about what we have achieved or failed to achieve after nearly five decades of national environmental protection policy. We focus here on a brief overview of trends in air quality, greenhouse gas emissions, water quality, toxic chemicals and hazardous wastes, and natural resources.[37]

Air Quality

Perhaps the best data on changes in the environment can be found for air quality, even if disagreement exists over which measures and time periods are most appropriate. The EPA estimates that, between 1970 and 2018, aggregate emissions of the six principal, or criteria, air pollutants decreased by 74 percent even while the nation's gross domestic product (GDP) grew by 275 percent, its population grew by 56 percent, vehicle miles traveled increased by 173 percent, and energy consumption grew by 44 percent, all of which would likely have increased air pollution without federal laws and regulations.[38]

Progress generally continues. For example, between 2000 and 2018, monitored levels of the six criteria pollutants (that is, ambient air concentrations) showed improvement, with all declining during this period by between 16 and 93 percent. Ozone concentrations (using the 8-hour standard) declined by 16 percent, particulate matter by 31 percent and fine particulates (which pose a greater health risk) by 39 percent, lead by 93 percent, nitrogen dioxide (annual measure) by 49 percent, carbon monoxide by 59 percent, and sulfur dioxide by 80 percent. The changes are far more modest for these pollutants for the period 2010–2018, indicating a substantial slowing of progress in recent years.

Consistent with these trends, the number of days with unhealthy air quality in major US cities generally has been trending downward since 1990, although it has leveled off over the past decade. The EPA celebrates these achievements, saying that the "air quality benefits will lead to improved health, longevity, and quality of life for all Americans."[39] Despite the history of impressive gains in air quality, as of 2018, over 137 million people (about 41 percent of the US population) lived in counties with pollution levels above the standards set for at least one of these criteria pollutants, particularly ozone and fine particulates. These figures vary substantially from year to year, reflecting changing economic activity and weather patterns, including wildfire-related air pollution. For example, air quality improved in many areas in the spring of 2020 due to a coronavirus-related decline in economic activity and vehicle traffic, and thus reduced use of fossil fuels.[40]

One of most significant remaining problems is toxic or hazardous air pollutants, which have been associated with cancer, respiratory diseases, and other chronic and acute illnesses. The EPA was extremely slow to regulate these pollutants and had established federal standards for only seven of them by 1989. Public and congressional concern over toxic emissions led Congress to mandate more aggressive action in the 1986 Superfund amendments as well as in the 1990 Clean Air Act Amendments. The former required manufacturers of more than 300 different chemicals (later increased by the EPA to over 650) to report annually to the agency and to the states in which they operate the amounts of those substances released to the air, water, or land. The EPA's TRI indicates that for the core chemicals from industry that have been reported in a consistent manner over time, total releases on- and off-site decreased by over 60 percent between 1988 and 2015, an impressive improvement in reducing public exposure to toxic chemicals. For a more recent period, 2007 through 2018, however, releases declined by only 9 percent.[41]

The annual TRI reports also tell us that industries continue to release very large quantities of toxic chemicals to the environment—3.8 billion pounds a year from about 21,000 facilities across the nation, based on the latest report. About 600 million pounds of the chemicals are released into the air, and those may pose a significant risk to public health.[42] It should be noted, however, that the TRI and related numbers on toxics do not present a full picture of public health risks. For instance, many chemicals and industries were added to TRI reporting requirements from the 1990s to the 2010s, complicating the determination of change over time. Using the original or core list of chemicals obviously doesn't account for those put on the list more recently. In addition to the TRI, under the 1990 Clean Air Act Amendments, the EPA regulates 188 listed air toxics, but nationwide monitoring of emissions is not standard.

Greenhouse Gas Emissions

The United States is making significant progress in addressing climate change despite fairly weak policies, largely because of improved energy efficiency, increased reliance on natural gas rather than coal for energy production, lower emissions from vehicles because of tougher fuel economy standards, and rapid growth in the use of renewable energy sources such as wind and solar power. The Obama administration advanced vehicle fuel economy standards and a Clean Power Plan designed to reduce reliance on coal-fired power plants. The Trump administration weakened both policies.[43] Even so, the new trends emerging in the nation's energy use are likely to continue.

According to the EPA's 2020 inventory of greenhouse gases, US emissions in 2018 totaled 6,678 million metric tons of CO_2 equivalent, a common way of accounting for emissions of all forms of greenhouse gases. Total US emissions of greenhouse gases peaked in 2007 at 7,414 million metric tons, and declined slowly after that, with some years showing a small increase or decrease. Emissions declined in 2015, 2016, and 2017, but increased by 3.1 percent in 2018, the last year covered in the report.[44] Global emissions, however, have been increasing, hitting a new high in 2019, even though emissions declined temporarily in early 2020 during the coronavirus pandemic.[45] The most notable declines in carbon dioxide releases in the United States have been in electricity generation even as there have been slight increases in emissions from transportation and industrial activity. Projections for the next several years depend on the pace of the nation's movement away from extensive reliance on coal for generating electricity and on its shift to cleaner-burning natural gas or renewable forms of energy, such as wind and solar power.

Despite this reduction in emissions, data from the National Oceanic and Atmospheric Administration's Annual Greenhouse Gas Index (based on highly precise atmospheric measurements) show the global concentration of the most important greenhouse gases continued to increase in recent years, reflecting a growth in "radiative forcing" or warming impact of nearly 60 percent since 1990.

Measurements of carbon dioxide concentrations at Hawaii's Mauna Loa Observatory, a long-time test site, show years of continuous increases, with levels reaching 417 parts per million in May 2020. In addition, the International Energy Agency (IEA) reported that global carbon dioxide emissions (that is, releases to the atmosphere) grew significantly in recent years, reflecting an increasing global economy, although emissions leveled off in 2019. Given what the IEA called a "continuing decoupling of emissions and economic activity," the pattern may change in coming years. The explanations for this trend include the growing reliance on renewable sources of energy, switching from coal to natural gas for electricity generation, and gains in energy efficiency.[46] In this context, it is worth adding that the United States remains by far the world's leading emitter of greenhouse gases by large nations on a per capita basis (see Chapter 12).

Water Quality

The nation's water quality has improved since passage of the Clean Water Act of 1972, although more slowly and more unevenly than has air quality. However, monitoring data are far less adequate for water quality than for air quality, making judgments about improvement difficult. For example, the best evidence for the state of water quality can be found in the EPA's consolidation of state reports (mandated by the Clean Water Act), which are accessible at the agency website. For the most recent reporting period (which often is not current due to lax state reporting), the states collectively assessed only 31 percent of the entire nation's rivers and streams; 45 percent of lakes, ponds, and reservoirs; 64 percent of estuaries and bays; and a mere 8 percent of coastal shorelines and 13 percent of oceans and near coastal areas.

Based on these very limited inventories, 47 percent of the surveyed river and stream miles were considered to be of good quality and 53 percent impaired. Some 71 percent of lakes, ponds, and reservoirs also were found to be impaired. A classification of impaired means that water bodies are not meeting or fully meeting the national minimum water quality criteria for "designated beneficial uses" such as swimming, fishing, drinking-water supply, and support of aquatic life. The same survey found that 80 percent of the nation's assessed estuaries and bays were impaired, as were 72 percent of assessed coastal shorelines and 90 percent of assessed oceans and near coastal waters.[47]

The latest data on water quality show very little improvement in recent years, and in some of the categories, a decline in quality. In the face of a growing population and strong economic growth, prevention of significant further degradation of water quality could be considered an important achievement. At the same time, water quality clearly falls short of the goals of federal clean water acts.

The causes of impaired waters today are well understood. The EPA reports that the leading identifiable sources of impairment of rivers and streams, for example, are agriculture, human modification of waterways, atmospheric deposition of chemicals, habitat modification, unspecified nonpoint sources, and municipal discharges (in that order). That is, the causes no longer are point sources of pollution, such as industrial discharges, which have been well controlled with regulation under the Clean Water Act. Rather, they are largely nonpoint sources that are much more difficult to control and will take longer to affect.

To date, little progress has been made in halting groundwater contamination despite passage of the Safe Drinking Water Act of 1974, the Resource Conservation and Recovery Act of 1976, and their later amendments. Heading the list of contaminant sources are leaking underground storage tanks, septic systems, landfills, spills, fertilizer applications, large industrial facilities, hazardous waste sites, and animal feedlots. With about half of the nation's urban population relying on groundwater for drinking water (99 percent in rural areas), far more remains to be done.[48]

The surge in natural gas drilling around the nation through hydraulic fracturing or fracking has sparked additional concerns over groundwater quality and its possible impacts on human health. Fracking involves the injection of massive amounts of water mixed with sand and various chemicals under high pressure to release natural gas from shale formations. There were over 1.7 million active wells in the nation in recent years, and they yielded about two-thirds of natural gas production; natural gas is now the leading source of the country's electricity generation. One consequence, however, is increasing citizen concern about the risks posed by fracking's possible contamination of groundwater. Fracking is regulated primarily by the states rather than the federal government.[49]

Toxic Chemicals and Hazardous Wastes

Progress in dealing with hazardous wastes and other toxic chemicals has been the least satisfactory of all pollution control programs. Implementation of the major laws has been extraordinarily slow due to the extent and complexity of the problems, scientific uncertainty, litigation by industry, public fear of siting treatment and storage facilities nearby, budgetary limitations, and poor management and lax enforcement by the EPA. As a result, gains have been modest when judged by the most common measures.

For example, the nation has yet to agree on how to dispose of high-level radioactive wastes from civilian nuclear power plants, despite several major national policies that date back decades. There also is the enormous task of cleaning up contaminated federal facilities, such as former nuclear weapons production plants, as well as tens of thousands of abandoned mines on federal lands in the West for which the federal government is liable for remediation. One comprehensive assessment in 2016 put the total federal environmental liability at over $400 billion, with the eventual cleanup cost likely to be much higher than that.[50]

One of the major federal programs aimed at toxic and hazardous chemicals is Superfund. For years, the government made painfully slow progress under the program in cleaning up the nation's worst hazardous waste sites. By the late 1990s, however, the pace of action improved significantly. Even so, by 2020, the EPA reported 1,335 sites remained on the program's National Priorities List or NPL; the NPL includes the top 1,600 or so sites out of some 40,000 Superfund sites in the country.[51] Many cleanup efforts have been successful to date, and those sites are removed from the list. Yet the pace of future cleanup remains in doubt given scarce federal funds for the program. In addition, some of the costliest cleanups, including chemical containments in rivers and bays, remain to be addressed. The EPA captured the challenge well in a report from several years ago. It noted that the "Superfund cleanup work EPA is doing today generally is more difficult, is more technically demanding, and consumes considerable resources at fewer sites than in the past."[52]

Historically, the EPA has set a sluggish pace in the related area of testing and acting on toxic chemicals, including pesticides. For example, under a 1972 law mandating control of pesticides and herbicides, only a handful of chemicals used to manufacture the fifty thousand pesticides in use in the United States had been fully tested or retested. The Food Quality Protection Act of 1996 required the EPA to undertake extensive assessment of the risks posed by new and existing pesticides. Following a lawsuit, the EPA began moving more quickly toward meeting the act's goal of protecting human health and the environment from these risks. The agency has said in recent years that it has begun a new program to reevaluate all pesticides in use on a regular basis, at least once every 15 years.[53]

Similarly, limited progress in implementing the relatively weak Toxics Substances Control Act of 1976 finally led to the act's amendment in 2016 with congressional approval of the Frank R. Lautenberg Chemical Safety for the 21st Century Act. That act mandated that the EPA evaluate existing chemicals with a new risk-based safety standard, that it do so with clear and enforceable deadlines, with increased transparency for chemical information, and with assurances that the agency would have the budgetary resources to carry out its responsibilities for chemical safety. A list of the first ten chemicals to be studied was released in late 2016. However, it remains unclear whether the Trump administration will accord the new law the priority that Congress intended, particularly in light of the budgetary and staff cuts that are likely to hinder the act's implementation.[54]

Natural Resources

Comparable indicators of environmental progress can be cited for natural resource use. As is the case with pollution control, however, interpretation of the data is problematic. We have few good measures of ecosystem health, and controversies continue about how best to value ecosystem services. Moreover, the usual information supplied in government reports details only the area of land set aside for recreational and aesthetic purposes rather than how well ecosystem functions are being protected.[55] Nonetheless, the trends in land conservation and wilderness protection suggest important progress since the 1960s.

For example, the national park system grew from about 26 million acres in 1960 to over 84 million acres by 2020, and the number of units (that is, parks) in the system doubled. Since adoption of the 1964 Wilderness Act, Congress has set aside more than 111 million acres of wilderness through the national wilderness preservation system. Since 1968, it has designated parts of 226 rivers in 41 states as wild and scenic, with over 13,400 river miles protected by 2020. The Fish and Wildlife Service manages more than 150 million acres in more than 560 units of the national wildlife refuge system in all fifty states, far in excess of the total acreage in the system in 1970; about 95 million acres of this total are set aside as wildlife habitat.[56]

Protection of biological diversity through the Endangered Species Act has produced some success as well, although far less than its supporters believe essential. By March 2020, 47 years after passage of the 1973 act, more than 1,661 US plant and animal species had been listed as either endangered or threatened, and more than 1,169 active recovery plans were in effect. Yet only a few endangered species have recovered fully. The Fish and Wildlife Service reported in 2008 (no similar assessments have been made in more recent years) that 41 percent of those listed were considered to be stable or improving, but that 34 percent were considered to be declining in status, and that for 23 percent their status was unknown. About 2 percent were presumed to be extinct.[57]

ASSESSING ENVIRONMENTAL PROGRESS

As the data reviewed in the preceding sections suggest, the nation made impressive gains between 1970 and 2020 in controlling many conventional pollutants and in expanding parks, wilderness areas, and other protected public lands. Despite some setbacks, progress on environmental quality continues, even if it is highly uneven from one period to the next. In the future, however, further advances will be more difficult, costly, and controversial. This is largely because the easy problems have already been addressed. At this point, marginal gains—for example, in air and water quality—are likely to cost more per unit of improvement than in the past. Moreover, second-generation environmental threats such as toxic chemicals, hazardous wastes, and nuclear wastes are proving even more difficult to regulate than the "bulk" air and water pollutants that were the main targets in the 1970s. In these cases, substantial progress may not be evident for years to come, and it may well be expensive.

The same is true for the third generation of environmental problems, such as global climate change and the protection of biodiversity. Solutions require an unprecedented degree of cooperation among nations and substantial improvement in institutional capacity for research, data collection, and analysis, as well as for policy development and implementation. Hence, success is likely to come slowly and will reflect the extent to which national and international commitments to environmental protection grow and capabilities improve.

Some long-standing problems, such as population growth, will continue to be addressed primarily within nation-states, even though the staggering effects on natural resources and environmental quality are felt worldwide. By mid-2020, the Earth's population of about 7.7 billion people was increasing at an estimated 1.1 percent (or about 85 million people) each year, with a middle-range projection for the year 2050 at about 9.8 billion. The US population had a natural rate of growth (that is, not counting immigration) of 0.3 percent a year, and middle-range projections by the Population Reference Bureau put the US population at about 388 million by 2050 (see Chapter 13).[58]

CONCLUSION

Since the 1970s, public concern and support for environmental protection have risen significantly, spurring the development of an expansive array of policies that substantially increased the government's responsibilities for the environment and natural resources, both domestically and internationally. The implementation of these policies, however, has been far more difficult and controversial than their supporters ever imagined. Moreover, the policies have not been entirely successful, particularly when measured by tangible improvements in environmental quality. Further progress will likely require the United States to search for more efficient and effective ways to achieve these goals, including the use of alternatives to conventional command-and-control regulation, such as the use of flexible regulation, market incentives, and information disclosure or public education.[59] Despite these qualifications, the record since the 1970s demonstrates convincingly that the US government is able to produce significant environmental gains through public policies. Unquestionably, the environment would be worse today if the policies enacted during the 1970s and 1980s, and since then, had not been in place.

Emerging environmental threats on the national and international agenda are even more formidable than the first generation of problems addressed by government in the

1970s and the second generation that dominated political debate in the 1980s. Responding to these threats will require creative new efforts to improve the performance of government and other social institutions, as well as effective leadership to design appropriate strategies to combat these threats, both within government and in society itself. Some of these strategies might include sustainable community initiatives and corporate social responsibility actions. This new policy agenda is addressed in Part IV of the book and in Chapter 15.

Government obviously is an important player in the environmental arena, and the federal government will continue to have unique responsibilities, as will the fifty states and the more than ninety thousand local governments across the nation. President Obama assembled an experienced and talented environmental policy team to address these challenges and, at the launch of his administration, vowed to make energy and environmental issues "a leading priority" of his presidency and a "defining test of our time."[60] The Donald Trump administration adopted a dramatically different environmental policy agenda and set of priorities, the effects of which will become apparent only over the next few years. Readers can judge for themselves how well recent presidents and their appointees have lived up to the promises they made as they peruse the chapters in this volume. It is equally clear, however, that government rarely can pursue forceful initiatives without broad public support. Ultimately, society's values and priorities will shape the government's response to a rapidly changing world environment that, in all probability, will involve major economic and social dislocations over the coming decades.

NOTES

1. See survey data reviewed in Chapter 3; Riley E. Dunlap, "Public Opinion and Environmental Policy," in *Environmental Politics and Policy: Theories and Evidence*, 2nd ed., ed. James P. Lester (Durham, NC: Duke University Press, 1995), 63–114; Riley E. Dunlap, George H. Gallup Jr., and Alec M. Gallup, "Of Global Concern: Results of the Health of the Planet Survey," *Environment* 35, no. 9 (1993): 7–15, 33–40; and David P. Daniels, Jon A. Krosnick, Michael P. Tichy, and Trevor Tompson, "Public Opinion on Environmental Policy in the United States," in *The Oxford Handbook of U.S. Environmental Policy*, ed. Sheldon Kamieniecki and Michael E. Kraft (New York: Oxford University Press, 2013), 461–86.

2. Norman J. Vig and Michael E. Kraft, eds., *Environmental Policy in the 1980s: Reagan's New Agenda* (Washington, DC: CQ Press, 1984).

3. See Robert Durant, Rosemary O'Leary, and Daniel Fiorino, eds., *Environmental Governance Reconsidered: Challenges, Choices, and Opportunities*, 2nd ed. (Cambridge, MA: MIT Press, 2017); Daniel Fiorino, *The New Environmental Regulation* (Cambridge, MA: MIT Press, 2006); Marc Allen Eisner, *Governing the Environment: The Transformation of Environmental Regulation* (Boulder, CO: Lynne Rienner, 2007); and Christopher McGrory Klyza and David Sousa, *American Environmental Policy: Beyond Gridlock*, updated and expanded edition (Cambridge, MA: MIT Press, 2013).

4. For example, see Daniel A. Mazmanian and Michael E. Kraft, eds., *Toward Sustainable Communities: Transition and Transformations in Environmental Policy*, 2nd ed. (Cambridge, MA: MIT Press, 2009); Durant, O'Leary, and Fiorino, *Environmental Governance Reconsidered*; Klyza and Sousa, *American Environmental Policy*; and Michael E. Kraft,

Mark Stephan, and Troy D. Abel, *Coming Clean: Information Disclosure and Environmental Performance* (Cambridge, MA: MIT Press, 2011).

5. Kent E. Portney, *Taking Sustainable Cities Seriously: Economic Development, the Environment, and Quality of Life in American Cities*, 2nd ed. (Cambridge, MA: MIT Press, 2013).

6. The California policy is discussed in Chapter 2. On Minnesota's multifaceted and very successful energy policy, see Michel Wines, "Without Much Straining, Minnesota Reins in Its Utilities' Carbon Emissions," *New York Times*, July 17, 2014. See also Roger Karapin, *Political Opportunities for Climate Policy: California, New York, and the Federal Government* (New York: Cambridge University Press, 2016); and Barry G. Rabe, *Statehouse and Greenhouse: The Emerging Politics of American Climate Change Policy* (Washington, DC: Brookings Institution Press, 2004).

7. See Walter A. Rosenbaum, "Science, Policy, and Politics at the EPA," in *Environmental Policy*, 8th ed., ed. Norman J. Vig and Michael E. Kraft (Thousand Oaks, CA: Sage, 2013), 158–84. See also National Academy of Public Administration (NAPA), *Setting Priorities, Getting Results: A New Direction for EPA* (Washington, DC: NAPA, 1995), 124–25.

8. See, for example, Glenn Thrush and Coral Davenport, "Donald Trump Budget Slashes Funds for E.P.A. and State Department," *New York Times*, March 15, 2017; and Rebecca Beitsch and Rachel Frazin, "Trump Budget Slashes EPA Funding, Environmental Programs," *The Hill*, February 10, 2020; and Jeffrey Mervic, "Trump's New Budget Cuts All But a Favored Few Science Programs," *Science* 367 (6479): February 14, 2020: 723–24.

9. See Karapin, *Political Opportunities for Climate Policy* and Klyza and Sousa, *American Environmental Policy*, Chapter 7.

10. John W. Kingdon, *Agendas, Alternatives, and Public Policies*, 2nd ed. (New York: HarperCollins, 1995); and Frank R. Baumgartner and Bryan D. Jones, *Agendas and Instability in American Politics* (Chicago: University of Chicago Press, 1993).

11. For a review of how this process works, see Deborah Lynn Guber and Christopher J. Bosso, "Issue Framing, Agenda Setting, and Environmental Discourse," in *The Oxford Handbook of U.S. Environmental Policy*, ed. Sheldon Kamieniecki and Michael E. Kraft (New York: Oxford University Press, 2013), 437–60.

12. Roger W. Cobb and Charles D. Elder, *Participation in American Politics: The Dynamics of Agenda-Building* (Boston: Allyn & Bacon, 1972). See also Thomas A. Birkland, *After Disaster: Agenda Setting, Public Policy, and Focusing Events* (Washington, DC: Georgetown University Press, 1997).

13. Kingdon, *Agendas*.

14. For a more thorough discussion of how the policy cycle model applies to environmental issues, see Michael E. Kraft, *Environmental Policy and Politics*, 7th ed. (New York: Routledge, 2018), Chapter 3. The general model is discussed at length in James E. Anderson, *Public Policymaking: An Introduction*, 8th ed. (Stamford, CT.: Cengage Learning, 2015), as well as in Thomas A. Birkland, *An Introduction to the Policy Process: Theories, Concepts, and Models of Public Policy Making*, 5th ed. (New York: Routledge, 2020).

15. Klyza and Sousa, *American Environmental Policy*. See also Frank J. Thompson, Kenneth K. Wong, and Barry G. Rabe, *Trump, the Administrative Presidency, and Federalism* (Washington, DC: Brookings Institution, 2020).

16. Dunlap, "Public Opinion and Environmental Policy"; Deborah Lynn Guber, *The Grassroots of a Green Revolution: Polling America on the Environment* (Cambridge, MA: MIT Press, 2003); and Daniels et al., "Public Opinion on Environmental Policy in the United States."

17. Paul J. Culhane, *Public Lands Politics: Interest Group Influence on the Forest Service and the Bureau of Land Management* (Baltimore: Johns Hopkins University Press, 1981). See also

Richard N. L. Andrews, *Managing the Environment, Managing Ourselves: A History of American Environmental Policy*, 3rd ed. (New Haven, CT: Yale University Press, 2020); and Sally K. Fairfax, Lauren Gwin, Mary Ann King, Leigh Raymond, and Laura A. Watt, *Buying Nature: The Limits of Land Acquisition as a Conservation Strategy: 1780–2004* (Cambridge, MA: MIT Press, 2005).

18. Andrews, *Managing the Environment*; Kraft, *Environmental Policy and Politics*, Chapter 4.
19. Michael E. Kraft, "Population Policy," in *Encyclopedia of Policy Studies*, 2nd ed., ed. Stuart S. Nagel (New York: Marcel Dekker, 1994), 617–42.
20. J. Clarence Davies III and Barbara S. Davies, *The Politics of Pollution*, 2nd ed. (Indianapolis, IN: Bobbs-Merrill, 1975).
21. Evan J. Ringquist, *Environmental Protection at the State Level: Politics and Progress in Controlling Pollution* (Armonk, NY: M. E. Sharpe, 1993), Chapter 2; Davies and Davies, *The Politics of Pollution*, Chapter 2. A much fuller history of the origins and development of modern environmental policy than is provided here can be found in Andrews, *Managing the Environment* and in Michael J. Lacey, ed., *Government and Environmental Politics: Essays on Historical Developments since World War Two* (Baltimore, MD: Johns Hopkins University Press, 1989).
22. Samuel P. Hays and Barbara D. Hays, *Beauty, Health, and Permanence: Environmental Politics in the United States, 1955–1985* (Cambridge: Cambridge University Press, 1987). See also Dunlap, "Public Opinion and Environmental Policy," and Robert Cameron Mitchell, "Public Opinion and Environmental Politics in the 1970s and 1980s," in *Environmental Policy in the 1980s*, ed. Norman J. Vig and Michael E. Kraft (Washington, DC: CQ Press, 1984), 51–74.
23. National Environmental Policy Act of 1969, Pub. L. No. 91–90 (42 USC 4321–4347), Sec. 101. See also Lynton Keith Caldwell, *The National Environmental Policy Act: An Agenda for the Future* (Bloomington: Indiana University Press, 1998).
24. Michael E. Kraft, "Congress and Environmental Policy," in *The Oxford Handbook of U.S. Environmental Policy*, ed. Kamieniecki and Kraft, 280–305; Amy Below, "Parties, Campaigns, and Elections," in *The Oxford Handbook of U.S. Environmental Policy*, ed. Kamieniecki and Kraft, 525–51; and Charles R. Shipan and William R. Lowry, "Environmental Policy and Party Divergence in Congress," *Political Research Quarterly* 54 (June 2001): 245–63.
25. Vig and Kraft, *Environmental Policy in the 1980s*.
26. See Riley E. Dunlap, "Public Opinion on the Environment in the Reagan Era," *Environment* 29 (July–August 1987): 6–11, 32–37; and Mitchell, "Public Opinion and Environmental Politics."
27. The changing membership numbers can be found in Kraft, *Environmental Policy and Politics*, Chapter 4. See also Christopher J. Bosso, *Environment, Inc.: From Grassroots to Beltway* (Lawrence: University Press of Kansas, 2005).
28. President's Council on Sustainable Development, *Sustainable America*.
29. Katharine Q. Seelye, "Bush Picks Industry Insiders to Fill Environmental Posts," *New York Times*, May 12, 2001.
30. See Natural Resources Defense Council, *Rewriting the Rules: The Bush Administration's First-Term Environmental Record* (New York: NRDC, 2005); Bruce Barcott, "Changing All the Rules," *New York Times Magazine*, April 4, 2004; and Margaret Kriz, "Vanishing Act," *National Journal*, April 12, 2008, 18–23.
31. Eric Pianin, "War Is Hell: The Environmental Agenda Takes a Back Seat to Fighting Terrorism," *Washington Post National Weekly Edition*, October 29–November 4, 2001, 12–13. See also Barcott, "Changing All the Rules," and Joel Brinkley, "Out of the Spotlight, Bush Overhauls U.S. Regulations," *New York Times*, August 14, 2004.

32. See Denise Lu and Armand Emamdjomeh, "Local Programs Get the Biggest Hit in Proposed EPA Budget," *Washington Post*, April 11, 2017; and Milman, "Trump Budget Would Gut EPA Programs Tackling Climate Change and Pollution." Current and historical EPA budgets and staff numbers can be found at https://www.epa.gov/planandbudget/budget.

33. On the staff numbers, see EPA, *FY 2015: EPA Budget in Brief*, Publication No. EPA-190-S-14-001 (Washington, DC: EPA, Office of the Chief Financial Officer, March 2014), available at http://www2.epa.gov/sites/production/files/2014-03/documents/fy15_bib.pdf. The overall EPA budget numbers come from the Obama administration's fiscal year 2017 budget. Annual budget numbers in EPA's *FY 2017* document differ somewhat from the budget authority figures in the historical tables of the fiscal year 2017 budget. For consistency, we use the latter for comparisons over time.

34. See James K. Conant and Peter J. Balint, *The Life Cycles of the Council on Environmental Quality and the Environmental Protection Agency: 1970–2035* (New York: Oxford University Press, 2016). The book reports in detail on EPA and Council on Environmental Quality personnel and budgetary support over time.

35. Robert V. Bartlett, "Evaluating Environmental Policy," in *Environmental Policy in the 1990s*, 2nd ed., eds. Norman J. Vig and Michael E. Kraft (Washington, DC: CQ Press, 1994), 167–87; Evan J. Ringquist, "Evaluating Environmental Policy Outcomes," in *Environmental Politics and Policy*, ed. James P. Lester (Durham, NC: Duke University Press, 1995), 303–27; and Gerrit J. Knaap and Tschangho John Kim, eds., *Environmental Program Evaluation: A Primer* (Champaign: University of Illinois Press, 1998).

36. One of the most thorough evaluations of environmental protection policies of this kind can be found in J. Clarence Davies and Jan Mazurek, *Pollution Control in the United States: Evaluating the System* (Washington, DC: NAPA, 1995). See also Daniel Press, *American Environmental Policy: The Failures of Compliance, Abatement and Mitigation* (Northampton, MA: Edward Elgar Publishing, 2015).

37. For a fuller account, see Kraft, *Environmental Policy and Politics*, Chapter 2.

38. US Environmental Protection Agency (EPA), "Our Nation's Air," available at https://gispub.epa.gov/air/trendsreport/2019/#home, accessed May 24, 2020. Other data in this section come from the EPA's report Air Trends, which is much more comprehensive and covers differing time periods: https://www.epa.gov/air-trends/air-quality-national-summary, accessed May 24, 2020.

39. US EPA, "Air Trends Report."

40. US EPA, "Air Trends Report."

41. The data are from the *Toxics Release Inventory: National Analysis* for the most recent year available: https://www.epa.gov/trinationalanalysis/releases-chemicals, accessed March 27, 2020.

42. EPA, *2018 Toxics Release Inventory: National Analysis*, https://www.epa.gov/trinationalanalysis/releases-chemicals, accessed March 27, 2020. The volume of releases refers only to TRI facilities that reported to the EPA that year. Facilities falling below a threshold level are not required to report, nor are many smaller facilities. To view TRI data for anywhere in the United States via an interactive map, see https://www.epa.gov/trinationalanalysis/releases-chemicals.

43. See, for example, Coral Davenport, "U.S. to Announce Rollback of Auto Pollution Rules, a Key Effort to Fight Climate Change," *New York Times*, March 30, 2020.

44. See EPA, *Inventory of U.S. Greenhouse Gas Emissions and Sinks: 1990–2018* (Washington, DC: EPA, 2020), available at https://www.epa.gov/ghgemissions/inventory-us-greenhouse-gas-emissions-and-sinks, accessed May 8, 2020.

45. See Brad Plumer, "Carbon Dioxide Emissions Hit a Record in 2019, Even as Coal Fades," *New York Times*, December 4, 2019; and Chris Mooney, Brady Dennis, and John Muyskens, "Global Emissions Plunged an Unprecedented 17 Percent During the Coronavirus Pandemic," *Washington Post*, May 19, 2020.

46. US Department of Commerce, National Oceanic and Atmospheric Administration, *The NOAA Annual Greenhouse Gas Index* (Boulder, CO: NOAA Earth System Research Laboratory), available at https://esrl.noaa.gov/gmd/aggi/aggi.html, accessed March 31, 2020. The IEA report can be found at https://www.iea.org/articles/global-co2-emissions-in-2019.

47. US EPA, "National Summary of State Information," available at https://ofmpub.epa.gov/waters10/attains_nation_cy.control#total_assessed_waters. The same page allows review of reports on each of the fifty states.

48. US EPA, *National Water Quality Inventory: 2000 Report to Congress* (Washington, DC: Office of Water, EPA, 2002). After this publication, the EPA no longer included an assessment of groundwater in these reports. The US Geological Survey has an extensive program of monitoring and assessing groundwater. See its website (https://www.usgs.gov). See also the EPA's page on groundwater and drinking water: http://water.epa.gov/drink/.

49. A major study by the EPA found that "hydraulic fracturing activities can impact drinking water resources under some circumstances," but also that data gaps and uncertainties limit its ability to characterize the severity of the impacts. See US EPA, *Hydraulic Fracturing for Oil and Gas: Impacts from the Hydraulic Fracturing Water Cycle on Drinking Water Resources in the United States: Final Report* EPA/600/R-16/236F (Washington, DC: Office of Research and Development, 2016), available at https://www.epa.gov/hfstudy.

50. See Government Accountability Office, *Numbers of Contaminated Federal Sites, Estimated Costs, and EPA's Oversight Role*, GAO-15-830T (Washington, DC: GAO, 2015) and Government Accountability Office, *U.S. Government's Environmental Liability* (Washington, DC: GAO, 2016). The latter report was published as part of the GAO's 2017 High-Risk Series: Government Accountability Office, *Progress on Many High-Risk Areas, While Substantial Efforts Needed on Others*, GAO-17-317 (Washington, DC: GAO, February 2017), 232–47.

51. The summary data can be found at https://www.epa.gov/superfund/npl-site-totals-status-and-milestone, accessed on March 31, 2020.

52. On the cleanup of rivers, see Anthony DePalma, "Superfund Cleanup Stirs Troubled Waters," *New York Times*, August 13, 2012. For the quotation, see US Environmental Protection Agency, *The Office of Solid Waste and Emergency Response: Fiscal 2010 End of the Year Report* (Washington, DC: USEPA, 2010), 12, https://archive.epa.gov/region3/ebytes/web/pdf/oswer_eoy_2010.pdf.

53. The pertinent documents can be found at the EPA's website for pesticide programs: https://www.epa.gov/pesticides/.

54. See Coral Davenport and Emmarie Huetteman, "Lawmakers Reach Deal to Expand Regulation of Toxic Chemicals," *New York Times*, May 19, 2016. For details about the new act, see the EPA webpage on it: https://www.epa.gov/assessing-and-managing-chemicals-under-tsca/frank-r-lautenberg-chemical-safety-21st-century-act.

55. Hallett J. Harris and Denise Scheberle, "Ode to the Miner's Canary: The Search for Environmental Indicators," in *Environmental Program Evaluation: A Primer*, eds. Gerrit J. Knaap and Tschangho John Kim (Champaign: University of Illinois Press, 1998), 176–200. See also Gretchen C. Daily, ed., *Nature's Services: Societal Dependence on Natural Ecosystems* (Washington, DC: Island Press, 1997); and Water Science and Technology Board, *Valuing Ecosystem Services: Toward Better Environmental Decision-Making* (Washington, DC: National Academies Press, 2004).

56. The numbers come from the various agency websites.

57. The Fish and Wildlife Service website provides extensive data on threatened and endangered species and habitat recovery plans through its ECOS system at https://ecos.fws.gov/ecp/. Look for the "listed species summary" tab. The figures on improving and declining species come from the US Fish and Wildlife Service, *Report to Congress on the Recovery of Threatened and Endangered Species: Fiscal Years 2009–2010* (Washington, DC: Fish and Wildlife Service, January 2012), available at https://www.fws.gov/endangered/esa-library/pdf/Recovery_Report_2010.pdf.

58. Population Reference Bureau, "2020 World Population Data Sheet," available at https://www.prb.org.

59. See Mazmanian and Kraft, *Toward Sustainable Communities*; Fiorino, *The New Environmental Regulation*; Eisner, *Governing the Environment*; and Kraft, Stephan, and Abel, *Coming Clean*. A number of the chapters in Kamieniecki and Kraft, eds., *The Oxford Handbook of U.S. Environmental Policy* also analyze the promise of new policy approaches, as do the chapters in Durant, Fiorino, and O'Leary, *Environmental Governance Reconsidered*.

60. The quotation is from John M. Broder and Andrew C. Revkin, "Hard Task for New Team on Energy and Climate," *New York Times*, December 16, 2008. See also David A. Fahrenthold, "Ready for Challenges: Obama's Environmental Team: No Radicals," *Washington Post National Weekly Edition*, December 22, 2008–January 4, 2009, 34.

2

RACING TO THE TOP, THE BOTTOM, OR THE MIDDLE OF THE PACK?

The Evolving State Government Role in Environmental Protection

Barry G. Rabe

> *The problem which all federalized nations have to solve is how to secure an efficient central government and preserve national unity, while allowing free scope for the diversities, and free play to the... members of the federation. It is... to keep the centrifugal and centripetal forces in equilibrium, so that neither the planet States shall fly off into space, nor the sun of the Central government draw them into its consuming fires.*

> Lord James Bryce, *The American Commonwealth*, 1888

Before the 1970s, the conventional wisdom on federalism viewed "the planet States" as sufficiently lethargic to require a powerful "Central government" in many areas of environmental policy. States were widely derided as mired in corruption, hostile to innovation, and unable to take a serious role in environmental policy out of fear of alienating key economic constituencies. If anything, they were seen as "racing to the bottom" among their neighbors, attempting to impose as few regulatory burdens as possible. In more recent times, the tables have turned—so much so that current conventional wisdom often berates an overheated federal government that squelches state creativity and capability to tailor environmental policies to local realities. The decentralization mantra of recent decades has endorsed an extended transfer of environmental policy resources and regulatory authority from Washington, DC, to states and localities.

Governors turned presidents, such as Ronald Reagan, Bill Clinton, and George W. Bush, extolled the wisdom of such a strategy, at least in their rhetoric. Many heads of the US Environmental Protection Agency (EPA), including Gina McCarthy in the Obama administration and Scott Pruitt under Donald Trump, took federal office after extended state government experience. They frequently endorsed the idea of shifting some authority back to statehouses, while differing on just what that meant in practice. Pruitt repeatedly invoked the phrase "cooperative federalism" during his 2017 confirmation hearings in embracing a state-centered model of federalism. His 2019 successor, Andrew Wheeler, employed similar rhetoric in advancing many Trump initiatives designed to soften federal authority over air, water, and climate in favor of expanded state discretion. Of course, such a transfer would pose a potentially formidable test of the thesis that more localized units know best and has faced major political hurdles.

What accounts for this sea change in our understanding of the role of states in environmental policy? How have states evolved in recent decades, and what types of functions do they assume most comfortably and effectively? Despite state resurgence, are there areas in which states fall short? How did states respond to efforts by the Trump administration to reduce federal engagement and shift many environmental protection responsibilities to them? Looking ahead, should regulatory authority devolve to the states, or are there better ways to sort out federal and state responsibilities?

This chapter addresses these questions, examining evidence of state performance in environmental policy. It provides both an overview of state evolution and a set of brief case studies that explore state strengths and limitations. These state-specific accounts are interwoven with assessments of the federal government's role, for good or ill, in the development of state environmental policy. Indeed, as political scientist John Kincaid has noted in a sophisticated analysis of power shifting between federal and state governments since the founding of the Republic, federal environmental policy powers have expanded markedly since 1970 and yet the American system remains quite balanced between federal and state authority.[1] In order to be truly effective, US environmental policy often needs to reflect constructive engagement across state and federal levels rather than acrimony.

THE STATES AS THE "NEW HEROES" OF AMERICAN FEDERALISM

Policy analysts are generally most adept at analyzing institutional foibles and policy failures. Indeed, much of the literature on environmental policy follows this pattern, with criticism particularly voluminous and potent when directed toward federal efforts in this area. By contrast, states have received much more favorable treatment. Many influential books and reports on state government and federalism portray states as dynamic and effective. Environmental policy is often depicted as a prime example of this general pattern of state performance. Some analysts routinely characterize states as the "new heroes" of American federalism, as having long since eclipsed a doddering federal government. According to this line of argument, states are consistently at the cutting edge of policy innovation, eager to find creative solutions to environmental problems, and "racing to the top" with a goal of national preeminence in the field. When the states fall short, an overzealous federal partner is often said to be at fault.

Such assertions have considerable empirical support. The vast majority of state governments have undergone fundamental changes since the first Earth Day in 1970. Many states have drafted new constitutions and gained access to unprecedented revenues through expanded taxing powers. These state powers have been further refined and expanded through highly active constitutional amendment processes.[2] In turn, many state bureaucracies have grown and become more professionalized, as have staff serving governors and legislatures. Expanded policy engagement was further stimulated by increasingly competitive two-party systems in many regions between 1980 and 2010, intensifying pressure on elected officials to deliver desired services. Heightened use of direct democracy provisions, such as the initiative and referendum, and increasing activism by state courts and coalitions of elected state attorneys general create alternative routes for policy adoption. On the whole in recent decades, public opinion data have consistently found that citizens have a considerably higher degree of "trust and confidence" in the public services and regulations dispensed from their state capitals than those generated from Washington. This pattern was evident during 2020 when governors received far higher marks than President Trump in their ability to address the profound challenges of the COVID-19 crisis. These factors have converged to expand state capacity and commitment to environmental protection.

This transformed state role is evident in virtually every area of environmental policy. States directly regulate approximately 20 percent of the total US economy, including many areas in which environmental concerns come into play.[3] States operate more than 90 percent of all federal environmental programs that can be delegated to them. Collectively, they approach that high level of engagement in the issuance of all environmental permits and the implementation of all environmental enforcement actions. Despite this expanded role, federal financial support to states in the form of grants to fund environmental protection efforts has generally declined since the early 1980s and plunged further during the Trump presidency. This increasingly forces states to find ways to fund most of their operations even as many face significant fiscal strains and face constitutional limits on their ability to run deficits.

Many areas of environmental policy remain clearly dominated by states, including most aspects of waste management, groundwater protection, land use management, transportation, energy production, and electricity regulation. This state-centric role is also reflected in rapidly emerging areas, such as environmental risks to air, water, and land linked to dramatic expansion in the exploration of shale gas and oil via hydraulic fracturing ("fracking") techniques. In many instances, state action represents "compensatory federalism," whereby Washington proves "hesitant, uncertain, distracted, and in disagreement about what to do," and states respond with a "step into the breach."[4] Even in policy areas with an established federal imprint, such as air and water quality, states often have considerable latitude to oversee implementation and move beyond federal standards if they so choose. In air quality alone, more than a dozen states routinely adopt policies to either exceed federal standards or fill federal regulatory gaps, often setting models for national consideration. Political scientists Christopher McGrory Klyza and David Sousa confirm that "the greater flexibility of state government can yield policy innovation, opening the way to the next generation of environmental policy."[5]

That flexibility and commitment are further reflected in the institutional arrangements established by states to address environmental problems. Many states maintain comprehensive agencies that gather most environmental responsibilities under a single organizational umbrella. These agencies have sweeping, cross-programmatic responsibilities and some take the lead on emerging issues such as climate change. In turn, many states have continued to experiment with new organizational arrangements to meet evolving challenges, including the use of informal networks, special task forces, and interstate compacts to facilitate cooperation among various departments and agencies.[6]

This expanded state commitment to environmental policy may be accelerated, not only by the broader factors introduced above but also by features somewhat unique to this policy area. First, many scholars contend that broad public support for environmental protection provides considerable impetus for more decentralized policy development tailored to salient local concerns. Such "civic environmentalism" stimulates numerous state and local stakeholders to take creative collective action independent of federal intervention. As opposed to top-down controls, game-theoretic analyses of efforts to protect so-called common-pool resources, such as river basins and forests, side decisively with local or regional approaches to resource protection. Much of the leading scholarly work of the late Elinor Ostrom, who in 2009 became the first political scientist to win the Nobel Prize in economics, actively embraced "bottom-up" or "polyarchic" environmental governance, including possible climate change applications.[7]

Second, the proliferation of environmental policy professionals in state agencies and legislative staff roles has created a sizable base of talent and ideas for state-level policy innovation. Contrary to conventional depictions of agency officials as shackled by elected "principals," an alternative view finds considerable policy innovation or "entrepreneurship" in state policymaking circles. This pattern is especially evident in environmental policy because numerous areas of specialization place a premium on expert ideas and allow for considerable innovation within agencies.[8] Recent scholarly work on state environmental agency performance gives generally high marks to officials for professionalism, constructive problem-solving, and increasing emphasis on improving environmental outcomes, albeit with considerable state-to-state variation.[9] Networks of state professionals, working in similar capacities but across jurisdictional boundaries, have become increasingly influential in recent decades. These networks facilitate information exchange, foster the diffusion of innovation, and pool resources to pursue joint initiatives. Such multistate groups as the Environmental Council of the States, the National Association of Clean Air Agencies, and the National Association of State Energy Officials also band together to influence the design of subsequent federal policies, seeking either latitude for expanded state experimentation or federal emulation of state "best practices." Other entities, such as the Northeast States for Coordinated Air Use Management, the Great Lakes Commission, and the Pacific Coast Collaborative, represent state interests in certain regions.

Third, environmental policy in many states is stimulated by direct democracy, unlike the federal level, through initiatives, referendums, and the recall of elected officials. In every state except Delaware, state constitutional amendments must be approved by voters via referendum. Thirty-one states and Washington, DC, also have some form of direct democracy for approving legislation, representing well over half the US population. Use of this policy tool has grown at an exponential rate to consider a wide array of state environmental policy options, including nuclear plant closure, disclosure of commercial product toxicity, and public land acquisition. In 2016 and 2018, Washington voters decisively rejected proposals to establish the first tax in the United States on carbon dioxide emissions, a policy already in place in parts of Canada and Europe, prompting Governor Jay Inslee to pursue other paths to climate policy. In prior years, however, Washington voters approved a ballot proposition requiring a steady increase in the amount of electricity derived from renewable sources, as was the case in Colorado and Missouri. Western states have generally made the greatest use of these provisions on environmental issues, particularly Oregon, California, and Colorado. In 2020, voters considered a wide range of environmental ballot propositions, including far-reaching expansion in Missouri and Nevada of earlier renewable energy policies, an Oregon proposal to heighten water protection from logging and pesticides, and a Colorado proposal to reintroduce the gray wolf to wildlife areas.

THE CUTTING EDGE OF POLICY: CASES OF STATE INNOVATION

The convergence of these various political forces has unleashed substantial new environmental policy at the state level. Various researchers have attempted to analyze some of this activity through ranking schemes that determine which states are most active and innovative, often tracking how policy ideas then diffuse across states. Such studies consistently conclude that certain states tend to take the lead in most areas of policy innovation, followed by an often uneven pattern of innovation diffusion across state and regional boundaries.[10] For example, the American Council for an Energy-Efficient Economy produces annual rankings of states on the basis of their adoption rates for a range of policies that offer environmental protection through more efficient energy use. In 2019, its researchers found that Massachusetts retained the top ranking for the ninth straight year, followed by California and other states located primarily on the East and West coasts. Maryland registered the biggest advances of any state from 2018, while Hawaii also made major strides. Minnesota maintained the highest ranking among Midwestern states, Colorado led Mountain West states, and Florida received top rating in the Southeast. Kansas, Louisiana, and North and South Dakota ranked at the very bottom.[11]

Additional analyses have attempted to examine which economic and political factors are most likely to influence the rigor of state policy or the level of resources devoted to it.[12] An important but less examined question concerns the relationships between environmental policy and both environmental quality and economic growth. Policy scholars Daniel Fiorino and Riordan Frost have created an "eco-efficiency index" that looks across multiple areas of environmental protection over time, ranking states according to the "stress on health and ecology required to generate a given unit of income."[13] These rankings generally parallel earlier studies that track rates of policy innovation and adoption, with higher scores reflecting greater eco-efficiency (see Table 2.1). In turn, a more established body of research suggests that a number of state innovations offer promising alternatives to prevailing approaches, often representing a direct response to local environmental crises and revealing shortcomings in existing policy design. Brief case studies that follow indicate the breadth and potential effectiveness of state innovation.

Anticipating Environmental Challenges

One of the greatest challenges facing US environmental policy is the need to shift from a pollution control mode that reacts after damage has occurred to one that anticipates potential problems and attempts to prevent or minimize them. Some states have launched serious planning processes in recent decades, attempting to pursue preventative strategies in an increasingly systematic and effective way. All fifty states have

TABLE 2.1 ■ State Air, Climate, and Energy (ACE) Index

State	Weighted Overall Score	State	Weighted Overall Score
New Jersey	47.04	Georgia	9.60
California	41.36	Minnesota	9.34
Connecticut	36.17	Utah	9.16
New York	36.03	Texas	8.84
Rhode Island	30.49	Arizona	8.54
Massachusetts	28.81	South Carolina	8.31
Delaware	26.57	Indiana	7.00
Maryland	26.18	Iowa	6.70
Washington	17.86	Idaho	6.63
Virginia	17.36	Missouri	6.56
New Hampshire	16.91	Nebraska	6.44
Hawaii	16.71	Kentucky	6.15
North Carolina	15.15	South Dakota	5.61
Pennsylvania	14.56	Alabama	5.03
Illinois	13.55	Kansas	4.93
Florida	13.10	West Virginia	4.82
Vermont	12.99	Oklahoma	4.60
Nevada	12.58	Louisiana	4.58
Oregon	11.38	Arkansas	4.53
Colorado	11.02	New Mexico	4.31
Tennessee	11.00	Mississippi	4.17
Ohio	10.28	Alaska	4.00
Michigan	10.07	Montana	3.38
Wisconsin	9.99	North Dakota	3.35
Maine	9.87	Wyoming	2.20

Source: Daniel Fiorino and Riordan Frost, "The Pilot Eco-Efficiency Index: A New State Environmental Ranking for Researchers and Government" (paper presented at the Association for Public Policy Analysis and Management's fall research conference, Washington, DC, November 4, 2016), www.researchgate. net/publication/315736703_The_Pilot_Eco-Efficiency_Index_A_New_State_Environmental_Ranking_for_ Researchers_and_Government.

adopted at least one pollution prevention program, and some have taken particularly bold approaches, cutting across conventional programmatic boundaries with various mandates and incentives to pursue prevention opportunities. Thirty-four states have adopted laws that move beyond federal standards in preventing risks from chemical exposure, such as bans of specific chemicals thought to pose health risks or comprehensive chemical management systems.[14] California has been particularly active in this area and heavily influenced the design of new federal legislation adopted in 2016.

Minnesota has long joined California as a national leader in this area. It requires hundreds of state firms to submit annual toxic pollution prevention plans and prioritize "chemicals of concern."[15] These plans must outline each firm's current use and release of a long list of toxic pollutants and establish formal goals for their reduction or elimination over specified periods of time. Firms have considerable latitude in determining how to attain these goals, contrary to the technology-forcing character of much federal regulation. But they must meet state-established reduction timetables and pay fees on releases. Minnesota was also one of the first two states to ban bisphenol A, a controversial chemical used in plastics. It moved quickly to address polyfluoroalkyl (PFAS) chemicals in water supplies, supported by an $850 million 2018 legal settlement with a major manufacturer of these chemicals. From these earlier efforts, Minnesota and other states have established multidisciplinary teams that attempt to forecast emerging environmental threats and respond before problems arise, including review of potential environmental risks from nanotechnology and its production of tiny particles that may improve product design but also harbor environmental risks.[16] Minnesota has also taken a pioneering role in measuring the environmental impacts of carbon dioxide emissions and attaching a price to them in statewide electricity planning.[17] In 2019, Governor Tim Walz elevated the role of climate change in long-term policy development, creating a climate "sub-cabinet" involving every Minnesota department and establishing benchmarks for future mitigation and adaptation.[18]

Colorado has taken a "race-to-the-top" approach to policy designed to anticipate and thereby minimize environmental risks from hydraulic fracturing practices. The state has a long-standing history in oil and gas extraction and has sought in recent years to temper all-out energy production by addressing environmental challenges in the fracking era. This featured pioneering steps in requiring public disclosure of chemicals used in drilling operations, water quality sampling, air quality standards, and property owner protections.[19] It emerged through a deliberative process orchestrated by former Governor John Hickenlooper to engage diverse stakeholders to take proactive steps to mitigate risks.[20] In 2019–2020, Colorado adopted a suite of additional laws that went even farther, imposing unusually strong regulatory standards on methane and other hydrocarbons and nitrogen oxides from energy production and transmission and requiring continuous emissions monitoring with cutting-edge technology. The state also took steps to give localities considerable authority to add their own regulatory oversight while transitioning for a less carbon-intensive future through sweeping new energy efficiency and renewable energy policies. Another major energy-producing state, New Mexico, began moving in similar directions in 2020, developing performance-based regulatory standards and penalties that offered firms incentives for emission release reductions that could be verified while intensifying oversight on laggard firms.[21]

Economic Incentives

Economists have long lamented the penchant for command-and-control rules and regulations in US environmental policy. Most would prefer to see a more economically sensitive set of policies, such as taxes on emissions to capture social costs or "negative externalities" and provide monetary incentives for good environmental performance.[22] The politics of imposing such costs has proven contentious at all governmental levels, although a growing number of states have begun to pursue some form of this approach in recent years. In all, the states have enacted hundreds of measures that can be characterized as "green taxes," including environmentally-related "surcharges" and "fees" that avoid the explicit use of the label "tax" but are functional equivalents.[23] Revenues from such programs are often used to cover costs of popular programs such as recycling, land conservation, and energy efficiency. A growing number of states have begun to revisit their general tax policies with an eye toward environmental purposes, including major tax incentives in many states to purchase hybrid and electric vehicles or invest in renewable energy. Many states and localities have also developed taxes on solid waste, often involving a direct fee for garbage pickup while offering free collection of recyclables.

One of the earliest and most visible economic incentive programs involves refundable taxes on beverage containers.[24] Ten states—covering one-third of the population—operate such programs. Deposit collections flow through a system that includes consumers, container redemption facilities such as grocery stores, and firms that reuse or recycle containers. Michigan's program is widely regarded as among the most successful of these state efforts and, similar to a number of others, is a product of direct democracy. Michigan's program places a dime deposit on containers—double the more conventional nickel—which may contribute to its unusually high redemption rate above 95 percent through the 2010s. The failure to adjust this policy for inflation or expand it to other beverages, however, served to erode its effectiveness over time. This type of state policy has diffused to other products, including scrap tires, used motor oil, pesticide containers, appliances with ozone-depleting substances, electronic waste such as used computers, and plastic bags.

States also have constitutional authority to tax all forms of energy, including transportation fuel and electricity. Increasing the price of energy in concert with its environmental damage would likely discourage consumption and related environmental damage, just as sustained tax increases have elevated the costs of smoking and driven down rates of tobacco use in recent decades. Many states have been highly reluctant to move beyond their traditional levels of taxation for fuels such as gasoline that are commonly used to maintain highways and bridges. But ten states have worked over the past decade to place a price on the release of carbon emissions through an auctioning process linked to an emissions cap that declines over time. Building on pioneering American work to reduce sulfur dioxide emissions, nine northeastern states have maintained the Regional Greenhouse Gas Initiative (RGGI) that requires purchase, through quarterly public auctions, of allowances to emit carbon. This pricing mechanism also provides revenue whereby RGGI states can support alternative energy projects or rebate consumer electricity bills. Political scientist Leigh Raymond has argued that RGGI offers a "new model" for climate policy that is already influencing other governments in America and internationally.[25] California operates its own version of this cap-and-trade system in collaboration with Canadian province Québec. During 2019–2020, New Jersey and Virginia formalized plans to join RGGI and significantly expand its reach. In turn, Pennsylvania Governor Thomas Wolf issued a 2019 executive order preparing the Keystone State for RGGI membership. This faced legislative opposition and potential state court scrutiny but, if advanced, would literally double RGGI's size in terms of carbon emission volume.

RGGI's successes inspired creation of a transportation sector replica. The Transportation Climate Initiative involves a partnership between twelve northeastern and Atlantic states, designed to create a regional "cap-and-invest" system. This would price oil and gas use, reallocating revenue to participating states to develop more environmentally friendly transportation systems. This region continues to be a hotbed for experimentation on environmental pricing strategies, further reflected in active exploration of carbon tax options in such states as Massachusetts and New York.

Filling the Federal Void: Reducing Greenhouse Gases

As the RGGI case demonstrates, states have proven unexpectedly active players in the fight to reduce greenhouse gas emissions to curb climate change. While most Congresses and some presidents have struggled to make any policy contribution to this problem, a number of states have attempted to fill some of the "policy gap" created by federal inaction.[26] This American "bottom-up" approach has also emerged in other federal or multilevel governmental systems, including Canada, Australia, and the European Union.[27] Many states are responsible for substantial amounts of greenhouse gas emissions, even by global standards. If all states were to secede and become independent nations, eighteen of them would rank among the top fifty nations in the world in terms of releases. In response, many states have adopted policies that promise to reduce their greenhouse gas releases, although they often tend to also pursue these policies for other environmental and economic reasons.

State-level engagement on climate policy has tended to peak during periods where federal engagement is lowest, thereby seizing opportunities that were being ignored or reversed nationally. This has certainly been evident since the advent of the Trump administration, including formation of a 25-state coalition that has pledged to meet Paris Climate Agreement emission reduction commitments within their boundaries. A 2019 study of subfederal climate policy concluded that American states could be divided into three tiers in terms of environmental commitment. At the top, 45 percent of the population and one-third of total emissions involved "first-mover" states actively engaged in policy. In contrast, 35 percent of the population and 47 percent of total emissions involved "slow-follower" states that lagged far behind all others. In the middle were "fast-follower" states seen as attempting to keep pace with first-movers, with 20 percent of the population and emissions. The study concluded that full implementation of all state policies adopted through 2018 alone could achieve approximately two-thirds of total American emission reductions that had been pledged under Paris.[28]

One common climate policy involves a clean electricity mandate designed to accelerate state transition away from fossil fuel sources. Twenty-nine states and Washington, DC, have established "renewable portfolio standards (RPS)," beginning with Iowa in 1991, and three have clean energy standards. Eight additional states have nonbinding renewable energy goals and two have comparable clean energy goals. These policies generally follow a similar structure, although they vary in terms of both the definition of eligible sources and the overall targets and timetables for expanding capacity. Seven states (California, Hawaii, Maine, New Mexico, New York, Virginia, and Washington) expanded earlier RPS commitments between 2018 and 2020, with legislation mandating that all state electricity emanate from nonfossil fuel sources between 2045 and 2050. Three others (Connecticut, New Jersey, and Wisconsin) took similar steps via gubernatorial executive orders.[29] Even without additional federal or state policies, these existing state efforts were projected to increase the share of electricity provided by renewables nationally to 26 percent by 2030. In turn, 20 states have adopted energy efficiency equivalents of an RPS, mandating an ongoing increase in overall energy efficiency that in some cases is integrated with renewable energy mandates. Seven states have adopted nonbinding versions of these policies, and numerous states have adopted more rigorous efficiency standards for new buildings or appliance purchases. These policies loom large among state pledges to honor Paris reduction targets.

Several states also developed policy to reduce climate damage related to greenhouse gases other than carbon dioxide, including accelerated transition away from hydrofluorocarbons (HFCs) in refrigerators and air conditioning systems. HFCs were devised decades ago as far friendlier to the ozone layer than previous chemicals that they replaced. But they have far greater global warming potential per molecule than carbon or even methane. Cost-effective chemical alternatives are available, and there was widespread international support even among industry for a 2016 phase-out treaty. However, the Trump administration's decision to withdraw the treaty from Senate consideration in 2017 prompted four states to adopt legislation phasing out HFCs and two more to develop regulatory provisions. These state actions prompted serious bipartisan Congressional consideration of a national version of this approach in 2019–2020, although this ultimately collapsed.

California has ranked among the world's most active governments in addressing climate change, developing cap-and-trade policies alongside renewable electricity and fuel standards and energy-efficiency provisions.[30] It has adopted a number of pioneering climate statutes in recent decades, including multiple bills designed to achieve aggressive statewide emission reductions. California attempts to attain those goals through an all-out policy assault on virtually every sector that generates greenhouse gases, including industry, electricity, transportation, agriculture, livestock, waste management, and residential activity, giving extraordinary authority to the formidable California Air Resources Board in overseeing implementation. But California's flagship climate initiative reflected repeated use of a unique waiver it holds under federal air legislation. On more than 100 occasions since 1968, California has established more rigorous tailpipe emission standards for cars and trucks than the rest of the nation, although its waiver frequently leads to other states joining a "bandwagon" that ultimately prompts adoption of a national standard reflecting California's lead. This policy has resulted in substantial statewide and national emission reductions per vehicle in past decades and took new form in 2009 when the Obama administration embraced a bold California waiver focused on carbon emissions as national policy, merging vehicle emission with fuel economy standards and setting bold targets that would reach an average of 54.5 miles per gallon in 2025. This reform reflected a unique situation whereby one state can innovate within its own boundaries but leverage national-level change in the process through power granted to it through federal legislation, although it would face an unprecedented challenge to continue use of this power in 2020 as noted below.[31] Governor Gavin Newsom pushed even farther in late 2020, issuing an executive order to ban all sales of gas-powered vehicles in California by 2035, raising added questions about constitutional authority in his state.

Taking It to the Federal Government

At the same time that states have eclipsed the federal government through new policies, they have also made increasingly aggressive use of litigation to attempt to force the federal government to take new steps or reconsider previous ones. The ability of state attorneys general from the party opposite the president to take unified and aggressive countermeasures against executive branch policy reached new heights under Barack Obama, where virtually every major climate, air, or water initiative was actively confronted, resulting in protracted litigation and some state efforts to refuse to comply with federal orders.[32] Unlike their federal counterpart, most state attorneys general are elected officials, with powers that have expanded significantly in recent decades. They do not necessarily work collaboratively with the sitting governor and often use their powers as a base from which to secure broader visibility and seek higher office, most commonly governorships.[33]

Collectively, these officials have increasingly become a force to be reckoned with, particularly as they expand their engagement through challenges brought into the federal courts. Whereas President Obama was routinely challenged by Republican attorneys general, President Trump was peppered with comparable responses to his environmental deregulation efforts from Democratic attorneys general. California's Xavier Becerra, Massachusetts' Maura Healey, and New York's Letitia James routinely led coalitions tailored to each particular issue and these expanded in 2019–2020 after more Democratic attorneys general were elected. A 2019 study identified some 300 separate actions, ranging from multistate lawsuits to comment letters, that these officials took to block Trump administration actions on issues such as climate, air and water quality, chemical accidents, and public lands and wildlife.[34] It appeared increasingly likely that this form of collective action by states led by officials of the party opposite the sitting president might become a permanent feature of political opposition to policies of a given federal executive.[35]

STATE LIMITS

Such a diverse set of policy initiatives would seem to augur well for the states' involvement in environmental policy. Any such enthusiasm must be tempered, however, by a continuing concern over how evenly that innovative vigor extends over the entire nation. One enduring rationale for giving the federal government so much environmental policy authority is that states appear to face inherent limitations. Rather than a consistent, across-the-board pattern of dynamism, we see a more uneven pattern of performance than decentralization advocates might anticipate. Just as some states consistently strive for national leadership, others appear to seek the middle or bottom of the pack, seemingly doing as little as possible and rarely taking innovative steps. This imbalance becomes particularly evident when environmental problems are not confined to a specific state's boundaries. Many environmental issues are, by definition, transboundary, raising important questions of interstate and interregional equity in allocating responsibility for environmental protection. These doubts about state capacity and commitment raise important concerns for any effort to shift more responsibility for environmental protection from federal to state governments, as was a central Trump administration emphasis.

Uneven State Performance

Many efforts to rank states according to their environmental regulatory rigor, institutional capacity, or general innovativeness find the same subset of states at the top of the list year after year. By contrast, a significant number of states consistently tend to fall much further down the list, somewhat consistent with their placement in Table 2.1 and characterization above as "slow followers," raising questions as to their overall policy capacity and commitment. As political scientist William R. Lowry notes, "Not all states are responding appropriately to policy needs within their borders.... If matching between need and response were always high and weak programs existed only where pollution was low, this would not be a problem. However, this is not the case."[36] A 2018 study on state policy adoption across multiple policy areas confirmed wide environmental policy disparities among states, concluding that "the most conservative states on the environment simply do not pass the major environmental laws that the 'green' states do."[37]

Given all the hoopla surrounding the newfound dynamism of states racing to the top in environmental policy, there has been remarkably little analysis of the performance of states that not only fail to crack top-ten rankings but may view racing to the bottom as an economic development strategy. Such a downward race may be particularly attractive during recessions, as was reflected in recent efforts in states such as North Carolina, Ohio, West Virginia, and Wisconsin to weaken dramatically the implementation of existing policies, efforts that had the express goal of promoting economic growth by creating a policy environment friendlier to industry.[38] What we know more generally about state policy commitment should surely give pause over any claims that state dynamism is truly national in scope. Despite considerable economic growth in formerly poor regions, such as the Southeast, substantial variation endures among state governments in their rates of public expenditure, including their total and per capita expenditures on environmental protection.[39] Such disparities are consistent with studies of state political culture and social capital, which indicate vast differences in probable state receptivity to governmental efforts to foster environmental improvement. These divides increasingly reflect deep partisan cleavage, unlike early periods between the 1990s and 2000s where it was more common to see Republicans and Democrats join forces behind state environmental policy adoption as is discussed in Chapter 1.

Although many states have unveiled exciting new programs, nearly half have established some formal restrictions that preclude their environmental agencies from adopting any regulations or standards that are more stringent than those of the federal government in such areas as air and water quality.[40] EPA Office of Inspector General reports and other external reviews generate serious questions about how effectively states handle core functions either delegated to them under federal programs or left exclusively to their oversight. Studies of water quality program implementation have found that states use highly variable water quality standards in areas such as sewage contamination, groundwater protection, nonpoint water pollution from diffuse sources, wetland preservation, fish advisories, and beach closures. Inconsistencies abound in reporting accuracy, suggesting that national assessments of water quality trends that rely on data from state reports may be highly suspect.[41] More than half of the states lack comprehensive water management and drought response plans, and several with such plans have not revised them in many years.[42] Even in many high-saliency cases, such as Everglades protection, states have sought a federal rescue rather than taking serious unilateral action that can be sustained over time.[43] Agricultural interests, particularly those promoting regional sugar production, have proven formidable opponents of major restoration, which would restrict their access to massive water volumes.[44]

Similar issues have arisen as states have struggled in recent years to formulate policies to reduce environmental risks linked to shale gas and oil development, with many racing in the opposite direction from Colorado's. Many have proven particularly lenient with methane releases, even though these represent the permanent loss of a nonrenewable energy source and pose significant air quality and climate concerns. Many states have long recognized either direct venting of methane or flaring into carbon dioxide as wasteful and dangerous practices, yet offer generous exemptions or exceptions to established regulations. Venting and flaring triggered particular concern in the booming Bakken region (North Dakota and Montana) and Permian Basin (Texas and New Mexico) as methane releases soared amid rapid expansion of production. State officials routinely acknowledge the problem as well as their chronic failure to prepare for long-term remediation of hundreds of thousands of idle or "orphan" wells after production ends. However, they remained highly reluctant to impose methane regulations, bonding requirements, or taxes on prominent energy producing firms, fearful that they might shift operations to other states with softer standards in response or accelerate their pursuit of bankruptcy protection during economic downswings.[45]

Comparable problems have emerged in state enforcement of air quality and waste management programs, including basic data collection and reporting. Despite efforts in some states to integrate and streamline permitting, many have extensive backlogs and lack reliable measures of facility compliance with various regulatory standards. Existing indicators confirm enormous variation among states, although we likely know less about such variation than in the 1990s, given that the EPA has lost funding and staff to maintain state-by-state data in many areas of environmental policy. State governments—alongside their local counterparts—have understandably claimed much of the credit for increasing solid waste recycling rates from a national average of 6.6 percent in 1970 to 16 percent in 1990 to 35.2 percent in 2017. At the same time, state recycling rates and policy design vary markedly, with some states formally restricting what local governments can do.

There was also growing indication in some states during the previous decade that environmental policy faced major challenges in cases where state leaders assumed that government could be managed similarly to business and industry. Michigan shifted to total Republican control of the executive and legislative branches in 2010, and former Republican Governor Rick Snyder won high marks nationally for his role in addressing fiscal concerns and assisting cities such as Detroit navigate bankruptcy en route to economic recovery.[46] But the use of state-appointed emergency financial managers to oversee fiscally-challenged municipalities backfired with tragic consequences in the case of Flint, a declining city that had once been an auto manufacturing hub.[47] The search for fiscal balance led to a 2015 decision to shift the source of Flint's water supply to save money and resulted in significant lead exposure for a city of nearly 100,000 residents. A set of state environmental and public health agencies ignored early warning signs and failed to respond to the emerging crisis, as did regional EPA authorities based in Chicago. This resulted in substantial lead contamination for Flint residents and has necessitated massive efforts to provide alternative water supplies and begin to replace damaged water infrastructure that continued more than 5 years after the initial incident. Research on water quality trends indicates that Flint is not alone among American localities in this regard, raising questions over state and local stewardship of drinking water quality in Michigan and nationally as well as long-term challenges that may require massive new investments in modern water infrastructure that few states have shown commitment to supporting.[48]

Enduring Federal Dependency

Many states have proven reluctant or unable to tap into their own revenues to support environmental protection efforts, thereby developing a deep dependency on federal grant funding to cover core programs or launch new initiatives. There are enormous differences between states in terms of their tax base and both capacity and willingness to produce significant revenues, compounded by a focus in some to only pursue policy when most of the operational costs are covered through inter-governmental financial transfers. Indeed, considerable innovative state-level activity has been at least partially underwritten through federal grants, which can serve to stimulate additional state environmental spending.[49] Although a number of states have developed fee systems to cover much of their operational costs, many continue to rely heavily on federal grants to fund some core environmental protection activities. States have continued to receive other important types of federal support, including grants and technical assistance to complete air and water quality management, wetlands program development, drinking water infrastructure, brownfields reclamation, and more. On the whole, states have annually received between one-fifth and one-third of their total environmental and natural resource program funding from federal grants in recent years, although a few states have remained more heavily reliant on federal dollars. The overall level of federal support has declined in recent decades and reduction accelerated during the Trump administration.

State dependence on federal funding has grown in the majority of states given widespread reluctance to expand agency funding and staff, even during periods of relative fiscal well-being for many states during the latter 2010s. A 2019 Environmental

Integrity Project report on state commitment to environmental programs found that "a majority of states have cut their pollution control spending and staffing over the last decade—often more drastically than EPA—even at times when overall state budgets have grown and environmental challenges have increased."[50] Some states pursued particularly far-reaching reductions. The Texas Commission on Environmental Quality, for example, cut its budget by 35 percent between 2008 and 2018, even though overall state spending increased by 41 percent. Other states with particularly deep cuts included Indiana, North Carolina, Pennsylvania, and Wisconsin. In contrast, 20 states reduced their dependency on federal dollars through spending and staff increases during this period.

Furthermore, for all the opprobrium heaped on the federal government in environmental policy, it has provided states with at least three other forms of valuable assistance, some of which has contributed directly to the resurgence and innovation of state environmental policy. First, federal development of the Toxics Release Inventory, modeled after programs initially attempted in Maryland and New Jersey, has emerged as an important component of many of the most promising state policy initiatives. This program has generated considerable data concerning toxic releases and provided states with a vital data source for exploring alternative regulatory approaches.[51] Many state pollution prevention programs would be unthinkable without such an annual information source. This program has also provided lessons for states to develop supplemental disclosure registries for greenhouse gases and chemical releases related to hydraulic fracturing.[52]

Second, many successful efforts to coordinate environmental protection on a multistate, regional basis have received substantial federal input and support. A series of initiatives in the Chesapeake Bay, the Great Lakes Basin, and New England have received considerable acclaim for tackling difficult issues and forging regional partnerships; federal collaboration—via grants, technical assistance, coordination, and efforts to unify regional standards—with states has proven useful in these cases.[53] One model for engagement was the Great Lakes Restoration Initiative championed by the Obama administration; intended to address pressing regional environmental challenges, it was successful in accelerating ecological recovery in several states with legacies of heavy toxic contamination, although funding to sustain operations lagged in the Trump administration. Aside from RGGI, other recent regional initiatives in the West and Midwest to reduce greenhouse gases struggled to endure in the absence of federal engagement or support.

Third, the EPA can constrain state innovativeness, but its oversight of state-level program implementation often looks more constructive when considering the role played by the agency's ten regional offices. Most state-level interaction with the EPA involves such regional offices, which employ approximately two-thirds of the total EPA workforce and regularly delegate enormous implementation authority to states.[54] Relations between state and regional officials are generally more cordial and constructive than those between state and central EPA officials, and such relations may even be, in some instances, characterized by high levels of mutual involvement and trust.[55] Surveys of state environmental officials confirm that they have a more positive relationship with regional rather than central agency staff.[56] Regional offices have played a key role in many promising state-level innovations, particularly during sympathetic presidential administrations. Their involvement may include formal advocacy on behalf of states with central headquarters, direct collaboration on meshing state initiatives with federal requirements, and special grant support or technical assistance.

The Interstate Environmental Balance of Trade

States may be structurally ill equipped to handle a large range of environmental concerns. In particular, they may be reluctant to invest significant energies to tackle problems that might literally migrate to another state or nation in the absence of intervention. The days of state agencies being captured securely in the hip pockets of major industries are probably long gone, reflecting fundamental changes in state government.[57] Nonetheless, state regulatory dynamism may be particularly likely to decline when cross-boundary transfer is likely.

The state imperative of economic development clearly contributes to this phenomenon. As states increasingly devise economic development strategies that resemble Asian and European industrial policies, a range of research has concluded they are far more deeply committed to strategies that promote investment or development than to those that involve social service provision or public health promotion.[58] A number of states routinely offer incentives of tens of thousands of dollars per new job to prospective developers and have intensified efforts to outbid neighboring states in the struggling manufacturing sector. Energy-producing states often maintain generous tax preferences and provide infrastructure to support extraction firms. Environmental protection can be eminently compatible with economic development goals, promoting overall quality of life and general environmental attractiveness that entices private investment. In many states, tourism and recreation industries have played active roles in seeking strong environmental programs designed to maintain natural assets. In some instances, states may be keen to take actions that could produce internal environmental benefits as long as these actions do not disrupt their economic growth. California and other states that have formally endorsed setting strict emissions standards from vehicles or even bans over time on purchase of gas-powered cars and trucks, for example, have very few jobs to lose in the vehicle manufacturing sector. They also see potential economic advantages if they can take a lead role nationally in developing alternative transportation technologies.

But much of what a state might undertake in environmental policy may largely benefit other states or regions, thereby reducing an individual state's incentive to take meaningful action. In fact, in many instances, states continue to pursue a "we make it, you take it" strategy. As political scientist William Gormley notes, sometimes "states can readily export their problems to other states," resulting in potentially serious environmental "balance of trade" problems.[59] In such situations, states may be inclined to export environmental contaminants to other jurisdictions while enjoying any economic benefits to be derived from the activity that generated the contamination. One careful study of state air quality enforcement found no evidence of reduced regulatory effort along state borders but a measurable decline in effort along state borders with Mexican states or Canadian provinces.[60]

Such cross-boundary transfers take many forms and may be particularly prevalent in environmental policy areas in which long-distance migration of pollutants is most likely. Air quality policy has long fit this pattern. Midwestern states, for example, have historically depended on burning massive quantities of coal to meet electricity demands. Prevailing winds invariably transfer pollutants from this activity to other regions, particularly New England, leading to serious concern about various contamination threats. Nationally, many states fail to meet federal air quality standards due to "interstate 'downwind' pollution." Despite some advances linked largely to federal air policy, air pollution remains responsible for estimated premature deaths of more than 100,000 people per year.[61] A 2020 study concluded that between 41 and 53 percent of premature mortality due to air pollution exposure resulted from a state's emissions that occurred outside its boundaries. It found that electricity sector emissions were particularly prominent but that releases from other forms of commercial and residential activity had grown in significance over recent decades. Fine particulate matter and ozone emerged as particularly large public health concerns in this research, whereas sulfur dioxide emissions have declined as a threat.[62] Cross-border transfers have also contributed to the growing problem of airborne toxics that ultimately pollute water or land in other regions, including chronic Lake Superior water quality problems linked to air contaminants generated outside the Great Lakes. Political scientist John Kincaid has noted that "externality mitigation across states" contributed to greater centralization of environmental policy in past decades.[63]

Interstate conflicts, often becoming protracted battles in the federal courts, have endured in recent decades as states allege they are recipients of such unwanted "imports." This has included prolonged political and legal combat over EPA's Cross-State Air Pollution Rule, the agency's "good neighbor" provision intended to restrict cross-border exports of nitrogen oxides and sulfur dioxide emissions from twenty-eight midwestern and southern states into the northeast. No region of the nation or environmental media appears immune from this kind of conflict. Prolonged battles between Alabama, Florida, and Georgia over access to waters from Lake Lanier and six rivers that cross their borders, for example, reached new intensity in recent years, resulting in extended mediation, litigation, and uncertainty about long-term approaches. Growing water scarcity linked to increased demand for water and extended drought in many regions continues to exacerbate these conflicts.

Perhaps nowhere is the problem of interstate transfer more evident than in the disposal of solid, hazardous, and nuclear wastes. States have generally retained enormous latitude to devise their own waste management and facility siting systems, working either independently or in concert with neighbors. Many states, including a number of those usually deemed among the most innovative and committed environmentally, continue to generate substantial quantities of waste and have struggled to establish comprehensive recycling, treatment, storage, and disposal capacity. Instead, out-of-state (and -region) export has been an increasingly common pattern, with a system that often resembles a shell game in which waste is ultimately deposited in the least resistant state or facility at any given moment. This pattern is repeated in emerging areas of waste management, such as the disposal of wastes generated by hydraulic fracturing procedures, and it is perhaps best illustrated in the migration of wastes generated in western Pennsylvania to deep-injection wells in eastern Ohio. This policy triggered considerable controversy in Ohio, especially following a significant expansion of earthquake activity near areas that accepted large amounts of out-of-state fracking wastes injected into vacated wells, ultimately forcing Pennsylvania to develop an alternative.

No area of waste management, however, is as contentious as nuclear waste disposal. In the case of so-called high-level wastes, intensely contaminated materials from nuclear power plants that require between 10,000 and 100,000 years of isolation, the federal government and the vast majority of states have supported a 35-year effort to transfer these wastes to a geological repository in Nevada. Ferocious resistance by Nevada officials and concerns among states who would host transfer shipments have continued to scuttle this approach, leaving each of the 57 commercial nuclear power plants with 95 reactors located in 29 states a de facto storage site. In the case of "low level" wastes, greater in volume but posing a less severe health threat, states have received considerable latitude from Washington for decades to develop a strategy for creating a series of regional sites, including access to funds to develop facilities. But subsequent siting efforts have been riddled with conflict, and no long-term plans have emerged.[64] One facility established for hazardous waste in western Texas has volunteered as a potential "host" for such waste, though it was not designed for nuclear materials, is thousands of miles away from the bulk of generated waste, and has triggered political opposition in the state. The Trump administration in 2017 proposed reopening active pursuit of the Nevada disposal option, triggering renewed opposition from that state, only to reverse that position in 2020 as the president prepared a reelection bid in which Nevada would be a contested state. This pivot included a 2021 budget proposal lacking any funding for continued Yucca site development.

RETHINKING ENVIRONMENTAL FEDERALISM

Federalism scholars and some political officials have explored models for the constructive sharing of authority in the American federal system, many of which attempt to build on the respective strengths of varied governmental levels and create a more functional intergovernmental partnership.[65] But it has generally proven difficult to translate these ideas into actual policy, particularly in the area of environmental policy. Perhaps the most ambitious effort to reallocate intergovernmental functions in environmental protection took place in the 1990s during the Clinton administration, under the National Environmental Performance Partnership System (NEPPS). This effort was linked to Clinton's broader attempts to "reinvent government," heralded by proponents as a way to give states substantially greater administrative flexibility over many federal environmental programs if they could demonstrate innovation and evidence of improved environmental outcomes.[66] NEPPS also offered Performance Partnership Grants that would allow participating states to concentrate resources on innovative projects that promised environmental performance improvements.

More than 40 states elected to participate in the NEPPS program, which required extensive negotiations between state and federal agency counterparts. Although a few promising examples of innovation can be noted, this initiative failed to approach its ambitious goals, and in the words of two scholarly analysts, "there have been few real gains."[67] NEPPS stemmed from an administrative action by a single president and thereby lacked the clout of legislation or resilient political support. In response, federal authorities often resisted altering established practices and failed to assume the innovative role anticipated by NEPPS proponents. In turn, states proved considerably less amenable to innovation than expected. They tended to balk at any possibility that the federal government might establish—and publicize—serious performance measures that would evaluate their effectiveness and environmental outcomes.

Ultimately, many NEPPS agreements were signed, especially in the waning years of the Clinton administration, and these generally remain in place. But the Bush administration never pursued NEPPS with enthusiasm, and neither the Obama nor Trump administrations made significant efforts to revitalize this program. It thereby remains a very modest test of the viability of accountable decentralization, whereby state autonomy is increased formally in exchange for demonstrable performance. No subsequent administration has attempted such a conceptual and far-ranging attempt to improve intergovernmental performance, as federal environmental policy has lurched back and forth across priorities of respective presidents.

Challenges to State Routines

The future role of states in environmental policy may be further shaped by three additional developments. First, the COVID-19 pandemic upended American life and politics, leaving potential long-term impacts that could influence future state capacity to pursue environmental policy. The severe economic contraction linked to the pandemic threatened tax revenue and budget solvency in all 50 states, with particularly severe threats in states that produce substantial amounts of oil and natural gas and depend on related tax dollars. Most states had begun to recover by 2020 from the Great Recession of the prior decade, although states such as Alaska, Connecticut, Illinois, and Wyoming remained in dire fiscal straits even prior to the pandemic. In turn, chronic pressures for expanded spending in certain domains, such as unfunded pensions and benefits for state employees and state health policy, only intensified during the economic downswing, further threatening restoration of state fiscal support for environmental protection efforts.[68]

Second, a sequence of elections during the 2010s reversed a long-standing pattern of divided, joint-party control of most state governments in favor of sweeping control by one party, with particularly strong gains among Republicans through 2016. However, state elections between 2017 and 2019 tilted more power back toward Democrats, including a shift in 10 governorships and three attorneys general as well as more than 300 state legislative seats. As of 2020, Republicans held "trifectas," controlling both legislative chambers and the governorship in 21 states. Of the remaining states, 15 featured exclusive Democratic control of the legislative and executive branches and 14 had divided partisan control. This gave both parties considerable strongholds and raised the possibility of continued divides among states over environmental policy given severe partisan splits on many issues.

Third, one early testing ground for potential environmental policy shifts was reflected in a flurry of new legislative proposals between 2010 and 2020 to either downsize or repeal many established state policies. These were most commonly introduced by Republican legislators, reflecting the growing partisan polarization between the parties during this period. Many such proposals focused on climate change and some reflected standardized legislative templates produced by the conservative American Legislative Exchange Council (ALEC), which offers luxurious conference venues and detailed policy advice for state legislators. One early theme in ALEC-supported repeal bills involved the reversal of state renewable portfolio standards. Kansas embraced repeal in 2015 and Ohio dramatically weakened its commitment to expanding renewables in 2018, with strong backing from electric utilities.[69]

One emerging theme in such bills in recent years was placing a tax on wind turbines, advanced by supporters of fossil fuel interests to slow the pace of wind adoption. These taxes faced intense opposition from representatives of wind-producing districts in states such as Montana, North Dakota, and Wyoming, although Oklahoma created a 0.5 cent per kilowatt-hour tax on wind and also repealed its tax credit for wind farm developers. Other types of state climate policies also failed to prove durable, most commonly when partisan control of a governorship or legislature shifted from Democrats to Republicans. While RGGI and California remained steadfast in their commitment to carbon cap-and-trade policies, 13 other states abandoned such initiatives after 2010, including Arizona, Illinois, New Mexico, Oregon, and Utah. The 2020 elections would prove particularly important for the next decade of state politics, as states generally redraw legislative district boundaries after each decennial census and dominant political parties can often manipulate control over this process to gain considerable advantage in subsequent elections.

LOOKING AHEAD

Amid the continued squabbling over the proper role of the federal government vis-à-vis the states in environmental policy, remarkably little effort has been made to sort out which functions might best be concentrated in Washington and which ones ought to be transferred to state capitals. Some former governors and federal legislators of both parties offered useful proposals during the 1990s that might allocate such responsibilities more constructively than at present. These proposals have been supplemented in later decades by thoughtful scholarly works by think tanks, political scientists, economists, and other policy analysts. Interestingly, many of these experts concur that environmental protection policy defies easy designation as warranting extreme centralization or decentralization. Instead, many observers endorse a process of selective decentralization, one leading to an appropriately balanced set of responsibilities across governmental levels.

Different presidents have attempted to advance a more functional form of environmental federalism that allowed for intergovernmental collaboration and played to the respective strengths of both federal and state partners. The NEPPS experiment in the Bill Clinton administration was one such example, as was an effort under George W. Bush to create more flexible state compliance paths for some contaminants under the Clean Air Act. In 2015, the Obama administration followed in this arena with the launch of the Clean Power Plan that established a national cap on carbon emissions from the electricity sector but offered states considerable latitude in achieving reduction targets that they were given.

One common theme across the Clinton, Bush, and Obama presidencies was the absence of congressional capacity to either adopt new environmental legislation or revise existing statutes as discussed in Chapter 5. This generated major incentives for respective presidents to take unilateral executive actions to achieve their environmental policy goals, whether through executive orders, regulatory revision processes, or other mechanisms as discussed in Chapter 4. It also created an opening for states, particularly through coalitions of elected attorneys general of the party opposite the president, in effect leading the direct challenge on those portions of a president's environmental agenda that affected states and their respective environmental policy roles. This rapid-fire response of coordinated litigation reached new intensity under President Obama, as Republican attorneys general attacked not only the Clean Power Plan but nearly every other regulatory initiative involving climate change and both air and water quality. These state efforts served to delay and in some instances thwart Obama from achieving his environmental policy goals.

This set the stage for the transition to the Donald Trump presidency and the question of how his approach to environmental governance might impact states. Trump said relatively little about federalism in the 2016 campaign but made clear his strong opposition to the Clean Power Plan, linking his plan to gut it with his vows to revive the fortunes of coal mining and coal use in electricity. Trump won 26 of the 27 most carbon-intensive states, whereas his opponent, Hillary Clinton, won the 14 least carbon-intensive states. He would not introduce any new environmental legislation but rather pursue a "search and destroy" strategy, involving a sequence of regulatory reversals or delays that would systematically attempt to undermine every regulatory effort on climate and air and water pollution taken by Obama, each of which would involve states in some respects.

Some of these Trump efforts would serve to *empower* states that had opposed these regulations and preferred to do far less in these areas. In the case of the Clean Power Plan, a multiyear regulatory revision effort led to its replacement with the Affordable Clean Energy rule. This alternative in essence eliminated any consequential pressures on states to transition their electricity generation sectors toward less reliance on fossil fuels, particularly coal and natural gas. This shift would be generally welcomed by states with the least aggressive climate policies and greatest production and use of fossil fuels. But it would be aggressively opposed by those states that had already made significant climate policy commitments and planned to do more. Similar patterns emerged as Trump rolled out the other elements of his deregulatory strategy for methane emissions and water pollution, empowering those states most opposed to new federal environmental policy. All of these steps would be countered not by Congress but rather coalitions of Democratic attorneys general, designed to derail their implementation and leave their future highly uncertain as they navigated the federal courts.

Other Trump initiatives would formally attempt to *constrain* states, particularly ones led by Democrats, from taking innovative environmental policy steps. Many of these focused on California's effort to adopt major new climate policy initiatives, in many cases securing allies from other states and even Canadian provinces. This included an unprecedented and direct assault on formal powers granted to California on vehicle emissions under the Clean Air Act for over a half-century, including the 2019 repeal of a state waiver approved 6 years earlier. This set the stage for the Trump administration to write California and allied states out of the decision process in producing very modest vehicle emission standards through 2026. In turn, the Trump Justice Department sued California for allegedly encroaching on federal treaty-making powers by establishing a carbon pricing partnership with Québec, launching a federal court review. Moreover, President Trump and his environmental officials demonstrated loathing of California through public rhetoric denouncing the state for its alleged policy failures and enduring environmental problems, while also threatening unprecedented steps of terminating federal grants to the state. This took traditional intergovernmental conflict to a new and visceral level, triggering an aggressive set of responses by California leaders and no clear end to the turmoil in sight.[70] Federal courts might ultimately resolve these issues, although a future president could reverse the Trump actions.

This episode raised the broader question of whether the United States was lurching toward a permanent partisan and federalism divide whereby presidents of one political party were destined to engage in protracted political and legal combat with attorneys general of the opposite party or whether alternatives might emerge involving greater collaboration across parties, states, and levels of government. As political scientist John Kincaid lamented, "Coercive federalism has ironically relegitimized states' rights as a consolation prize for whichever party is out of power in Washington, D.C."[71] Under a Democrat like Barack Obama, Republicans found merit in expanding state resistance to new federal policies. Under a Republican like Donald Trump, Democrats found new virtues in "progressive federalism" and expanded state authority. In short, was this an inevitable new feature of American politics and federalism?

Scholars turned increasingly to other models and examples, some drawn from beyond American boundaries, in beginning to envision alternatives. Political scientist Donald Kettl released a major new book on the "Divided States of America" in early 2020 just as the COVID-19 pandemic began. He noted that "the great public policy challenges of the 21st Century are health and climate. Both of them raise questions that require redistributive answers, and the redistributive answers inevitably require a robust federal role, in leadership, policy, and funding."[72] Kettl invoked the ideas of Alexander Hamilton in proposing a new federal compact, with the federal government in a lead role in addressing profound problems of inequality across states. However, he further emphasized the need to allow states to play central roles in policy implementation and pursue innovation that can offer national models. Finally, he called for an expanded role for local governments in a revised American intergovernmental system.

There were no immediate prospects to transition to such a system, although a growing number of reform proposals with these qualities began to surface in 2019–2020. During 2019 House climate policy hearings, environmental policy scholar Tim Profeta unveiled a "comprehensive state-federal partnership" approach, whereby the federal government would establish enforceable greenhouse gas emission reduction targets but allow each state to decide how to achieve those reductions, permitting any revenues generated through a state carbon price to be kept within that state.[73] This approach would also allow all existing state climate policy efforts to continue and contribute to meeting emission reduction targets while providing flexibility to laggard states in devising their own plans. Enforcement would follow some of the traditional federalism patterns established under air and water quality statutes.

Americans could actually look across their northern border to Canada to see how such a system works in practice. Canada is a federal system much like the United States, although somewhat more decentralized in environmental policy through Constitutional design that gives considerable authority to provinces and territories to oversee their natural resources.[74] Canada also is a major producer of fossil fuels, particularly in Western provinces such as Alberta and Saskatchewan, and struggled for decades to devise a climate policy that was politically feasible and durable given the many fault lines among different jurisdictions and their leaders. Instead, a frequent Canadian climate policy approach involved leaders of multiple parties making bold declaration of their climate concern at international gatherings but then failing to deliver through policy given deep divides between regions and parties over possible next steps.

Justin Trudeau's 2015 election as Prime Minister gave his Liberal Party yet another chance to explore climate policy design and what emerged was the 2018 Greenhouse Gas Pollution Pricing Act. This built directly on the most innovative provinces, most notably British Columbia and its bold carbon tax design that reduced emissions, had no adverse economic impacts, and expanded political support across parties over elections over its first decade.[75] The emerging Pan-Canadian Framework used this case as a model and called upon all provinces and territories to develop some form of a carbon price that would rise steadily and reach $50 per ton (Canadian) by 2022. But the federal government offered these jurisdictions enormous latitude in policy design as long as they delivered results.[76] For provinces like British Columbia, compliance was automatic since they were already committed to such policy. For provinces willing to negotiate a plan with the federal government, they could design and implement an approved proposal, keeping any revenue and using it as they saw fit. For provinces unwilling to work cooperatively, the federal government would impose a carbon tax rising to $50 per ton upon them, collect the revenue, and return all of it to every citizen of that province through an annual dividend check, thereby denying provincial leaders any input into deciding how to spend the money. This triggered many controversies and yet moved relatively smoothly into implementation. Late 2019 elections not only reelected Trudeau but 62 percent of Canadians voted for one of the multiple national parties that had endorsed the strategy. Even amid the COVID-19 crisis in March 2020, Trudeau announced that Canada's climate policy would continue to move forward.

Perhaps Canada provides a model of sorts for American consideration. In turn, there were at least a few signs in the early 2020s that there might be some ways to develop more collaborative relations across states and, possibly, between state and federal governments. Political scientists Paul Nolette and Colin Provost concluded that there were at least a few policy areas where states were developing multistate partnerships and even partisan attorneys general were able to work across divides. In particular, they noted that issues such as opioids, tobacco use reduction, sex trafficking, and elder abuse garnered unexpected ability to build coalitions involving both Republican and Democratic states and even partnerships between rural and urban jurisdictions.[77] Some energy policy analysts began to note that some of the greatest growth in renewable energy deployment was occurring in the very states least likely to have adopted major climate or environmental policies.[78] This included states such as Texas, which produces considerably more wind energy than any other state. Wyoming, North Dakota, and Nevada are the top-ranked states in terms of renewable energy generated per capita.

Wyoming's Chokecherry and Sierra Madre Wind Energy Project, scheduled to open in 2026, will be the United States' largest wind farm and one of the world's biggest, with plans to sell power to California.[79]

Gaining political support for siting renewable technologies and needed transmission capacity may be easier in states with open vistas, a history of energy production and policies amenable to new development, and established systems of royalties that give property owners financial incentives to welcome developers. In contrast, states such as California and New York that are most likely politically to adopt bold renewable energy standards and related climate policies may face the greatest political opposition to actually siting these facilities within their boundaries. This raises questions of whether they force this development upon reluctant in-state localities or import such energy from other states where siting is easier. It also prompts the broader issue of the conditions under which disparate states might find common cause moving forward, particularly in the aftermath of the COVID-19 pandemic and lessons emerging from it on ways to improve the performance of both federal and state governments in responding to public policy challenges. Much as Lord Bryce pondered centuries ago, it is possible, at least in theory, to envision a political system in which multiple levels of government work toward the common good on such issues as environmental protection.

SUGGESTED WEBSITES

Environmental Council of the States (www.ecos.org) The Environmental Council of the States represents the lead environmental protection agencies of all fifty states. The site contains access to state environmental data and periodic "Green Reports" on major issues.

Georgetown Climate Center (www.georgetownclimate.org) The Georgetown Climate Center provides extensive data bases and reports on different dimensions of state climate and energy policy. This includes a State Energy Analysis Tool that provides highly detailed information on state energy sources and usage and updates on the Transportation Climate Initiative, where it has played a central convening role among participating states.

National Conference of State Legislatures (www.ncsl.org) The National Conference of State Legislatures conducts extensive research on a wide range of environmental, energy, and natural resource issues for its primary constituency of state legislators, as well as for the general citizenry. The organization offers an extensive set of publications, including specialized reports and monthly review of state policy developments.

National Governors Association (www.nga.org) The National Governors Association maintains an active research program concerning state environmental protection, natural resources, and energy concerns. It has placed special emphasis on maintaining a database on state "best practices," which it uses to promote diffusion of promising innovations and to demonstrate state government capacity in federal policy deliberations.

State Energy & Environmental Impact Center, New York University School of Law (www.law.nyu/centers/state-impact.edu) The State Energy & Environmental Impact Center keeps close tabs on the actions of state attorneys general and state governments in environmental policy. This includes databases on multistate litigation and other strategies in instances where states choose to challenge federal environmental policy decisions.

NOTES

1. John Kincaid, "Dynamic De/Centralization in the United States, 1970–2010," *Publius: The Journal of Federalism* 49 (October 2018): 166–93.
2. John Dinan, *State Constitutional Politics: Governing by Amendment in the American States* (Chicago: University of Chicago Press, 2018).
3. Paul Teske, *Regulation in the States* (Washington, DC: Brookings Institution Press, 2004), 9.
4. Martha Derthick, "Compensatory Federalism," in *Greenhouse Governance*, ed. Barry Rabe (Washington, DC: Brookings Institution Press, 2010), 66.
5. Christopher McGrory Klyza and David Sousa, *American Environmental Policy: Beyond Gridlock,* updated and expanded edition (Cambridge, MA: MIT Press, 2013), 247.
6. Stephen Goldsmith and Donald F. Kettl, eds., *Unlocking the Power of Networks* (Washington, DC: Brookings Institution Press, 2009).
7. Elinor Ostrom, *Governing the Commons: The Evolution of Institutions for Collective Action* (New York: Cambridge University Press, 1990); Elinor Ostrom, *A Polycentric Approach to Climate Change* (Washington, DC: World Bank, 2009).
8. Barry G. Rabe, *Statehouse and Greenhouse: The Emerging Politics of American Climate Change Policy* (Washington, DC: Brookings Institution Press, 2004).
9. Michelle Pautz and Sara Rinfret, *The Lilliputians of Environmental Regulation: The Perspective of State Regulators* (New York: Routledge, 2013).
10. Andrew Karch, *Democratic Laboratories: Policy Diffusion Among the American States* (Ann Arbor: University of Michigan Press, 2007).
11. American Council for an Energy-Efficient Economy, *2019 State Energy Efficiency Scorecard* (Washington, DC: ACEEE, 2019).
12. Evan J. Ringquist, *Environmental Protection at the State Level: Politics and Progress in Controlling Pollution* (Armonk, NY: M. E. Sharpe, 1993).
13. Daniel Fiorino and Riordan Frost, "The Pilot Eco-Efficiency Index: A New State Environmental Ranking for Researchers and Government" (paper presented at the Association for Public Policy Analysis and Management's fall research conference, Washington, DC, November 4, 2016), www.researchgate.net/publication/3157 36703_The_Pilot_Eco-Efficiency_Index_A_New_State_Environmental_Ranking_for_ Researchers_and_Government.
14. Kathy Kinsey, "Neither of These Bills Address the Law's Failings," *Environmental Forum* 31 (May–June 2014): 48.
15. Linda Breggin, "Broad State Efforts on Toxic Controls," *Environmental Forum* 28 (March–April 2011): 10.
16. Christopher Bosso, ed., *Governing Uncertainty: Environmental Regulation in the Age of Nanotechnology* (Washington, DC: Resources for the Future, 2010), 105–30.
17. Barry G. Rabe, *Can We Price Carbon?* (Cambridge, MA: MIT Press, 2018), 239–40.
18. Glen Stubbe, "Walz Announces Climate Change Subcabinet," *Minnesota Star-Tribune* (December 2, 2019).

19. Andrew Kear, "Natural Gas Policy Path—Built to Boom," *Journal of Policy History* 30 (2018): 334–65.

20. Jonathan M. Fisk, *The Fracking Debate: Intergovernmental Politics of the Oil and Gas Renaissance* (New York: Routledge, 2018).

21. Nathan Hultman et al., *Accelerating America's Pledge: Going All-In to Build a Prosperous, Low-Carbon Economy for the United States* (New York: Bloomberg Philanthropies, 2019), Chapter 3.

22. Gilbert E. Metcalf, *Paying for Pollution: Why a Carbon Tax is Good for America* (New York: Oxford University Press, 2019).

23. Barry G. Rabe and Christopher Borick, "Carbon Taxation and Policy Labeling: Experience from American States and Canadian Provinces," *Review of Policy Research* 29 (May 2012): 358–82.

24. Samantha McBride, *Recycling Reconsidered: The Present Failure and Future Promise of Environmental Action in the United States* (Cambridge, MA: MIT Press, 2012).

25. Leigh Raymond, *Reclaiming the Atmospheric Commons: The Regional Greenhouse Gas Initiative and a New Model of Emissions Trading* (Cambridge, MA: MIT Press, 2016).

26. Roger Karapin, *Political Opportunities for Climate Policy: California, New York, and the Federal Government* (New York: Cambridge University Press, 2016).

27. Vivian E. Thomson, *Sophisticated Interdependence in Climate Policy: Federalism in the United States, Brazil, and Germany* (London: Anthem Press, 2014).

28. Nathan Hultman et al., *Accelerating America's Pledge*.

29. Kelly Trumbull, Colleen Callahan, Sarah Goldmuntz, and Michelle Einstein, *Progress Toward 100% Clean Energy in Cities & States Across the US* (Los Angeles: UCLA Luskin Center for Innovation, 2019).

30. David Vogel, *California Greenin': How the Golden State Became an Environmental Leader* (Princeton: Princeton University Press, 2018).

31. Barry G. Rabe, "Leveraged Federalism and the Clean Air Act: The Case of Vehicle Emissions Control," in *Lessons from the Clean Air Act: Building Durability and Adaptability into U.S. Climate and Energy Policy*, ed. Ann Carlson and Dallas Burtraw (Cambridge: Cambridge University Press, 2019), Chapter 4.

32. Ben Merelman, *Conservative Innovators: How States are Challenging Federal Power* (Chicago: University of Chicago Press, 2019); Paul Nolette and Colin Provost, "Change and Continuity in the Role of State Attorneys General in the Obama and Trump Administrations," *Publius: The Journal of Federalism* 48 (Summer 2018): 469–94.

33. Paul Nolette, *Federalism on Trial: State Attorneys General and National Policymaking in Contemporary America* (Lawrence, KS: University Press of Kansas, 2015).

34. David J. Hayes et al., *300 and Counting: State Attorneys General Lead the Fight for Health and the Environment* (New York: State Energy & Environmental Impact Center, 2019).

35. Frank J. Thompson, Kenneth K. Wong, and Barry G. Rabe, *Trump, the Administrative Presidency, and Federalism* (Washington, DC: Brookings Institution Press, 2020).

36. William R. Lowry, *The Dimensions of Federalism: State Governments and Pollution Control Policies*, rev. ed (Durham, NC: Duke University Press, 1997), 125.

37. Jacob M. Grumbach, "From Backwaters to Major Policymakers: Policy Polarization in the States, 1970–2014," *Perspectives in Politics* 16 (June 2018): 416–35; Rebecca Bromley-Trujillo and Mirya R. Holman, "Climate Change Policymaking in the States: A View at 2020," *Publius: The Journal of Federalism* 50 (Summer 2020): 446–72.

38. In Wisconsin, for example, Governor Scott Walker oversaw significant reductions in state environmental staff and research capacity, with a primary focus on accelerating approval time for proposed development. See Steven Verberg, "DNR to Alter Handling of Pollution, Parks, Enforcement," *Wisconsin State Journal* (December 1, 2016); Evan Osnos, "Chemical Valley," *New Yorker* (April 7, 2014), 38–49; and Trip Gabriel, "Ash Spill Shows How Watchdog Was Defanged," *New York Times* (February 28, 2014).

39. R. Steven Brown, *Status of State Environmental Agency Budgets* (Washington, DC: Environmental Council of the States, 2012).

40. Linda K. Breggin, "Stringency Laws Widely Adopted," *Environmental Forum* 32, no. 3 (May–June 2015): 18.

41. John A. Hoornbeek, "The Promise and Pitfalls of Devolution: Water Pollution Policies in the American States," *Publius: The Journal of Federalism* 35 (Winter 2005): 87–114.

42. Shama Gamkhar and J. Mitchell Pickerill, "The State of American Federalism 2011–2012," *Publius: The Journal of Federalism* 42 (Summer 2012): 357–86.

43. Sheldon Kamieniecki, *Corporate America and Environmental Policy* (Palo Alto, CA: Stanford University Press, 2006), 253.

44. William Lowry, *Repairing Paradise: The Restoration of Nature in America's National Parks* (Washington, DC: Brookings Institution Press, 2010), Chapter 4.

45. Barry G. Rabe, Claire Kaliban, and Isabel Englehart, "Taxing Flaring and the Politics of State Methane Release Policy," *Review of Policy Research* 37 (2020): 6–38.

46. Nathan Borney, *Detroit Resurrected: To Bankruptcy and Back* (New York: W.W. Norton, 2017).

47. Anna Clark, *The Poisoned City: Flint's Water and the American Urban Tragedy* (New York: Metropolitan Books, 2018).

48. David Switzer and Manuel P. Teodoro, "The Color of Drinking Water: Class, Race, Ethnicity and Safe Drinking Water Act Compliance," *Journal of American Water Works Association* 108, no. 8 (2017): E416–E424; Sara Hughes, "Flint, Michigan, and the Politics of Safe Drinking Water in the United States," *Perspectives in Politics* 18 (July 2020): 1–14.

49. Benjamin Y. Clark and Andrew B. Whitford, "Does More Federal Environmental Funding Increase or Decrease States' Efforts?" *Journal of Policy Analysis and Management* 30 (Winter 2010): 136–52.

50. Keene Kelderman et al., *The Thin Green Line: Cuts in State Pollution Control Agencies Threaten Public Health* (Washington, DC: Environmental Integrity Project, 2019).

51. Michael E. Kraft, Mark Stephan, and Troy D. Abel, *Coming Clean: Information Disclosure and Environmental Performance* (Cambridge, MA: MIT Press, 2011).

52. Matthew J. Hoffmann, *Climate Governance at the Crossroads* (New York: Oxford University Press, 2011); Michael E. Kraft, *Using Information Disclosure to Achieve Policy Goals: How Experience with the Toxics Release Inventory Can Inform Action on Natural Gas Fracturing*, Issues in Energy and Environmental Policy No. 6 (Ann Arbor, MI: Center for Local, State, and Urban Policy, March 2014).

53. Paul Posner, "Networks in the Shadow of Government: The Chesapeake Bay Program," in *Unlocking the Power of Networks*, ed. Stephen Goldsmith and Donald F. Kettl (Washington, DC: Brookings Institution Press, 2009), Chapter 4; Barry G. Rabe and Marc Gaden, "Sustainability in a Regional Context: The Case of the Great Lakes Basin," in *Toward Sustainable Communities: Transition and Transformations in Environmental Policy*, ed. Daniel A. Mazmanian and Michael E. Kraft, 2nd ed (Cambridge, MA: MIT Press, 2009), 266–69.

54. Donald F. Kettl, *The Divided States of America: Why Federalism Doesn't Work* (Princeton: Princeton University Press, 2020), 119.

55. Denise Scheberle, *Federalism and Environmental Policy: Trust and the Politics of Implementation*, rev. ed (Washington, DC: Georgetown University Press, 2004), Chapter 7.

56. Michelle Pautz and Sarah Rinfret, *The Lilliputians of Environmental Regulations*, 50–51.

57. Paul Teske, *Regulation in the States*; Pautz and Rinfret, *The Lilliputians of Environmental Regulation*.

58. John D. Donahue, *Disunited States: What's at Stake as Washington Fades and the States Take the Lead* (New York: Basic Books, 1997); Paul E. Peterson, *The Price of Federalism* (Washington, DC: Brookings Institution Press, 1995), Chapter 4.

59. William T. Gormley Jr., "Intergovernmental Conflict on Environmental Policy: The Attitudinal Connection," *Western Political Quarterly* 40 (1987): 298–99.

60. David M. Konisky and Neal D. Woods, "Exporting Air Pollution? Regulatory Enforcement and Environmental Free Riding in the United States," *Political Research Quarterly* 63 (2010): 771–82.

61. William Boyd, "The Clean Air Act's National Ambient Air Quality Standards," in *Lessons from the Clean Air Act*, ed. Ann Carlson and Dallas Burtraw (Cambridge: Cambridge University Press, 2019), 15–18.

62. Irene C. Dedoussi, Sebastian D. Eastham, Erwan Monier, and Steven R.H. Barrett, "Premature Mortality Related to United States Cross-State Air Pollution," *Nature* 587 (February 13, 2020): 261–68.

63. Kincaid, "Dynamic De/Centralization in the United States, 1790–2010," 180.

64. Daniel J. Sherman, *Not Here, Not There, Not Anywhere* (Washington, DC: Resources for the Future, 2011).

65. Alice Rivlin, "Rethinking Federalism for More Effective Governance," *Publius: The Journal of Federalism* 42 (Summer 2012): 357–86; Jenna Bednar, *The Robust Federation* (New York: Cambridge University Press, 2008); R. Daniel Keleman, *The Rules of Federalism* (Cambridge, MA: Harvard University Press, 2004).

66. John Buntin, "25 Years Later, What Happened to 'Reinventing Government'?" *Governing* (September 2016).

67. Klyza and Sousa, *American Environmental Policy, 1990–2006*, 253.

68. Alan Greenblatt, "State Budget Fallout: A Hurricane That Hits All Over the Country," *Governing* (April 9, 2020).

69. Leah Stokes, *Short Circuiting Policy: Interest Groups and the Battle Over Clean Energy and Climate Policy in the American States* (New York: Oxford University Press, 2020).

70. Thompson, Wong, and Rabe, *Trump, the Administrative Presidency, and Federalism*, 96–104.

71. John Kincaid, "Introduction: The Trump Interlude and the States of American Federalism," *State and Local Government Review* 49 (September 2017): 165.

72. Kettl, *The Divided States of America*, 200.

73. David Obey, "House Panel Showcases Federalism Framework for Broad GHG Cuts," *Inside EPA/Climate* (December 6, 2019); Tim Profeta, "Using the Old to Solve the New—Creating a Federal/State Partnership to Fight Climate Change," *Duke Nicholas Institute for Environmental Policy Solutions Policy Brief* (October 2019).

74. Andrea Olive, *The Canadian Environment in Political Context* (Toronto: University of Toronto Press, 2016).

75. Rabe, *Can We Price Carbon?*, 88–119.

76. Brendan Boyd and Barry Rabe, "Whither Canadian Climate Policy in the Trump Era?" in *Canada–U.S. Relations: Sovereignty of Shared Institutions?* ed. David Carment and Christopher Sands (New York: Palgrave Macmillan, 2018), 239–60.

77. Nolette and Provost, "Change and Continuity in the Role of State Attorneys General in the Obama and Trump Administrations," 483–87.

78. Miriam Fischlein et al., "Which Way Does the Wind Blow? Analysing the State Context for Renewable Energy Development in the United States," *Environmental Policy and Governance* 24 (2014): 169–87; "A Renewable-Energy Boom is Changing the Politics of Global Warming," *The Economist* (March 14, 2020); Elizabeth Weise, "Wind Energy Gives American Farmers a New Crop to Sell in Tough Times," *USA Today* (February 16, 2020).

79. Daniel Fiorino, Carley Weted, and Sabina Blanco Vecchi, "The Politics of Energy Efficiency: Explaining Policy Variations in the American States" (paper presented at the 2019 Conference of the International Public Policy Association, Montreal).

POLITICS, PRICES, AND PROOF

American Public Opinion on Environmental Policy

Christopher Borick and Erick Lachapelle

As the second decade of the twenty-first century drew to a close, climate change and other environmental matters had become some of the American public's most significant areas of concern. Surveys during this period showed a majority of Americans accepted the existence of climate change, were concerned about the impacts of this phenomenon, and wanted their government to take action to mitigate the warming planet. Beyond climate matters, Americans were indicating increased concern for an array of other environmental issues, including drinking water safety and local air quality conditions.[1]

In a representative democracy, such as the United States, it seems reasonable to expect that these public concerns would be translated into significant efforts on the part of the federal government to address them. However, this expectation would not be supported by an examination of national-level environmental policy throughout this era. In fact, as Americans increasingly demonstrated concerns regarding environmental matters between 2017 and 2020, the federal government under the Trump presidency was aggressively scaling back major US climate regulations and an array of other environmental policies and programs (see Chapter 4).[2]

The recent disconnect between the American public's views on the environment and the policies and actions of the federal government in the area of environmental protection raises a number of important questions. Is this absence of alignment between opinion and policy in the area of the environment fairly normal, or is the Trump era an outlier? If public opinion has varied effects on environmental policy adoption, are there some dimensions of public opinion that have greater impact on policy efforts than others? Finally, if public opinion can affect policy adoption under certain conditions, what factors determine the public's views on the environment? In this chapter we will attempt to answer these questions and to provide insight into causes, forms, and impact of American attitudes, beliefs, and policy preferences regarding the environment.

THE RELATIONSHIP BETWEEN PUBLIC OPINION AND ENVIRONMENTAL POLICY

Does public opinion regarding the environment affect environmental policy? The opening vignette would seem to suggest that it doesn't. Of course, any single period or example may not be indicative of broader relationships. While many factors influence the adoption of environmental policies in the United States, as earlier chapters in this volume have shown, the role of public opinion is considered by scholars to be a crucial determinant of what issues governments will act upon.[3] This relationship was on full display during the 1970s when the US government significantly expanded its role in environmental protection. As the 1970s dawned, surveys showed an increasing number of Americans were concerned with environmental issues such as air and water pollution, and that a majority were calling for government action to address these environmental threats.[4] The increased public awareness and support for government intervention culminated in an unparalleled array of legislation emerging from Congress in the early 1970s, including landmark laws such as the National Environmental Policy Act (1970), Clean Air Act (1970), Clean Water Act (1972), and Endangered Species Act (1973) (see Chapter 1).

While the 1970s appear to provide evidence of congruence between opinion and policy in the realm of environmental matters, and the recent Trump era demonstrates a high degree of opinion–policy incongruence, much of the twenty-first century shows a more complicated and nuanced relationship between what Americans think about the environment and what government does in terms of environmental policy. One complicating factor is that the public's views on environmental issues may send mixed messages regarding preferences for action. The more complex nature of American views on environmental protection was on display during the opening decade of the twenty-first century. Between 2000 and 2010, Gallup found a dramatic 30-point increase in the percentage of Americans who believed economic growth should be given priority, even if the environment suffers to some extent.[5] But as Americans were increasingly willing to sacrifice a degree of environmental quality to attain economic growth during the early 2000s, a majority continued to claim that the US government was doing too little to protect the environment.

Opinion regarding environmental matters can also shift quickly as government is wrestling with policy actions. Such was the case during the early stages of the Obama administration. In the decade leading up to the 2008 election Americans had increasingly indicated that they thought global warming was occurring. But as the Obama administration and Congress worked on passing major climate legislation, public opinion shifted substantially, with acceptance of the underlying problem becoming much more contested. The Pew Research Center found that the percentage of Americans who believed there was solid evidence of global warming dropped from 71 percent in April 2008 to only 57 percent in October of 2009.[6] The decline in public acknowledgment of climate change came as the United States was experiencing its worst economic crisis since the Great Depression of the 1930s, and a number of scholars found the worsening economic conditions directly contributed to the shift in opinion.[7] Other research suggests that efforts by fossil fuel industries, Republican congressional leaders, and conservative media figures helped to raise skepticism about climate science among portions of the American public.[8]

During the second decade of the twenty-first century, as the nation climbed out of a deep recession to a period of record economic growth, Americans increasingly expressed worries about environmental quality, with concern regarding a number of environmental issues growing significantly.[9] As can be seen in Figure 3.1, since 2011 there was fairly steady growth in the number of Americans who believe protection of the environment should be given priority, even at the risk of curbing economic growth.

FIGURE 3.1 ■ Environmental Protection versus Economic Growth

Environmental Protection vs. Economic Growth

With which one of these statements do you most agree -- protection of the environment should be given priority, even at the risk of curbing economic growth (or) economic growth should be given priority, even if the environment suffers to some extent?

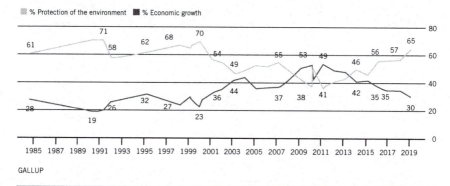

GALLUP

Source: Gallup – https://news.gallup.com/poll/248243/preference-environment-economy-largest-2000.aspx

During Obama's second term government policy responded to the increased public support for environmental protection efforts with executive actions such as the creation of the Clean Power Plan and entry into the Paris Climate Agreement. Yet as we discussed in the introduction, despite strong public concerns with environmental matters and desire to have government take greater actions to protect the environment, the Trump administration undertook a historic rollback of existing environmental rules and dismantling of Obama-era climate policy.

The inconsistent relationship between opinion and policy in the area of the environment leads us to consider what might explain such variation. So far, we have discussed public opinion as a single concept, but of course opinion is multidimensional, and it is in those dimensions that varied impacts on policy adoption may rest. Indeed, much scholarship has explored aspects of public opinion and the impact of those aspects on government engagement, particularly on environmental matters.[10] In this next section we seek to more fully explore the dimensions of public opinion regarding environmental issues, and how those dimensions may affect environmental policy adoption in the United States. In particular we examine aspects of public opinion including: (1) awareness and concern; (2) issue saliency; and (3) policy preferences.

Awareness and Concern: What do Americans know about the environment, and how concerned is the public about environmental issues? For years opinion researchers have tried to assess the public's awareness of environmental issues and to determine the degree to which individuals worry about environmental matters. The results of this research paint a complex picture of American perceptions of the natural environment. As noted in the previous section, Americans became increasingly aware and concerned about environmental matters in the 1960s and 1970s, particularly with issues such as air and water pollution. In numerous surveys during this period Americans reported that they believed the nation's air and water quality had substantially diminished over time, and that they were concerned about the impacts of pollution on human health.[11] Despite economic downturns and the energy crises of the 1970s, environmental awareness and concern levels had become so consistent by the 1980s that many analysts described pollution issues as "enduring concerns" of the American public.[12]

But as awareness of core environmental issues such as air and water pollution became established during the 1970s and 1980s, new environmental issues and challenges began to emerge. In particular, during the 1980s, environmental concerns such as depletion of the ozone layer, rainforest loss, acid rain, and global warming were identified by scientists as significant threats. Polling showed fairly widespread public awareness and concern about all these issues in the United States. For example, a Gallup poll in 1989 found a majority of Americans (51 percent) worried a great deal about ozone depletion, with these relatively high levels of concern persisting throughout the remainder of the twentieth century. Concern regarding acid rain and rainforest destruction also became widespread and fairly consistent throughout the late 1980s and 1990s.[13]

As perhaps the most significant environmental threat to emerge in the 1980s, global warming began to receive greater media attention late in that decade. This attention arrived particularly after the record warmth and severe drought of 1988, and the high-profile congressional testimony of NASA researcher James Hanson regarding the scientific evidence of this phenomenon. When first surveyed about this issue in 1986, just 39 percent of Americans indicated that they had heard or read about the greenhouse effect (as global warming or climate change was commonly known as at the time), but by September 1988, 58 percent were aware of this matter. During the 1990s, global warming awareness levels would rise to over 80 percent of the American public, and by the first decade of the twenty-first century awareness would top the 90 percent mark.[14]

While most Americans were aware of the issue of global warming by the late 1980s, acceptance of the presence of the problem and concern about the impacts of a warming planet were far less consistent. In fact, there have been significant shifts in American acceptance of evidence of global warming since the issue emerged in the 1980s. For example, between 1997 and 2008, Gallup found the percentage of Americans that believed global warming had already begun increased from 48 percent to 61 percent. Yet by 2011 the percentage of Americans who thought global warming was happening had dropped back to 49 percent, or almost where it was in 1997. It would not be until 2017 that at least 61 percent of Americans (the 2008 level) would report that global warming was already going on.

Public concern regarding global warming has also shifted substantially since Americans became aware of the issue over three decades ago. When Gallup first asked about concern regarding global warming in 1989, about one in three Americans indicated that they worried "a great deal" about this issue. Over the course of the next two decades there would be modest shifts in concern levels, with a high of 40 percent concerned "a great deal" in 2007, and a low of only 25 percent greatly concerned in 2011. However, the percentage of Americans who worry "a great deal" about global warming has risen steadily since 2011, reaching a mark of 44 percent in 2019.

Notably, as concerns about climate change have increased, public worries about air pollution have significantly declined during the last thirty years. In 1989, 63 percent of

Americans indicated that they worried a great deal about air pollution, compared with 43 percent that maintained the same level of concern in 2019. Concerns with drinking water quality have also declined in the twenty-first century, dropping from 72 percent of Americans that worried a great deal about this matter in 2000, to only 48 percent maintaining this level of concern in 2012. But since the record low levels of concern found in 2012, concerns have increased, peaking at 63 percent in 2017 in the wake of the Flint, Michigan, drinking water crisis that emerged in 2015.[15]

Saliency: With a majority of Americans aware of an array of environmental challenges, and large percentages concerned about environmental threats, it seems reasonable to expect government to respond with policy efforts to address those concerns. As emphasized earlier in the chapter, the wave of environmental policies during the 1970s aligned with the growing concerns of the American public regarding the environment, thus indicating a general alignment of opinion and policy. But in terms of significant policy effort to address global warming, public concerns have not necessarily translated into major policy efforts, particularly at the federal level. While numerous factors may have contributed to limited federal climate policy action, one aspect of public opinion seems likely to be affecting the situation—issue saliency. Salience refers to the importance of an issue, particularly as it applies to the constituents of elected officials.

Government officials in democratic systems may maximize the likelihood of reelection by attempting to address the most salient concerns of their constituents. The public may hold and express concerns for many issues, but given the limited time and resources available to officials, there is an increased likelihood that only the most salient concerns will be acted on.[16] As noted earlier, Americans generally express concern for environmental issues, but how do those concerns compare with other issues that are in play at the same time? To help answer this question we turn to a series of polls conducted by the Pew Center over the last twenty-five years that examine the relative importance, or saliency, of issues for Americans (Pew 2020).

The Pew findings demonstrate that while large portions of the American public think the environment and climate change should be a top priority for government, even larger portions of the public think other issues should be a top priority. During the period between 2010 and 2020, an annual average of 35 percent of Americans identified dealing with global climate change as a top priority for the federal government, with 48 percent identifying protection of the environment as a top priority for the US government. While those results certainly indicate that many Americans believe environmental issues should be a top priority for the national government, other issues are rated much higher. For example, during the 2010–2020 period, defending the country against terrorist attacks (74 percent), improving the job situation (69 percent), and reducing health-care costs (63 percent) were among the issues that more Americans identified as a top priority for the federal government than environmentally related options. In fact, in most years during this period, dealing with climate change placed 18th out of 19 options that Pew presented to respondents.[17]

Of course, in an era where party polarization is elevated, as it has increasingly been in the twenty-first century,[18] issue saliency may be quite varied for individuals across party divides. According to a 2020 Pew Research Center study, the divide in issue saliency between Democrats and Republicans is the largest on environmental issues, and particularly climate change. As can be seen in Figure 3.2, while 78 percent of

FIGURE 3.2   ■   American Views on Top Priorities for the Federal Government by Political Party

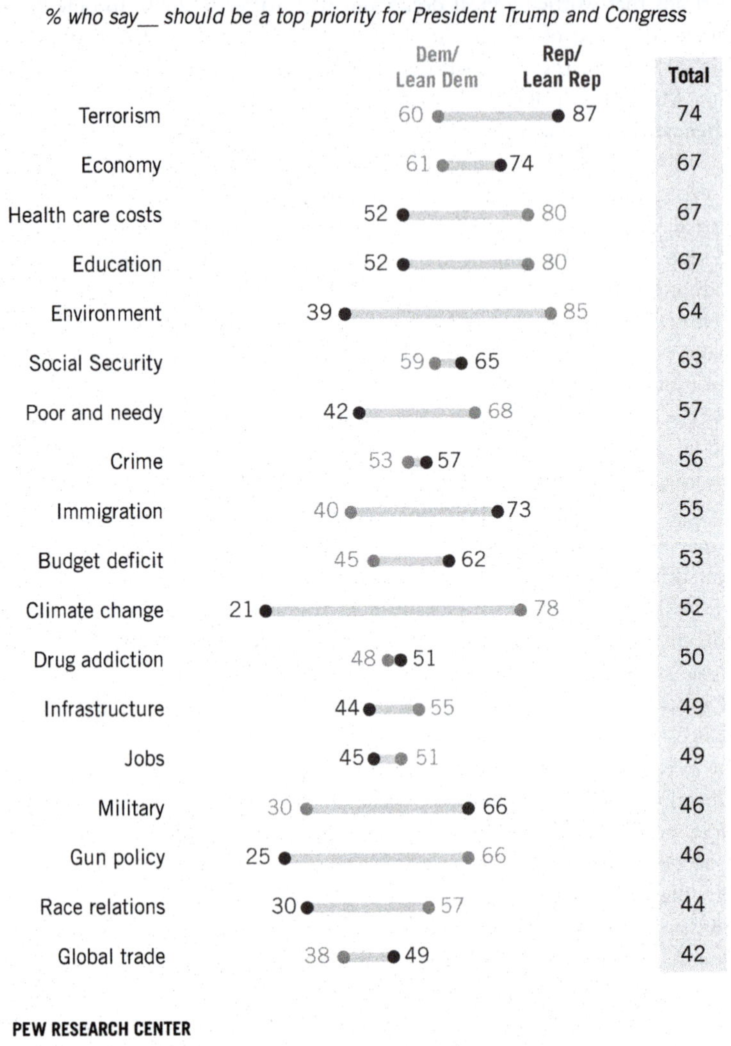

% who say___ should be a top priority for President Trump and Congress

	Dem/ Lean Dem	Rep/ Lean Rep	Total
Terrorism	60	87	74
Economy	61	74	67
Health care costs	52	80	67
Education	52	80	67
Environment	39	85	64
Social Security	59	65	63
Poor and needy	42	68	57
Crime	53	57	56
Immigration	40	73	55
Budget deficit	45	62	53
Climate change	21	78	52
Drug addiction	48	51	50
Infrastructure	44	55	49
Jobs	45	51	49
Military	30	66	46
Gun policy	25	66	46
Race relations	30	57	44
Global trade	38	49	42

PEW RESEARCH CENTER

Source: "Wide partisan gaps on climate change, environment, guns and stronger military." Pew Research Center, Washington, DC (February 13, 2020), https://www.people-press.org/2020/02/13/as-economic-concerns-recede-environmental-protection-rises-on-the-publics-policy-agenda/pp_2020-02-13_political-priorities_0-02/

Democrats thought that climate change should be a top priority for the federal government in 2020, only 21 percent of Republicans maintained the same view. The 57-point gap regarding climate change is far and away the highest among the 18 issues Pew tested, with the 46 percent gap on the "environment" the second largest divide.

While these differences in saliency by themselves may not explain the Trump administration's decision to dismantle federal climate and environmental regulations, they certainly help provide insight into why the connection between aggregate measures of public support for greater environmental protection may not result in policies that reflect these preferences.

Differences in issue saliency can help explain a degree of the differences in state environmental policy outcomes that were discussed in the federalism chapter in the first section of the book. While many factors contribute to the stark differences in state environmental policies that Barry Rabe notes in Chapter 2, a number of scholarly studies have shown that issue saliency is one of the determinants of environmental policy adoption at the state level. In particular, a study by Rebecca Bromley-Trujillo and John Poe[19] finds that general concern among state residents regarding climate change is a fairly weak predictor of state climate policy adoption, but when issue salience is considered, public opinion plays a much greater role in explaining the strength of state efforts to address climate change.

Policy Preferences: As the previous discussion has shown, higher levels of awareness, concern, and most importantly salience help increase the likelihood that governments at both the national and subnational levels engage in policy development. Yet the presence of these conditions can't necessarily explain the types of policy actions that are selected. For any environmental problem there are often numerous policy options available for policymakers to select from.

The wave of environmental policy action in the 1970s generally applied a traditional "command and control" approach to environmental problems. Under this approach, governments set firm environmental standards, monitor to make sure those standards are being met, and if not, enforce the law with some form of penalty (e.g., fines). As noted in Chapter 10, command and control policies have been criticized for their lack of flexibility and high administrative costs, although their political attractiveness is enhanced by their ability to hide costs from consumers. In contrast, alternative policies that focus on economic incentives and disincentives as the primary mechanisms are likely to prove far more cost-effective, but can face greater political challenges because of their direct imposition of costs on consumers. Among the tools most commonly applied under the "market based" approach are taxes and the establishment of markets for the trade of pollution credits. Examples of economic-oriented approaches range from a city charging residents a fee on garbage disposal that is determined by how much waste they produce, to the creation of the Regional Greenhouse Gas Initiative (RGGI) by northeastern states that oversees a market to trade carbon dioxide credits and a similar one operated by California and the Canadian province of Québec (see Chapter 2).

Given the varied policy tools that can be applied to environmental problems, does the public have clear preferences for which approaches are best? The National Surveys on Energy and the Environment (NSEE) have regularly asked Americans about their energy policy preferences since 2008, with the results showing strong support for regulatory tools, but more mixed support for market-oriented approaches.[20] Substantial majorities of Americans have supported increasing federal fuel efficiency standards for automobiles during both the Obama and Trump administrations, and at the state level policies that require the use of renewable energy have been consistently popular, even across partisan divides. Americans were also largely opposed to the Trump administration's proposal to scale back existing automobile fuel efficiency standards in 2018.

While Americans strongly support the use of tax incentives for individuals to purchase electric and hybrid vehicles, they are less supportive of marked-based policies such as energy taxes. For example, they are generally opposed to increasing gas taxes as a means of reducing fossil fuel use. In a series of NSEE polls between 2008 and 2017, support for increasing the gas tax as a means of reducing fossil fuel emissions never surpassed 30 percent. A broader tax on carbon also receives lower rates of support than regulatory options though it elicits more nuanced reactions from the American public. A carbon tax, which places a fee on all fossil fuels, has become a much discussed policy option in climate policy debates.[21] As with a gas tax, a majority of Americans have opposed a carbon tax since the concept has been tested in polls in the twenty-first century.[22] However, the scope and design of the carbon tax, and the use of the revenue generated from this tool, have significant impact on public support levels. Studies have found that the way the carbon tax is described to individuals can significantly enhance the level of support that this policy option garners.[23] Additionally, using the revenues from a carbon tax to enhance the development of renewable energy, or to offset other taxes by returning the revenue to taxpayers, helps to significantly increase public support for this policy tool among survey respondents.[24]

American views on cap-and-trade systems have also been fairly mixed as this policy alternative has emerged over the last three decades. Originally heralded by economists as a more efficient approach to address environmental problems such as air pollution, cap-and-trade systems became popular among conservatives in the 1980s, culminating with the establishment of a sulfur dioxide market as part of the 1990 Clean Air Act Amendments.[25] The success of the sulfur dioxide market helped make a carbon dioxide cap-and-trade system a key component of 2009 national climate policy efforts in Congress. But as the debates evolved, Republican opposition to the legislation, including the branding of cap-and-trade as "cap-and-tax," led to the demise of the efforts.[26] Indeed, public concerns with the costs of cap-and-trade were significant at the time, with majority support (53 percent) for the general concept of cap-and-trade falling to only 42 percent when a $15 a month cost was affixed to the policy, and 22 percent when the cost rose to $50 a month.[27]

While efforts to establish a national cap-and-trade system for greenhouse gases may have received mixed reviews from the American public, the largest existing market for greenhouse gases has drawn more positive public reactions. The RGGI noted earlier has remained fairly popular among citizens of the northeast states where it is centered, with a 2016 poll showing a substantial majority of residents in that region supporting the continuation of this market for greenhouse gases.[28] This may reflect its unique pattern of revenue use among its participating states, which has built public support, as well as its relatively modest price to date, which has served to minimize possible public opposition over costs.

One final perspective on public opinion regarding environmental policies relates to the levels of government that individuals want engaged on environmental matters. Once again, the federal system in the United States allows for varied levels of government to be involved in environmental protection. David Konisky finds that Americans generally prefer the federal government take the lead in addressing most environmental issues, and in particular those that relate to pollution, and have a national or global scale (e.g., global warming).[29] At the same time, the public prefers that state and local governments take the lead on handling local or regional issues (e.g., managing urban sprawl). These results suggest a desire among many in the public to match governmental policy assignment with the geographic scale of the problem. However, the public might not always view the roles of government in a policy realm as exclusive to one level. NSEE studies over the course of the last decade indicate that in the realm of climate change, most Americans believe that all levels of government have responsibility to address the warming planet. For example, in 2017 the NSEE found that approximately 7 out of 10 Americans thought that federal, state, and local governments had responsibility to address global warming. In essence, government engagement on climate change does not have to be thought about as a zero-sum situation or as a dual federalism framework in which actions at one level preclude actions at another.[30]

We now move our discussion of public opinion regarding the environment to an examination of the determinants of public views on environmental matters. A key question for scholars, and for those interested in environmental politics, is how to understand differences in environmental public opinion across people, places, and over time. To this end, a number of determinants have been identified as key for understanding differences and change in public opinion. These can usefully be categorized in terms of either "aggregate" or "individual" level determinants. Whereas aggregate data—as the name suggests—are compiled or summed to answer questions about groups or populations (usually among geographical units like countries, counties, or states), individual-level data are measured at the micro-level among units (usually people). We examine aggregate level and individual determinants here.

AGGREGATE-LEVEL DETERMINANTS

What explains shifts in environmental concern over time? A long-standing debate in environmental politics concerns the extent to which the environment and economy are substitutes. This line of reasoning can be traced back to Abraham Maslow's hierarchy of needs theory[31] which suggests the environment is a luxury good—people may become concerned about environmental degradation, but only after more basic needs for food, shelter, and security are met.[32] From this perspective, it is common to assume that the public is more likely to support environmental protection under more favorable economic conditions, while they are likely to prioritize their immediate economic needs over environmental protection during economic hard times. Empirical support for this idea is found in the literature—at least at the aggregate level.

For example, Matthew Kahn and Matthew Kotchen find a dip in Google searches for "global warming" following an increase in state unemployment rates, and that unemployment rates are also associated with lower belief in global warming and support for climate policy as measured in national opinion surveys.[33] Analysis of aggregate data from national samples obtained in the United States between 2002 and 2010 finds that structural economic factors are among the best predictors (along with elite cues) of shifting public concern over climate change.[34] Similarly, Lyle Scruggs and Salil Benegal find substantial evidence to support the claim that changes in macroeconomic conditions affect the priority given to and beliefs in global warming.[35] However, individual-level evidence from panel survey data has challenged this general conclusion, suggesting that some caution is warranted when drawing conclusions about micro-level opinion change from aggregate economic trends.[36]

Geography: Though useful for analyzing longer-term dynamics and trends, a focus on national-level aggregate data is akin to looking at public opinion from 20,000 feet. It allows one to see the forest, but it can also mask important differences among the trees (i.e., at more local levels). Focusing attention on national-level aggregate data also ignores the potential importance of context, effectively treating public opinion as if it comes from nowhere in particular. However, there are good reasons to expect that differences in environmental and economic contexts shape opinion. For example, some of the earliest literature[37] expected to find differences in environmental concern across urban population centers (which may be more polluted, overcrowded, or home to more highly educated professionals) and rural settings (where many of the "extractive" industries like farming, logging, and mining are found). Since then, it is common to find place-based variation in levels of environmental concern, though these relationships are usually modest and can vary depending on the issues studied, measures used, and the areas considered.

Analyzing data from a large number of surveys administered at the national and regional levels, Lawrence Hamilton and colleagues document substantial place-to-place variation at the county level and find that such variation helps predict individual climate beliefs.[38] Other research has taken national-level opinion data and "downscaled" them to develop opinion estimates at the congressional district and county level.[39] These estimates document the geographic distribution of public opinion on climate change in the United States, providing an important resource for students, scholars, policymakers, and practitioners interested in local climate opinions. The variation found in these data have both a regional component (with greater concern along the coasts relative to inland) as well as an urban–rural divide, and distribution of opinions in these data at the state level strongly correlate to the number of climate policies enacted.[40] The extent to which these regional differences are a result of varying levels of vulnerability to environmental problems, however, remains an open question.

Experience: The question of whether or not, and the extent to which, experience with local environmental problems helps shape levels of environmental concern is a long-standing one in the field. Findings from this research have generally shown that while objective environmental conditions can influence the public's level of environmental concern, these effects tend to be modest and short-lived. Some of the earlier work in this area explored attitudes toward nuclear energy and found—somewhat surprisingly—that proximity to nuclear energy facilities was associated with higher levels of support for nuclear energy.[41] In a more recent study, Bradford Bishop finds that residence in counties that are more economically dependent on the oil industry was not associated with greater support for offshore oil drilling before the *Deepwater Horizon* incident, but that people living in these counties became more supportive following the spill.[42] However, the study found no relationship between living in a county affected by the spill and opposition to offshore oil drilling, suggesting that local economic benefits trump local environmental risks in the formation of public attitudes.

In the area of climate change, Konisky and colleagues merge survey data with micro-level geospatial data on extreme weather events and find a modest relationship between experience with extreme weather and the level of climate change concern, though these impacts are short-lived.[43] Though it might be tempting to conclude from these results that the public is largely unresponsive to changes in objective environmental conditions, it remains possible that this might change in the future as the consequences of major environmental problems like climate change become more severe, frequent, and widespread across geographies that are less socially adapted to these impacts.

INDIVIDUAL-LEVEL DETERMINANTS

Compared to aggregate-level determinants, the research on micro-level predictors of environmental concern is considerably larger. This reflects a seismic shift among social scientists in the 1960s and 1970s who began to eschew grand theorizing and sought instead to uncover micro-level explanations of social phenomenon. Along these lines, environmental scholarship in the 1970s and 1980s took a great interest in exploring the sociodemographic characteristics that could explain differences in environmental concern. This early work found that younger and more educated Americans tended to be more concerned about environmental quality and more supportive of measures to remedy environmental problems. Since then, decades of research have established robust patterns over time, with greater environmental concern found among younger, female, liberal, and better educated Americans relative to their older, male, conservative, and less educated counterparts. Despite their importance, however, the impact of demographic determinants of environmental public opinion has come to be dwarfed and sometimes moderated by political partisanship. This is especially—though not exclusively—true in the area of climate change public opinion.

Demographics: Some of the earliest work looking at environmental public opinion focused on age, gender, and education, both to highlight the potential implications of these group-level differences for policymakers, as well as to get a better sense of what types of people are more and less likely to show heightened concern for environmental problems[44]

Age: Among the demographic determinants of environmental opinion, one of the most important is age. Relative to older people, younger Americans are generally more concerned about environmental problems, while also being more supportive of many of the environmental policy solutions discussed above. The negative association commonly found between age and environmental concern has led to some debate over the precise mechanism at play. One interpretation, rooted in Ronald Inglehart's post-materialism thesis, suggests that values have changed over time from emphasizing material economic and security conditions associated with survival to postmaterial values emphasizing autonomy and self-expression.[45] This thesis relies on the idea of generational replacement, whereby highly salient historical events that occur during one's adolescent and young adult life can have lasting consequences as younger generations replace older cohorts.

However, to the extent that solving current environmental problems may require substantial change to current institutions and lifestyles, and given that younger people are less invested in the dominant social order, an alternative interpretation is that younger people are more concerned about the environment precisely because they are young. From this perspective, people's priorities change as they move along the life cycle, such that more short-term economic needs gradually displace concern over the environment. These different possibilities are directly relevant for thinking about the current youth generation's increasingly visible action on environmental issues (including climate strikes around the world), as they imply very different consequences for the prospects for the continuation of heightened public pressure for greater environmental protection (Table 3.1).

Ethnicity and Gender: Unlike age differences, which are generally consistent in terms of their predictive power over time, whether or not a "gender gap" exists in environmental concern has been less conclusive. Initially, it was not uncommon to find weak, mixed, or null effects of gender on environmental attitudes.[46] However, more recent work suggests that women are more risk averse than men.[47] Moreover, in the area of climate change, women tend to display consistently higher levels of belief and concern than do their male counterparts.[48] Though usually included in studies as a

TABLE 3.1 ■ Americans' Attitudes About Global Warming, by Age				
	18 to 34	**35 to 54**	**55 and Older**	**Age Gap (18 to 34 minus 55+)**
	%	%	%	pct. pts.
Think global warming will pose a serious threat in your lifetime	51	47	29	22
Think global warming is caused by human activities	75	62	55	20
Think problem of global warming is underestimated in the news	48	38	31	17
Think most scientists believe global warming is occurring	73	69	58	15
Worry a great deal/fair amount about global warming	70	63	56	14
Think effects of global warming have already begun	62	63	54	8
Understand global warming issue very/fairly well	82	80	76	6

Source: Gallup, May 11, 2018. https://news.gallup.com/poll/234314/global-warming-age-gap-younger-americans-worried.aspx. Results are based on aggregated telephone interviews from four separate Gallup polls conducted from 2015 through 2018 with a random sample of 4,103 adults, aged 18 and older.

control, the persistent differences found between men and women have prompted more research to explain their different levels of concern. Of the two most common accounts, the idea that gendered social roles (e.g., homemaker status, parenthood) lead to different levels of concern has received little empirical support. However, an alternative "gender socialization" model has been found to be more compelling.[49] This model suggests that different values and social expectations taught to boys and girls at a young age (e.g., competitiveness vs. cooperation, unemotional vs. compassionate) can have long-lasting consequences in terms of environmental problem perceptions between women and men.

Yet another explanation for the gender gap is found in the "differential vulnerability hypothesis," which suggests that some demographic groups (like women and non-whites) are more directly exposed to environmental risks and are thus more aware and concerned. While nonwhite Americans (especially Latinos) have been found to report higher levels of belief in and concern for climate change—a racial spillover that seemed to increase during the Obama presidency,[50] these kinds of ethnicity–environment relationships may be complicated by other confounding factors, such as differences in income.[51] Thus, in addition to highlighting the importance of things like race and gender, this research points to the importance of intersectionality, or the study of overlapping and interrelated power structures that discriminate and disadvantage different social groups.[52]

Education: Given the complexity of environmental problems, education, knowledge, and overall scientific literacy are commonly thought to play an important role in shaping public opinion. Take the example of climate change; a complex problem, involving the concentration of invisible gases that alter the Earth's energy balance, raising the *global average* temperature, with wide-ranging effects spread out over the short, medium, and long term. In light of its inherent complexity, and given mounting scientific evidence regarding the existence of climate change and its human causes, it is reasonable to expect that people with higher education are more likely to be exposed to and to be motivated to use their analytical tools to develop a better understanding of this complex problem, to better understand the risks involved, and thus become more concerned about the issue. The expected relationship between higher education and greater environmental concern, however, has proven to be much more complicated.

To be sure, education has performed well as a predictor of general environmental concern both in terms of its consistency, and at certain times, rivaling the predictive power of ideology.[53] However, more recent studies suggest that the straightforward relationship between education and environmental concern has changed. Particularly in the area of climate opinion, education can be inconsistently related to concern (i.e., having both a positive and negative association) depending on ideology, worldview, and partisanship. Several studies have shown that higher education or self-reported knowledge of climate change increases concern among Democrats, while among Republicans, these associations were found to be weak, negative, and in some cases, null.[54] To explain this inconsistent relationship, scholars have identified the role of (affective) elite cues and media campaigns (like Al Gore's *Inconvenient Truth* and the conservative climate change countermovement) whose discourse is more likely to reach more educated (attentive) and strongly identified partisans.[55]

Another influential approach that helps explain the inconsistent effect of education or knowledge on climate change beliefs and attitudes points to the role of alternative cognitive frameworks. Based on a key distinction in modes of reasoning between "system 1" (fast, instinctive, and emotional) and "system 2" (slower, deliberative, more logical), scholars have identified the role of cultural and identity-based cues that can dominate information processing in the opinion formation process.[56] Echoing the studies mentioned earlier that show differential effects of knowledge and education conditional on ideology and partisanship, Dan Kahan and colleagues[57] have shown that rather than bridging the gap, scientific literacy can polarize people with opposing cultural commitments, such that even when people are perfectly capable of thinking analytically about complex problems (i.e., they score high on a battery of science literacy measures), they may instinctively express opinions based on their underlying cultural commitments and worldviews. This is because, for issues like climate change which are imbued with cultural significance, individuals are motivated to deflect threats to their underlying identities and values. To the extent that climate change—or environmental protection more generally—poses a fundamental challenge to the prevailing capitalist order, for instance, people with more hierarchical and individualistic worldviews are likely to intuitively downplay the risks of climate change, since accepting these risks would invite more government regulation of commerce and industry. Meanwhile, people with more egalitarian and communitarian values—for whom the prevailing capitalist order is the source of rising inequality and a threat to core values—might be motivated to embrace the notion that climate change and environmental degradation pose severe risks.

Ideology, Partisanship, and Polarization: Political divisions along the traditional liberal–conservative spectrum also map closely onto the societal debate between those who criticize and those who defend the industrial capitalist system.[58] As a result, it should come as no surprise that, for decades, political ideology has consistently been an important predictor of environmental concern.[59] More recent studies confirm that political ideology continues to be among the most important determinants of individual opinion decades later.[60] Conservatives tend to support business and industry and are thus more likely to oppose environmental reforms, which usually entail government interventions to which conservatives are opposed.

In contrast to ideology, partisanship has not always been an important predictor of environmental attitudes. Four decades ago, initial studies found very weak associations between partisanship and environmental public opinion. This prompted the conclusion that partisanship was "not a crucial variable" for explaining differences in environmental concern among the general public (though at this time partisan differences were more pronounced among political elites and the educated public) and led some to anticipate that environmental concern would continue to broaden across partisan lines.[61] Over time, however, political partisanship emerged as a key determinant of environmental public opinion, as a burgeoning literature in political science would soon show.

The importance of partisanship as a determinant of environmental attitudes can be illustrated by looking at public opinion on climate change. In this area, one of the most robust generalizations is that, relative to Republicans, Democrats are more likely to perceive human-caused global warming as real, be concerned about its effects, and support greenhouse gas reduction policies.[62] Typically, when partisanship is considered, its association dominates other relationships (save for ideology, to which it is strongly correlated). Moreover, partisanship moderates the influence of other variables, like knowledge and education, as well as the impact of alternative issue labeling (e.g., "global warming" vs. "climate change"), which can further widen the partisan gap.[63] Looking at data from the NSEE (see Figure 3.3), it is not uncommon to find a 30 percent gap in levels of belief on the basic question of whether or not the Earth is warming,[64] while on a range of measures, the size of the partisan gap appears to have increased over time.[65]

In fact, a report by the Pew Research Center suggests that partisan polarization over climate change was the widest ever in 2020, with a 52 percent gap between Republicans and Democrats on the question of whether climate change should be a top priority for President Trump and Congress.[66] At the same time, emerging research suggests the ground may be shifting, with increasing numbers of Americans categorized as being concerned or alarmed about global warming.[67] Important work from Hank Jenkins-Smith and colleagues, echoing evidence highlighted by Alexandre Morin-Chassé and Erick Lachapelle, suggests that Republican attitudes may be more variable than expected, especially among younger Republicans during recent years.[68]

Much research has gone into understanding how environmentalism and climate change in particular became such polarized topics. Here, it is important to note that political polarization was much less prevalent before the 1990s. Moreover, there is strong evidence from congressional roll call votes that polarization among partisan elites preceded polarization among the public.[69] The general consensus is that political partisans in the general population responded to elite cues over this period, a phenomenon which continues to this day. Political campaigns like the climate denial countermovement, amplified by partisan media, and reinforced through relatively closed information networks (or "echo chambers"), have made the environment and especially climate change an identity marker.

FIGURE 3.3 ■ Percent of Americans Who Say There Is Solid Evidence of Global Warming, by Party Affiliation

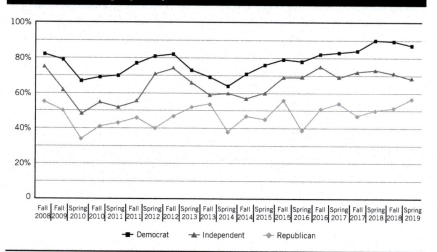

Source: The National Surveys on Energy and The Environment: (October 19, 2019), http://closup.umich.edu/files/ieep-nsee-10-year-beliefs.pdf

Partisans are exposed to very different types of information and have differential incentives to assimilate or discard information based on its consistency with their underlying identity and values. Though hardly unique to the environmental issue area, this elite-driven process of partisan polarization is key to understanding the divergent beliefs and policy preferences of Americans on environmental issues, as well as the current political difficulty in enacting policies to address climate change. The past has shown that the environment need not be a partisan issue. However, the extent to which recent dynamics foreshadow change on the horizon, partly in response to changing environmental conditions and partly due to the attitudes of younger generational cohorts, remains to be seen.

A BOLD NEW ERA OF PUBLIC OPINION?

While this chapter has argued that public opinion is important, numerous trends in the way individuals communicate have posed a challenge to conventional theories of opinion formation and how opinion itself is measured. Indeed, the rise of social media platforms as an important (if often polluted) medium for communication and exchange about climate change has contributed to the further fragmentation of the media environment and to increased polarization. This new reality, where people are exposed to very different kinds of information about climate change, is emerging as an essential component of understanding why Americans hold vastly different attitudes and preferences with respect to climate change beyond the traditional types of aggregate and individual level determinants discussed in this chapter. Meanwhile, the evolution of communication technologies is also changing the way scholars measure public opinion. For instance, on-line data collection drawn from social media and on-line panels have become increasingly popular, replacing traditional modes of data collection involving telephone and mail-back surveys.

The evolution in mode of data collection has helped solve certain problems with declining response rates over the telephone (as people are less inclined to want to answer questions over the phone, or even own a landline), but this has also created new challenges for researchers relying on crowdsourcing and voluntary opt-in panels (like Amazon's Mechanical Turk), which raise new questions (and potentially create new problems) related to the types of people answering questions, their level of attention, and the generalizability of survey results more generally.[70] At the same time, new approaches to collecting public opinion and behavior over popular social media like Facebook and Twitter have also opened up new research opportunities for public sentiment analysis, sampling, and learning about the influence of bots while also creating further challenges in terms of data comparability over different modes.[71] While the relative merits and drawbacks of these different approaches continue to be the subject of close scrutiny, the emerging generation of scholars have at their disposition a variety of tools they can use to measure dynamics in environmental public opinion.

CONCLUSION

This chapter has provided an overview of the dimensions of American public opinion towards the environment, the factors that shape such opinions, and the impact of these views on policymaking in the United States. While by itself public opinion may not determine environmental policy adoption, the public's perceptions of environmental matters and their policy preferences have played an important role in shaping expanded environmental policy engagement over the last 50 years. But as was shown in this chapter, when environmental issues are less salient to the American public, government officials may not prioritize efforts in this area. Such has been the case in the area of climate policy, where despite widespread and growing public acknowledgment of the problem over the last decade, the federal government has not expanded its climate policy efforts, but instead retreated from earlier commitments.

Notably, as the nation heads into the new decade, the saliency of climate change and environmental protection in general are rising. In 2020, the National Surveys on Energy and the Environment found that a record 75 percent of adult Americans believe that there is clear evidence that temperatures are rising on the planet. In that same year, a study by the Pew Research Center found that for the first time in the history of their survey, nearly the same amount of Americans (64 percent) said protecting the environment should be a top national priority as they think strengthening the economy (62 percent) should be a top priority for the Congress and President. And for the first time Pew found a majority of Americans (52 percent) think that dealing with climate change should be a top policy priority for the federal government. Such shifts in opinion may act as a catalyst for substantial climate and environmental policy engagement in Washington, DC, during the decade of the 2020s. But as we discussed in this chapter, opinions regarding the environment can shift quickly, and the increased prominence of environmental, and in particular climate concerns, can retreat behind other priorities. Certainly, the tremendous economic disruptions caused by the COVID-19 pandemic in 2020 will serve as a great test of the most recent elevated saliency of environmental concerns in the United States. The Great Recession a decade earlier helped erode an emerging American consensus on the need to address climate change, and the even greater scope and scale of the economic challenges in the wake of the pandemic may once again alter how Americans think about the environment and the government's role in protecting it.

SUGGESTED WEBSITES

Gallup (www.gallup.com) Gallup is one of the oldest and most established public opinion research organizations in the United States. Gallup conducted some of the earliest surveys on public opinion about environmental issues in the late 1960s and 1970s and maintains some of the longest time series opinion data on American views about environmental matters.

National Surveys on Energy and the Environment (closup.umich.edu/national-surveys-on-energy-and-environment/) The National Surveys on Energy and Environment (NSEE) is a biannual national opinion survey on energy and climate issues.

Pew Research Center (www.pewresearch.org) The Pew Research Center is a nonpartisan fact tank that informs the public about the issues, attitudes, and trends shaping the world, with regular focus on environmental and energy topics.

Resources for the Future (www.rff.org/climatesurvey/) RFF's *Surveying American Attitudes toward Climate Change and Clean Energy* project is a polling partnership with Stanford University.

Roper Center for Public Opinion Research (www.ropercenter.cornell.edu) The Roper Center collects, preserves, and disseminates public opinion data, including many international, national, and state surveys on environmental matters.

Yale Program on Climate Communication (www.climatecommunication.yale.edu) The Yale Program on Climate Communication (YPCC) conducts scientific research on public climate change knowledge, attitudes, policy preferences, and behavior at the global, national, and local scales.

NOTES

1. Justin McCarthy, "In U.S., Water Pollution Worries Highest Since 2001," *The Gallup Organization Social and Policy Issues*, March 31, 2017.
2. Nadja Popovich, Livia Albeck-Ripk, and Kendra Pierre-Louis, "The Trump Administration Is Reversing Nearly 100 Environmental Rules. Here's the Full List," *New York Times*, May 6, 2020.
3. John W. Kingdon, *Agendas, Alternatives, and Public Policies*, 2nd ed. (New York, NY: HarperCollins, 2002).
4. Richard Anthony, "Polls, Pollution and Politics: Trends in Public Opinion on the Environment," *Environment* 24 (1982): 14–20.
5. Lydia Saad, "Preference for Environment Over Economy Largest Since 2000," April 4, 2019, https://news.gallup.com/poll/248243/preference-environment-economy-largest-2000.aspx.
6. Tom Rosenteil, "Fewer Americans See Solid Evidence of Global Warming," *Pew Research Center*, October 22, 2009, https://www.pewresearch.org/2009/10/22/fewer-americans-see-solid-evidence-of-global-warming/.
7. Lyle Scruggs and Salil Benegal, "Declining Public Concern About Climate Change: Can We Blame the Great Recession?" *Environmental Change* (2012). doi:10.1016/j.gloen vcha.2012.01.002.
8. Robert Brulle, Jason Carmichael, and J. Craig Jenkins, "Shifting Public Opinion on Climate Change: An Empirical Assessment of Factors in Influencing Concern Over Climate Change in the U.S., 2002–2010," *Climatic Change* 114, no. 2 (February 2012): 169–88. doi:10.1007/s10584-012-0403-y.
9. Lydia Saad and Jeffery Jones, "U.S. Concern About Global Warming at Eight Year High," March 16, 2016, https://news.gallup.com/poll/190010/concern-global-warming-eight-year-high.aspx.
10. Benjamin Page and Robert Shapiro, "The Effects of Public Opinion on Policy," *American Political Science Review* 77 (March 1983): 175–90; Martin Johnson, Paul Brace, and Kevin Arceneaux, "Public Opinion and Dynamic Representation in the American States: The Case of Environmental Attitudes," *Social Science Quarterly* 86 (February 2005): 87–108; Rebecca Bromley-Trujillo and John Poe, "The Importance of Salience: Public Opinion and State Policy Action on Climate Change," *Journal of Public Policy* 40, no. 2 (June 2020): 280–304.
11. Riley Dunlap, "Trends in Public Opinion Toward Environmental Issues: 1965–1990," *Society and Natural Resources* 4 (1991): 285–312.
12. Everett Carl Ladd, "Cleaning the Air: Public Opinion and Public Policy on the Environment," *Public Opinion* 5 (February/March 1982): 16–20.

13. Frank Newport and Lydia Saad, "Americans Consider Global Warming Real, But Not Alarming," April 9, 2001, https://news.gallup.com/poll/1822/americans-consider-global-warming-real-alarming.aspx.

14. Matthew Nisbet and Teresa Myers, "Twenty Years of Public Opinion about Global Warming," *The Public Opinion Quarterly* 71, no. 3 (Autumn 2007): 444–70.

15. Perri Ruckart, Adrienne Ettinger, Mona Hanna-Attisha, Nicole Jones, Stephanie David, and Patrick Breysse, "The Flint Water Crisis: A Coordinated Public Health Emergency Response and Recovery Initiative," *Journal of Public Health Management Practice* 25 (January/February 2019): S84–90.

16. Robert Dahl, *A Preface to Democratic Theory* (Chicago, IL: The University of Chicago Press, 1956).

17. Pew Research Center, "As Economic Concerns Recede, Environmental Protection Rises on the Public's Policy Agenda," February 2020, www.people-press.org.

18. Pew Research Center, "Partisan Antipathy: More Intense, More Personal," October 10, 2019, www.people-press.org/2019/10/10/partisan-antipathy-more-intense-more-personal/?utm_source=link_newsv9&utm_campaign=item_268982&utm_medium=copy.

19. Rebecca Bromley-Trujillo and John Poe, "The Importance of Salience."

20. Sarah Mills, Natalie Fitzpatrick, Barry Rabe, Christopher Borick, and Erick Lachapelle, "Fuel Economy, Electric Vehicle Rebates, and Gas Taxes: 10 Years of Transportation Policies in the NSEE," *Issues in Energy and Environmental Policy, 38*. (Ann Arbor, MI: Center for Local, State, and Urban Policy at the Gerald R. Ford School of Public Policy, University of Michigan, July 2018).

21. Erick Lachapelle, "Communicating about Carbon Taxes and Emissions Trading Programs," *Oxford Research Encyclopedia of Climate Science* (2017). doi:10.1093/acrefore/9780190228620.013.431.

22. Steffano Carattini, Maria Carvalho, and Sam Fankhauser, "Overcoming public resistance to carbon taxes," *Wiley Interdisciplinary Reviews. Climate Change* 9, no. 5 (2018): e531.

23. Andrea Baranzini and Stefano Carattini, "Effectiveness, Earmarking and Labeling: Testing the Acceptability of Carbon Taxes with Survey Data," *Environmental Economics and Policy Studies* 19, no. 1 (2017): 197–227.

24. Natalie Fitzpatrick, Barry Rabe, Sarah Mills, Christopher Borick, and Erick Lachapelle, "American Opinions on Carbon Taxes and Cap-and-Trade," *Issues in Energy and Environmental Policy, 35*. (Ann Arbor, MI: Center for Local, State, and Urban Policy at the Gerald R. Ford School of Public Policy, University of Michigan, 2018).

25. Tom Tietenberg, "Cap-and-Trade: The Evolution of an Economic Idea," *Agriculture and Resource Economics Review* 39, no. 3 (2010): 359–67.

26. John Broder, "Cap-and-Trade Loses Its Standing as the Energy Policy of Choice," *The New York Times*, March 25, 2010.

27. Barry Rabe and Christopher Borick, "The Climate of Belief: American Public Opinion on Climate Change," *Issues in Governance Studies*, Number 31 (January 2010).

28. Hart Research Associates and Chesapeake Beach Consulting, *Findings from a Survey in States Participating in RGGI*, August 9, 2016, www.Sierraclub.org.

29. David Konisky, "Preferences for Environmental Policy Responsibility," *Publius: The Journal of Federalism* 41 (2011): 76–100.

30. Sarah Mills, Natalie Fitzpatrick, Barry Rabe, Christopher Borick, and Erick Lachapelle, "Should State and Local Governments Address Climate Change: 10 Years of Climate Federalism in the NSEE. *Issues in Energy and Environmental Policy, 34*. (Ann Arbor, MI: Center for Local, State, and Urban Policy at the Gerald R. Ford School of Public Policy, University of Michigan, March 2018).

31. Abraham Maslow, *Motivation and Personality*, 2nd ed. (New York, NY: Viking Press, 1970).

32. Kent Van Liere and Riley Dunlap, "The Social Bases of Environmental Concern: A Review of Hypotheses, Explanations, and Empirical Evidence," *Public Opinion Quarterly* 44, no. 2 (1980): 181–97.

33. Matthew Kahn and Matthew J. Kotchen, "Business Cycle Effects on Concern About Climate Change: The Chilling Effect of Recession," *Climate Change Economics* 2, no. 3 (2011): 257–73.

34. Robert Brulle, Jason Carmichael, and J. Craig Jenkins, "Shifting Public Opinion on Climate Change."

35. Lyle Scruggs and Salil Benegal, "Declining Public Concern About Climate Change."

36. Matto Mildenberger and Anthony Leiserowitz, "Public Opinion on Climate Change: Is There an Economy-Environment Tradeoff?" *Environmental Politics* 26, no. 5 (2017): 801–24.

37. Kent Van Liere and Riley Dunlap, "The Social Bases of Environmental Concern."

38. Lawrence Hamilton, Joel Hartter, Mary Lemcke-Stampone, David Moore, and Thomas Safford, "Tracking Beliefs about Anthropogenic Climate Change. *PLoS One* 10, no. 9 (2015): e0138208. doi:10.1371/journal.pone.0138208.

39. Peter Howe, Matto Mildenberger, Jennifer Marlon, and Anthony Leiserowitz, "Geographic Variation in Opinions on Climate Change at State and Local Scales in the USA," *Nature Climate Change* 5 (2015): 596–603.

40. Patrick Egan and Megan Mullin, "Climate Change: US Public Opinion," *Annual Review of Political Science* 20 (2017): 209–27.

41. Eric R. A. N. Smith, *Energy, the Environment, and Public Opinion* (Lanham, MD: Rowman & Littlefield, 2002).

42. Bradford Bishop, "Focusing Events and Public Opinion: Evidence from the Deepwater Horizon Disaster," *Political Behavior* 36 (2014): 1–22.

43. David Konisky, Llewellyn Hughes, and Charles Taylor, "Extreme Weather Events and Climate Change Concern," *Climatic Change* 134 (2016): 533–47.

44. Kent Van Liere and Riley Dunlap, "The Social Bases of Environmental Concern."

45. Ronald Inglehart, "Public Support for Environmental Protection: Objective Problems and Subjective Values in 43 Societies," *PS: Political Science and Politics* 28, no. 1 (1995): 57–72.

46. Kent Van Liere and Riley Dunlap, "The Social Bases of Environmental Concern"; Robert Jones and Riley Dunlap, "The Social Bases of Environmental Concern: Have They Changed Over Time," *Rural Sociology* 57, no. 1 (1992): 28–47.

47. Paul Slovic, "Trust, Emotion, Sex, Politics, and Science: Surveying the Risk-assessment Battlefield," *Risk Analysis* 19 (1999): 689–701.

48. Patrick Egan and Megan Mullin, "Climate Change: US Public Opinion."

49. Chenyang Xiao, and Aaron McCright, "Explaining Gender Differences in Concern About Environmental Problems in the United States," *Society and Natural Resources* 25 (2012): 1067–84.

50. Salil D. Benegal, "The Spillover of Race and Racial Attitudes into Public Opinion About Climate Change," *Environmental Politics* 27 (2018): 4.

51. Adam R. Pearson, Matthew T. Ballew, Sarah Naiman, and Jonathan Schuldt, "Race Class, Gender and Climate Change Communication," *Climate Science* (April 2017).

52. Julia Robertson Hathaway, "Climate Change, the Intersectional Imperative, and the Opportunity of the Green New Deal," *Journal of Environmental Communication* 14, no. 1 (2020): 13–22.

53. Robert Jones and Riley Dunlap, "The Social Bases of Environmental Concern"; Aaron McCright, Chenyang Xiao, and Riley Dunlap, "Political Polarization on Support for Government Spending on Environmental Protection in the USA, 1974-2012, *Social Science Research* 48 (2014): 251–60.

54. Lawrence Hamilton, "Education, Politics, and Opinions About Climate Change Evidence for Interaction Effects," *Climatic Change* 104 (2011): 231–42.
55. Aaron McCright, "Political Orientation Moderates Americans' Beliefs and Concern About Climate Change: An Editorial Comment," *Climatic Change* 104 (2011): 243–53; Alexandre Morin-Chassé and Erick Lachapelle, "Partisan Strength and the Politicization of Global Climate Change: A Re-examination of Schuldt, Roh, and Schwarz," *Journal of Environmental Studies and Sciences* 10 (2020): 31–40.
56. Daniel Kahneman, *Thinking Fast and Slow* (New York, NY: Farrar, Straus and Giroux, 2012).
57. Dan Kahan, Ellen Peters, Maggie Wittlin, Paul Slovic, Lisa Larrimore Ouellette, Donald Braman, and Gregory Mandel, "The Polarizing Impact of Science Literacy and Numercy on Perceived Climate Change Risks," *Nature Climate Change* 2 (2012): 732–5.
58. Aaron McCright and Riley Dunlap, "Anti-Reflexivity: The American Conservative Movement's Success in Undermining Climate Science and Policy," *Theory, Culture & Society* 27, no. 2–3 (2010): 100–33.
59. Kent Van Liere and Riley Dunlap, "The Social Bases of Environmental Concern."
60. Aaron McCright, Sandra Marquart-Pyatt, Rachael Shwom, Steven Brechin, and Summer Allen, "Ideology, Capitalism, and Climate: Explaining Public Views About Climate Change in the United States," *Energy Research & Social Science* 21 (2016): 180–9.
61. Kent Van Liere and Riley Dunlap, "The Social Bases of Environmental Concern."
62. Aaron McCright, Sandra Marquart-Pyatt, Rachael Shwom, Steven Brechin, and Summer Allen, "Ideology, Capitalism, and Climate: Explaining Public Views About Climate Change in the United States," *Energy Research & Social Science* 21 (2016): 180–9.
63. Jonathan Schuldt, Sarah Roh, and Norbert Schwarz, "'Global Warming' or 'Climate Change'? Whether the Planet Is Warming Depends on Question Wording," *Public Opinion Quarterly* 75, no. 1 (2011): 115–24.
64. Christopher Borick, Natalie Fitzpatrick, Sarah Mills, Erick Lachapelle, and Barry Rabe, "Belief (and Disbelief) in Global Warming: 10 Years of Attitudes About Climate Change in the NSEE," *Issues in Energy and Environmental Policy, 43* (Ann Arbor, MI: Center for Local, State, and Urban Policy at the Gerald R. Ford School of Public Policy, University of Michigan, October, 2019).
65. Riley Dunlap, Aaron McCright, and Jerrod Yarosh, "The Political Divide on Climate Change: Partisan Polarization Widens in the U.S." *Environment: Science and Policy for Sustainable Development* 58, no. 5 (2016): 4–23; Riley Dunlap, "Partisan Polarization on the Environment Grows Under Trump," *Gallup Blog*, April 5, 2019, https://news.gallup.com/opinion/gallup/248294/partisan-polarization-environment-grows-trump.aspx.
66. Nadja Popovich, "Climate Change Rises as a Public Priority, But It's More Partisan than Ever," *The New York Times*, February 20, 2020.
67. Matthew Goldberg, Abel Gustafson, Seth Rosenthal, John Kotcher, Edward Maibach, and Anthony Leiserowitz, *For the First Time, the Alarmed Are Now the Largest of Global Warming's Six Americas* (New Haven, CT: Yale University and George Mason University. Yale Program on Climate Change Communication, January 16, 2020).
68. Alexandre Morin-Chassé and Erick Lachapelle, "Partisan Strength and the Politicization of Global Climate Change"; Lawrence Hamilton, Joel Hartter, and Erin Bell, "Generation Gaps in US Public Opinion on Renewable Energy and Climate Change," *PLoS One* 14, no. 7 (2019): e0217608.
69. Patrick Egan and Megan Mullin, "Climate Change: US Public Opinion."
70. Robert Groves, "Three Eras of Survey Research," *Public Opinion Quarterly* 75, no. 5 (2011): 861–71; Michael Chmielewski and Sarah Kucker, "An MTurk Crisis: Shifts of Data Quality and the Impact on Study Results," *Social Psychology and Personality Science* 11, no. 4 (2020): 464–73.

71. Emily Cody, Andrew Reagan, Lewis Mitchell, Peter Sheridan Dodds, and Christopher Danforth, "Climate Change Sentiment on Twitter: An Unsolicited Public Opinion Poll," *PLoS One* 10, no. 8 (August 20, 2015): e0136092; S. Mo Jang and P. Sol Hart, "Polarized Frames on 'Climate Change' and 'Global Warming' Across Countries and States: Evidence from Twitter Big Data" *Global Environmental Change* 32 (May 2015): 11–17; Baobao Zhang, Matto Mildenberger, Peter Howe, Jennifer Marlon, Seth Rosenthal, and Anthony Leiserowitz, "Quota Sampling Using Facebook Advertisement," *Political Science Research Methods* (2018): 1–7. doi:10.1017/psrm.2018.49; Oliver Milman, "Revealed: Quarter of All Tweets About Climate Crisis Produced by Bots," *The Guardian* (February 2020), www.theguardian.com/technology/2020/feb/21/climate-tweets-twitter-bots-analysis.

FEDERAL INSTITUTIONS AND POLICY CHANGE

4 PRESIDENTIAL POWERS AND ENVIRONMENTAL POLICY

Norman J. Vig

I have an Article II where I have the right to do whatever I want as president.
President Donald J. Trump, July 23, 2019

Thanks to our bold regulatory campaign, the United States has become the number-one producer of oil and natural gas anywhere in the world, by far.
President Donald J. Trump, State of the Union Address, February 4, 2020

In early September of 2020, with the presidential campaign in full swing, President Donald J. Trump gave a speech near Palm Beach, Florida, to burnish his environmental credentials. He announced that he was about to sign an executive order that would extend a 10-year moratorium on oil drilling in the eastern half of the Gulf of Mexico and off the Atlantic coasts of Florida, Georgia and South Carolina, thus reversing an unpopular policy he had announced two years earlier (he later extended the moratorium to North Carolina as well). In addition to the drilling moratorium, Trump claimed several environmental achievements including signing of the Great American Outdoors Act, an omnibus conservation bill recently passed by a rare bipartisan majority in Congress after years of effort (Chapter 5). He proclaimed the Act "the most significant investment in our national parks in over a century" and went on to call himself "number one since Teddy Roosevelt."[1] While these claims could be written off as campaign hyperbole, they reflected the fact that environmental policy had become an important issue in the presidential election. Trump, who was widely regarded as the most anti-environmental president in memory, thus felt it necessary to address a growing political backlash over his environment record, at least in potential swing states.

The event also highlighted two dominant trends in environmental policy in recent decades. The first is the expansion of presidential powers in defining national environmental priorities and policies. In the absence of congressional action, what has been labeled "the administrative presidency" has become the default mode for environmental governance—that is, the use of executive rather than legislative powers to control both the direction and the implementation of these policies.[2] The second trend is the growing partisan divide over climate change and other environmental issues. Although there are some signs of moderation, it appears that opinion over environmental issues is more deeply polarized than at any time in recent history (see Chapter 3).[3] The result has been sharp policy oscillations depending on which party wins the presidency.

Such presidential policy "cycling" has been especially obvious in the case of environmental policy under Barack Obama and Donald Trump. President Obama used expansive executive powers to launch a series of domestic and international initiatives to attack climate change for the first time and to extend the reach of previous environmental laws through new rules and regulations. During the 2016 election campaign, Mr. Trump repeatedly called climate change a "hoax" and threatened to reverse Obama's priorities: he promised to repeal Obama's Clean Power Plan and other climate change regulations; to "cancel" the Paris Agreement on Climate Change; to dismantle much of the Environmental Protection Agency (EPA); and to remove barriers to the rapid expansion of oil, gas, and coal production as part of his "America First" energy plan. The result was a stark reversal of nearly all of Obama's achievements and, in some cases, a weakening of basic environmental legislation dating back to the 1970s. During the 2020 campaign, the Democratic presidential nominee, Joseph R. Biden Jr., pledged to reverse all of these changes and to enact environmental policies going well beyond those of the Obama administration in which he had served as vice-president (Chapter 15).

Article II of the Constitution does indeed grant the president broad executive authority to conduct foreign and domestic policy, but President Trump's view of unfettered presidential power was unprecedented.[4] In this chapter I will compare the environmental records of presidents since 1980. But first it is necessary to take a closer look at the powers of the presidency as traditionally understood and the criteria for comparing presidents' environmental policies.

PRESIDENTIAL POWERS AND CONSTRAINTS

The formal *roles* of the president have been summarized as commander in chief of the armed forces, chief diplomat, chief executive, legislative leader, and opinion/party leader.[5] If we look only at environmental policy, the president's role as chief executive has been most important. But some presidents, including Teddy Roosevelt, Franklin Roosevelt, and Richard Nixon, have played a leading role in enacting environmental legislation and in using the "bully pulpit" to rally public opinion behind new environmental policies. The role of chief diplomat has also become more important as many environmental problems have required international solutions; for example, Ronald Reagan supported and signed the landmark Montreal Protocol on Substances that Deplete the Ozone Layer in 1987, and Barack Obama played a leading role in negotiating the Paris Climate Agreement of 2015. Finally, as commander in chief of the armed forces, the president also deals with a growing range of environmental issues, including the destabilization of regions and governments affected by climate change and threats posed by melting Arctic ice and rising sea levels.[6]

Some of the president's *powers* are "contextual"; that is, they shape the general context of policymaking and the broad directions of the administration. Presidents can draw attention to issues and frame the political agenda through speeches, press conferences, and, as in the case of Mr. Trump's "twitter presidency," the new social media;[7] they can propose legislation and budgets; they nominate and appoint cabinet members and other key officials; and they can reorganize their staff and departments to better implement their policies. Other powers of the president are more "unilateral" and allow the president to influence policy decisions directly. These include the power to veto legislation; to issue executive orders, directives, and proclamations; to make executive agreements; and to monitor and control regulatory processes.[8] One unique power is the authority to designate national monuments to protect areas of exceptional scenic, scientific, cultural, or historical value under the Antiquities Act of 1906. Trump has also claimed the authority to drastically alter monuments designated by his predecessors.

Article II of the Constitution directs the president to "take Care that the Laws be faithfully executed." Most legislation passed by Congress is quite broad, however, and leaves much to the discretion of the president and the implementing agencies. The EPA, for example, is charged by the Clean Air Act to formulate and issue detailed standards, rules, and regulations to control the emission of pollutants to ensure healthy air quality (see Chapter 7). These rules and regulations are adopted through lengthy administrative proceedings, and once finalized, they have the full force of law. Presidents therefore try to centralize and control rulemaking to ensure that it reflects their policy agendas. All presidents since Nixon have required that important regulations be cleared by the Office of Management and Budget (OMB) before they are proposed by the EPA and other regulatory agencies. Presidents can thus have a major impact on policy decisions through controlling the bureaucracy rather than passing new legislation.[9]

Ultimately, however, presidents cannot govern alone; they are part of a government of "separated powers."[10] Their policies must be grounded in statutes passed by Congress, and Congress also determines the budgets and spending limits of all executive agencies. The Senate can refuse to confirm the president's nominees for office and deny the ratification of treaties. Even with a majority in both houses, the president may not have sufficient support to pass new legislation. Without such legislation, executive actions may be difficult to carry out. There can be considerable resistance to policy change within the bureaucracy itself.[11] Moreover, most major rules and regulations, as well as many executive orders and other presidential actions, are challenged in the courts by affected parties, often tying up policies in litigation for years (see Chapter 6). Executive orders can also be modified or rescinded by subsequent presidents, so they are rarely permanent. Finally, in our federal system, most environmental laws depend heavily on the states for implementation. In recent years groups of states have frequently challenged federal actions through lawsuits filed by their attorneys general (see Chapter 2).

COMPARING ENVIRONMENTAL PRESIDENCIES

Presidential leadership on environmental policy can be analyzed and compared in different ways. We can examine how each president performs the five basic *roles* noted in the previous section as they pertain to the environment. Two of the political scientists who developed this approach to presidents' environmental records have extended their analysis to twelve past presidents, focusing on four differentiating factors: *political communication, legislative leadership, administrative actions,* and *environmental diplomacy.*[12] Presidents were then ranked according to their positive or negative impacts on the environment, yielding a "continuum of greenness." Seven presidents are classified as having had a positive impact (Franklin Roosevelt, Harry Truman, John Kennedy, Lyndon Johnson, Richard Nixon, Jimmy Carter, and Bill Clinton), three as having mixed impacts (Dwight Eisenhower, Gerald Ford, and George H. W. Bush), and two as having negative impacts (Ronald Reagan and George W. Bush). Roosevelt and Nixon rank as the "greenest" by these criteria.[13] Roosevelt greatly expanded the conservation programs pioneered by his cousin, Teddy Roosevelt; and Nixon proclaimed the 1970s the "environmental decade," created the EPA by executive action, and signed most of the landmark environmental legislation of our era.[14]

For our purposes, a president's overall environmental performance can be evaluated by examining a few basic indicators: (1) the president's environmental *agenda* as expressed in campaign statements, policy documents, and other pronouncements; (2) presidential *appointments* to key positions in government departments and agencies and to the White House staff; (3) the relative priority given to environmental programs in the president's proposed *budgets*; (4) presidential *legislative success*; (5) *executive orders and other unilateral actions* by the president; and (6) presidential support for or opposition to *international environmental agreements*. By these criteria, some presidents have been much more favorable to the environment than others.

Space does not allow a full comparison of all recent administrations. I will briefly summarize the environmental records of four presidents since 1980: Reagan, G.H.W. Bush, Clinton, and G.W. Bush. We can see many contrasts in presidential leadership from these case studies—as well as some of the dynamics of policy cycles over time. Fuller attention will be paid to the Obama presidency's environmental policy and the actions of the Trump administration to undo it.

THE REAGAN REVOLUTION: ENVIRONMENTAL BACKLASH

The "environmental decade" of the 1970s came to an abrupt halt with Reagan's inauguration in 1981. Although the environment was not a major issue in the election, Reagan was the first president to come to office with an avowedly antienvironmental agenda. Influenced by the Sagebrush Rebellion—an attempt by several western states to claim ownership of federal lands—as well as by his years of public relations work for corporate and conservative causes, Reagan viewed environmental conservation as fundamentally at odds with economic growth and prosperity. He saw environmental regulation as a barrier to "supply side" economics and sought to reverse or weaken many of the policies of the previous decade.[15] Although only partially successful, Reagan's agenda laid the groundwork for renewed attacks on environmental policy in later decades.

Because it came after a period of economic decline, Reagan's landslide victory appeared to reflect a strong mandate for policy change. And with a new Republican majority in the Senate, he was able to gain congressional support for the Economic Recovery Tax Act of 1981, which embodied much of his program. He reduced income taxes by nearly 25 percent and deeply cut spending for environmental and social programs. Despite this initial victory, however, Reagan faced a Congress that was divided on most issues and did not support his broader environmental goals. On the contrary, the bipartisan majority that had enacted most of the environmental legislation of the 1970s remained largely intact.

Faced with this situation, Reagan turned to what has been termed an "administrative presidency."[16] Essentially, this involved an attempt to change federal policies by maximizing control of policy implementation within the executive branch. The administrative strategy initially had four major components: (1) careful screening of all appointees to environmental and other agencies to ensure compliance with Reagan's ideological goals, (2) tight policy coordination through cabinet councils and White House staff, (3) deep cuts in the budgets of environmental agencies and programs, and (4) an enhanced form of regulatory oversight to eliminate or revise regulations considered burdensome by industry.

Reagan's appointment of officials who were overtly hostile to the mission of their agencies aroused strong opposition from the environmental community. His selection of Anne Gorsuch to head the EPA and James Watt as secretary of the interior provoked controversy from the beginning because both were attorneys who had spent long years litigating against environmental regulation. Both made it clear that they intended to rewrite the rules and procedures of their agencies to accommodate industries such as mining, logging, and oil and gas. Watt was also designated head of the new cabinet council to coordinate policies in all the environmental and natural resource agencies, and to bring these policies in line with the president's agenda.

In the White House, Reagan lost no time in changing the policy machinery to accomplish the same goal. He all but eliminated his environmental advisers and instead appointed Vice President George H. W. Bush to head a new Task Force on Regulatory Relief, to identify and modify or rescind regulations targeted by business and industry. More important, in February 1981 Reagan issued Executive Order 12291, which carried White House control of the regulatory process to a new level. All major regulations (costing over $100 million) were now to undergo prepublication review by the Office of Information and Regulatory Affairs (OIRA) in the OMB. They were to be accompanied by rigorous benefit–cost analyses demonstrating that benefits exceeded costs, and they had to include evaluation of alternatives to ensure that net social benefits were maximized. OIRA consequently held up, reviewed, and revised hundreds of EPA and other regulations to reduce their effect on industry. The number of new environmental rules declined sharply in the Reagan years.[17] At the same time, Reagan's budget cuts had major effects on the capacity of environmental agencies to implement their growing policy mandates. The EPA lost approximately one-third of its operating budget and one-fifth of its personnel in the early 1980s. The White House Council on Environmental Quality (CEQ) lost most of its staff and barely continued to function. In the Interior Department and elsewhere, funds were shifted from environmental to development programs.[18]

Not surprisingly, Congress responded by investigating OIRA procedures and other activities of Reagan appointees, especially Gorsuch and Watt. Gorsuch came under heavy attack for confidential dealings with business and political interests that allegedly led to sweetheart deals on matters such as Superfund cleanups. After refusing to disclose documents, she was found in contempt of Congress and forced to resign (along with twenty other high-level EPA officials) in March 1983. Watt was pilloried in Congress for his efforts to open virtually all public lands (including wilderness areas) and offshore coastal areas to mining and oil and gas development, and he left office later in 1983.[19] William Ruckelshaus, the original EPA administrator, was brought back in to run the EPA and gradually restored the funding and credibility of the agency.

Because of these embarrassments and widespread public and congressional opposition to weakening environmental protection, Reagan's deregulatory campaign was thus largely spent by the end of his first term. Recognizing that his policies had backfired, the president took few new initiatives during his second term. Congress passed a series of amendments to existing laws such as Superfund, the Safe Drinking Water Act, and the Clean Water Act over Reagan's opposition (see Appendix 1). These laws mandated stricter regulatory timetables and enforcement and were intended to reduce the discretionary authority of the EPA and other executive agencies. Indeed, the backlash *against* Reagan *strengthened* rather than weakened key environmental statutes, thus reinforcing these policy cycles.

PRESIDENT GEORGE H. W. BUSH: CLEAN AIR LEGISLATION

George H. W. Bush's presidency thus returned to a more moderate tradition of Republican leadership, particularly in the first 2 years. While promising to "stay the course" on Reagan's economic policies, Bush also pledged a "kinder and gentler" America. Although his domestic policy agenda was the most limited of any recent president, it included action on the environment. Indeed, during the campaign, Bush declared himself a "conservationist" in the tradition of Teddy Roosevelt and promised to be an "environmental president."

If Bush surprised almost everyone by seizing the initiative on what most assumed was a strong issue for the Democrats, he impressed environmentalists even more by soliciting their advice and by appointing environmental leaders to his administration. William Reilly, the highly respected president of the World Wildlife Fund and the Conservation Foundation, became EPA administrator; and Michael Deland, formerly New England director of the EPA, became chairman of the CEQ.

Bush pursued a bipartisan strategy in passing the Clean Air Act Amendments of 1990, arguably the single most important legislative achievement of his presidency. His draft bill, sent to Congress on July 21, 1989, had three major goals: to control acid rain by reducing by nearly half sulfur dioxide emissions from coal-burning power plants by 2000, to reduce air pollution in eighty urban areas that still had not met 1977 air quality standards, and to lower emissions of nearly two hundred airborne toxic chemicals by 75–90 percent by 2000. To reach the acid precipitation goals—to which the White House devoted most of its attention—Bush proposed a cap-and-trade system rather than command-and-control regulation, so as to achieve emissions reductions more efficiently.[20] The act also prohibited the use of CFCs and other ozone-depleting chemicals by 2000. Bush also signed the Intermodal Surface Transportation Efficiency Act (ISTEA) of 1991 which provided major funding for mass transportation planning and development.

However, during his reelection campaign, Bush declared a moratorium on further environmental regulation and retreated to a more traditional Republican stance. He also threatened to boycott the UN Conference on Environment and Development (the Earth Summit) in June 1992 until he had ensured that the climate change convention to be signed would contain no binding targets for carbon dioxide reduction. He further alienated the environmental community by refusing to sign the Convention on

Biological Diversity, despite efforts by his delegation chief, William Reilly, to seek a last-minute compromise. Thus, despite Bush's other foreign policy accomplishments, the United States failed to lead in environmental diplomacy (see Chapter 12).

THE CLINTON PRESIDENCY: FRUSTRATED AMBITIONS

President Bill Clinton entered office in 1993 with high expectations from environmentalists. His campaign promises included many environmental pledges: to raise the Corporate Average Fuel Economy (CAFE) standard for automobiles, support renewable energy research and development, limit US carbon dioxide emissions to 1990 levels by 2000, create a new solid waste reduction program and provide other incentives for recycling, pass a new Clean Water Act with standards for nonpoint sources, reform the Superfund program and tighten the enforcement of toxic waste laws, protect ancient forests and wetlands, preserve the Arctic National Wildlife Refuge, sign the biodiversity convention, and restore funding to UN population programs.[21]

Clinton's early actions indicated that he intended to deliver on his environmental agenda. The environmental community largely applauded his appointments to key environmental positions. Perhaps most important, Vice President Al Gore was given the lead responsibility for formulating and coordinating environmental policy in the White House. Several of his former Senate aides were also appointed to high positions, including EPA administrator Carol Browner. Other appointments to the cabinet and executive office staffs were largely proenvironmental. The most notable environmental leader was Bruce Babbitt, a former Arizona governor and president of the League of Conservation Voters, who became secretary of the interior. In contrast to his predecessors in the Reagan and Bush administrations, Babbitt came to office with a strong reform agenda for western public lands management.

Although Clinton entered office with an expansive agenda and Democratic majorities in both houses of Congress, his environmental agenda quickly got bogged down. Two events early in the term gave the administration an appearance of environmental policy failure. Babbitt promptly launched a campaign to "revolutionize" western land use policies; it included a proposal in Clinton's first budget to raise grazing fees on public lands closer to private market levels (something natural resource economists had advocated for many years). The predictable result was a furious outcry from cattle ranchers and their representatives in Congress; Clinton backed down and removed the proposal from the bill. Much the same thing happened on the so-called BTU tax. This was a proposal to levy a broad-based tax on the energy content of fuels as a means of promoting energy conservation and addressing climate change. Originally included in the president's budget package at Gore's request, it was eventually dropped in favor of a much smaller increase in the federal gasoline tax (4.3 cents per gallon) in the face of fierce opposition from members of both parties in Congress. Clinton was, however, able to get bipartisan support for the passage of two relatively uncontroversial bills in 1996, the Food Quality Protection Act, which established a new basis for regulating pesticide uses, and the Safe Drinking Water Act Amendments (Appendix 1).

Clinton also relied heavily on his executive powers to pursue his environmental agenda. A "reinventing environmental regulation" initiative launched in 1995 created some fifty new EPA programs to encourage voluntary pollution reduction and to reward states and companies that exceeded regulatory requirements.[22] EPA administrator Browner also strengthened existing regulations and enforcement. For example, in 1997, she issued tighter ambient air quality standards for ozone and small particulate matter. In the final year of the Clinton administration, the EPA proposed a series of new regulations tightening standards on other forms of pollution, including diesel emissions from trucks and buses and arsenic in drinking water. It also established a legal basis for future regulation of carbon emissions from cars and light trucks (see Chapter 7).

In addition to strengthening EPA regulations, the Clinton administration took numerous executive actions to protect public lands and endangered species. For example, it helped to broker agreements to protect old-growth forests in the Pacific Northwest, the Florida Everglades, and Yellowstone National Park. The White House actively promoted voluntary agreements to establish habitat conservation plans to protect wildlife throughout the country.[23] Clinton also used his authority under the Antiquities Act to establish or enlarge twenty-two national monuments covering more than six million acres. Finally, in January 2001, Clinton issued a long-awaited executive order protecting nearly sixty million acres of "roadless" areas in national forests from future road construction and hence from logging and development. He could thus claim to have preserved more public land in the contiguous United States than any president since Theodore Roosevelt.[24]

However, the Clinton administration largely failed to develop an effective response to the greatest challenge of the new century: climate change. After defeat of the BTU tax in 1993, Clinton's climate change proposals called for only voluntary actions; and the administration refused to commit the United States to binding reductions prior to the Kyoto treaty negotiations in December 1997. By then, the Senate had gone on record opposing any agreement limiting US emissions.[25] Ultimately, Clinton authorized Vice President Gore to break the deadlock at Kyoto with an offer to reduce US emissions to 7 percent below 1990 levels by 2008–2012, and the United States signed the Kyoto Protocol in 1998. However, Congress made it clear that it would not ratify the agreement and prohibited all efforts to implement it.

PRESIDENT GEORGE W. BUSH: REGULATORY STALEMATE

George W. Bush took office in 2001 with a weak mandate to govern. He had lost the popular vote to Al Gore and had been declared the Electoral College winner only after several weeks of wrangling over contested Florida ballots, a dispute culminating in the Supreme Court's intervention. However, in the wake of the September 11, 2001, terrorist attacks on New York City and Washington, DC, his powers were greatly enlarged. Like Reagan, Bush used the executive powers of the presidency to advance an antiregulatory, probusiness agenda, though not to the same degree. He exercised the powers of appointment, budget, regulatory oversight, and rulemaking to weaken environmental policies.[26] Vice President Dick Cheney played a leading role in selecting cabinet appointees and, like previous vice presidents, in shaping energy and environmental policies.

With the exception of Christine Todd Whitman, the former governor of New Jersey who was picked to head the EPA, Bush's initial appointments to environmental and natural resource agencies were largely drawn from business corporations or from conservative interest groups, law firms, and think tanks. Among the more controversial of these appointees were secretary of the interior Gale Norton, a protégée of James Watt and a strong advocate of resource development; J. Steven Griles, her deputy secretary and a longtime coal and oil industry lobbyist; Julie MacDonald, deputy assistant secretary of interior for fish and wildlife (responsible for the Endangered Species Act); and Mark Rey, a timber industry lobbyist, as undersecretary of agriculture for natural resources and environment (including the US Forest Service). All of these officials left office under a cloud of investigation after ignoring numerous statutory and regulatory limits on resource exploitation and were replaced by more seasoned administrators in the second term.

President Bush's first budget proposal, for fiscal year 2002, called for a modest 4 percent increase in overall domestic discretionary spending, but an 8 percent reduction in funding for natural resource and environmental programs (the largest cut for any sector). The EPA's budget was to be slashed by nearly $500 million, or 6.4 percent, and the Interior Department budget was slated for a 3.5 percent cut.[27] Congress, however, did not approve these budget cuts. Total EPA spending remained near $8 billion through 2008 but declined somewhat when adjusted for inflation. On the other hand, federal outlays for the Departments of Energy and the Interior increased significantly, and overall spending for natural resources and the environment rose by about 25 percent between 2001 and 2008.[28] Thus, environmental agencies did not suffer the budget cuts they did during the Reagan administration.

After suspending or rejecting many of Clinton's last-minute regulations, the Bush White House reestablished the Reagan-era rules for regulatory review. OIRA carried out an extensive analysis of proposed regulations, demanding that agencies justify all new rules by strict cost–benefit analysis. At the behest of business groups, existing rules were also reviewed in order to reduce the burden of regulation wherever possible.[29] Going one better than Reagan, in January 2007, the president issued a new executive order requiring that each agency must have a regulatory policy office run by a political appointee to manage the regulatory review process.[30] The order also granted OIRA new authority to review and edit "agency guidance documents," including scientific reports and memoranda, and to hold up proposed regulations indefinitely. The result was a highly politicized form of administration in which the political interests of the president and his supporters frequently overrode scientific and technical considerations in the bureaucracy.

President Bush's energy and environmental agenda was quickly shaped after he took office. During spring 2001, a national energy plan was drafted in secrecy by a task force appointed by Vice President Cheney. The plan called for major increases in future energy supplies, including domestic oil, gas, nuclear, and "clean coal" development, and for streamlining environmental regulations to accelerate new energy production. A bill incorporating these and other aspects of the Bush–Cheney plan quickly passed the House of Representatives in 2001 but later stalled in the Senate when authorization to drill in the Arctic National Wildlife Refuge was defeated. Eventually, an energy bill was passed in 2005 providing large subsidies, loan guarantees, and other incentives to conventional energy producers (see Appendix 1).

The administration's "Clear Skies" bill, introduced in Congress in 2002, incorporated many of industries' suggestions for scaling back the pollution control requirements of the Clean Air Act. When this legislation went nowhere, Bush proceeded to issue executive orders that, in effect, implemented similar rules. For example, one of the more controversial rules relaxed requirements for the installation of new pollution equipment when power plants and oil refineries expanded or increased production (see Chapter 7). A broader Clean Air Interstate Rule issued in 2005 set standards for conventional air pollutants such as sulfur dioxide and ozone in twenty-eight eastern states, while another rule regulated mercury emissions from coal-fired power plants. These rules raised current standards but were less rigorous than those recommended by EPA scientists and, in most cases, were overturned by the courts.[31]

Perhaps President Bush's most significant executive action was his rejection of the Kyoto Protocol on climate change in 2001. Calling the treaty "fatally flawed," the president officially withdrew US participation in the international regime for regulating greenhouse gases. This was part of a larger shift away from international treaty obligations in the Bush administration, but it presaged an 8-year effort to block any mandatory requirements for controlling greenhouse gases (despite Bush's campaign promises to regulate carbon dioxide). Instead, Bush supported continuing research programs on climate science and technological development, including new efforts to develop hydrogen energy and other alternative fuels.[32] The Energy Independence and Security Act of 2007 incorporated many of these proposals and also modestly raised auto fuel efficiency standards for the first time in decades.[33] But despite a landmark US Supreme Court decision, *Massachusetts v. EPA*, in April 2007 holding that the EPA could regulate greenhouse gases under the Clean Air Act, the White House refused to allow the agency to develop a regulatory strategy for climate change or to grant California a waiver to regulate carbon dioxide emissions from motor vehicles (see Chapters 6 and 7).

PRESIDENT BARACK OBAMA: CLIMATE BREAKTHROUGH

Barack Obama took office in January 2009 with what appeared to be a strong electoral mandate. Although he inherited the multiple economic and foreign policy crises left by the Bush administration, his campaign based on messages of "hope" and "change we can believe in" seemed to provide an opening for far-reaching reforms. Obama had also endorsed a strong environmental agenda. Among other things, he promised to create a cap-and-trade program for reducing US greenhouse gas emissions, to make massive investments in renewable energy to create millions of new jobs, to double auto fuel economy standards by 2025, to tighten air pollution standards for mercury and other pollutants from power plants, to make major improvements in energy efficiency standards for buildings and in the national electricity grid, and to reverse Bush administration policies on mining and the protection of roadless areas.[34] The Democratic Party also gained the largest majorities in Congress in decades, making a legislative strategy appear feasible.

Obama's climate change bill, the American Clean Energy and Security Act of 2009, would have established a national cap-and-trade system to reduce carbon dioxide emissions throughout the US economy. The bill passed by a close, largely partisan, vote in the House, but failed in the Senate after lengthy compromise negotiations (Chapter 5). Once the Republicans regained a majority in the House of Representatives in 2010, it became impossible to enact any environmental legislation. Thus, like his predecessors, Obama turned increasingly to an administrative strategy to carry out his agenda. Following his reelection in 2012, he made it clear that he would act with or without the support of Congress.[35]

President Obama's choices for cabinet and top White House staff positions were well regarded by environmentalists. His first-term "green team" included Lisa Jackson, a chemical engineer who had served in the EPA and as commissioner of the New Jersey Department of Environmental Protection, as EPA administrator; Steven Chu, a Nobel Prize–winning physicist who directed the Lawrence Berkeley National Laboratory, as energy secretary; and Sen. Ken Salazar, D-Colorado, as interior secretary. Obama also appointed several other top scientists to the administration, including John Holdren, a Harvard physics professor, as White House science adviser.[36] Obama's second-term appointments were considered strong environmentalists but pragmatic administrators as well. Gina McCarthy, the assistant administrator in charge of air and radiation at the EPA, became the new EPA head; Ernest J. Moniz, a distinguished physicist from the Massachusetts Institute of Technology, replaced Steven Chu at the Energy Department; and Sally Jewell, an oil geologist and businesswoman, became secretary of the interior. These and other scientists played a more prominent role in decision-making than in any recent presidency.

Despite taking office amidst the worst recession since the 1930s and strong opposition from Republicans to raising the national debt, President Obama presided over a significant increase in federal spending for energy and the environment. Many of his goals for developing new energy technologies were incorporated into his emergency "stimulus bill" (the American Recovery and Reinvestment Act of 2009), which included some $80 billion in new spending, tax incentives, and loan guarantees to promote energy efficiency, renewable energy sources, fuel-efficient cars, mass transit, and cleaner fuel technologies. He also called for large increases in the budget of the EPA and in the budgets of other environmental agencies. His first budget (for fiscal year 2010) requested $10.5 billion for the EPA (48 percent more than requested by President Bush in his final budget), and Congress approved $10.2 billion. Actual EPA spending rose from $8 billion in 2009 to over $12.7 billion in 2012 and then fell back to an estimated $8.3 billion in 2016. Overall spending for natural resources and the environment increased from $35.5 billion in 2009 to a peak of $45.9 billion in 2011 before settling back to an estimated $39.5 billion in 2016.[37]

Given his experience as a senator, his criticism of George W. Bush's style of leadership, and his preference for legislative solutions, it is not surprising that President Obama was slower to develop an administrative presidency. Nevertheless, he utilized some executive powers from the beginning. Upon taking office, he suspended or revoked a number of Bush's executive orders and regulations, including those on California's request for a waiver to regulate greenhouse gas emissions from automobiles, oil and gas leasing in potential wilderness areas, and the political direction of regulatory review. In March 2009, he issued a "Presidential Memorandum on Scientific Integrity" to heads of agencies and departments to prevent political misuse of scientific research and information such as was alleged to have occurred during his predecessor's tenure.[38]

Obama took a relatively cautious approach to the use of unilateral powers during his first term. Nevertheless, among other regulations, the EPA issued ones to control mercury and other toxic emissions from industrial boilers, incinerators, and power plants; to tighten standards for the emission of sulfur dioxide, nitrogen oxides, and particulates that drift downwind from twenty-eight eastern states and the District of Columbia (the Cross-State Air Pollution Rule); and to limit some carbon emissions from large industries. In August 2012, following negotiations with the auto industry, environmentalists, and energy experts, Obama raised fuel economy standards for cars and light trucks to 54.5 miles per gallon by 2025.[39]

At the Copenhagen climate change conference in December 2009, President Obama pledged to reduce US greenhouse gas emissions by 17 percent from 2005 levels by 2020, roughly the reduction expected from passage of the climate change bill then pending in Congress. Despite the failure of the bill, Obama continued to espouse this goal and in his second term made its achievement a top priority.

The president outlined his strategy for climate change in a speech at Georgetown University in June 2013.[40] The plan, which contained some seventy-five specific proposals, relied heavily on executive action and expansive interpretations of existing law. It included the regulation of carbon dioxide emissions from new and existing power plants and a continuing shift to cleaner fuels such as natural gas; a redoubling of wind and solar energy production and opening public lands and military bases to renewable energy construction; a new goal for the federal government to consume 20 percent of its electricity from renewable sources by 2020; increased preparedness for the impacts of climate change through support of infrastructure improvements; and the reassertion of American international leadership on climate change. As part of reclaiming global environmental leadership, the United States would negotiate cooperative agreements with rapidly developing countries such as China, end US aid and support for the construction of coal-fired power plants abroad, and seek a new global treaty to cut carbon pollution.[41]

In September 2013, the EPA proposed a carbon dioxide emission standard for all *new* electric power plants that would make future construction of coal-fired plants prohibitively expensive; and in June 2014, it issued a draft rule that would require *existing* coal plants to reduce their carbon emissions by approximately 30 percent from 2005 levels by 2030. After a year of public comment and negotiation, the "Clean Power Plan" rule was finalized in August 2015.[42] The regulations would give considerable flexibility to the states in designing their implementation plans and allowed states up to 15 years to comply, but it was nevertheless denounced by the Republican Party—and some Democrats—as a "war on coal" during the 2014 congressional elections.

Despite losing the Senate in the election, President Obama followed through on his promise to pursue a new international climate change agreement. He announced a landmark executive agreement with China a week after the elections. Under the agreement, China made a commitment to limit its use of fossil fuels for the first time by pledging that its carbon emissions will peak no later than 2030. On its side, the United States pledged to reduce its carbon emissions by 26–28 percent by 2025, compared to 2005. This joint leadership by the United States and China, as well as the Clean Power Plan to limit domestic emissions, paved the way for the Paris Agreement in December 2015 (see Chapter 12).[43] The United States and China also brokered an international agreement in 2016 to phase out use of hydrofluorocarbons (HFCs), which are widely used as refrigerants throughout the world but are potent greenhouse gases.[44]

However, Republican majorities in Congress continued to oppose all actions to limit climate change. The Senate voted to block the power plant regulations after Senate Majority Leader Mitch McConnell urged states not to comply with them.[45] Attorneys general from twenty-nine Republican states together with dozens of corporations and industry groups filed suit in the District of Columbia Circuit Court to block them. After the court refused to grant a preliminary injunction, the US Supreme Court, in an unusual move, intervened in early 2016 to order an indefinite stay pending full review by the appellate court.[46] Several other rules, including EPA regulations limiting emissions of mercury and other toxic pollutants from power plants and regulations to control oil and gas fracking on public lands, were also held up by the courts.

President Obama continued to battle Congress on other fronts as well. He delayed a decision on the controversial Keystone XL oil pipeline, vetoed a bill passed by Congress to approve it, and finally rejected the pipeline entirely on grounds that it would contribute disproportionately to carbon emissions.[47] After weeks of protests by Native American and environmental groups, he also held up the Dakota Access pipeline, which would transport oil from North Dakota to Illinois. In the closing months of his administration, several other important environmental decisions were promulgated, including EPA rules to reduce carbon emissions from heavy-duty trucks and tractor-trailers and to reaffirm fuel economy standards for cars and light trucks through 2025 models; Interior Department regulations to limit methane emissions from oil and gas operations on public lands and to protect streams from mountain coal-mining discharges; and a presidential order under the 1953 Outer Continental Shelf Lands Act to permanently ban oil drilling along the Atlantic Seaboard and in much of the Arctic Ocean.[48] Republicans in Congress vowed to reverse most if not all of these new regulations.

Finally, President Obama left a strong legacy of public lands conservation by designating 29 new national monuments and expanding several others. Among the most significant of the new protected areas were the 1.35-million-acre Bears Ears National Monument in Utah and the Gold Butte National Monument in the Mojave Desert in Nevada. Altogether, Obama designated more monuments under the Antiquities Act of 1906 than any previous president.[49]

PRESIDENT DONALD J. TRUMP: POLICY REVERSAL

Donald Trump came to the presidency with the most antienvironmental agenda since Ronald Reagan. During the election campaign, he had repeatedly attacked Barack Obama's environmental initiatives, especially Obama's actions to limit climate change. He promised to roll back these policies and to dismantle the EPA "in almost every form."[50] Echoing Reagan—and most Republicans—he argued that environmental regulations were "job-killers" impeding economic growth and declared that he would remove all rules and regulations deemed "unnecessary" and impose a temporary moratorium on new ones. He called for the repeal of Obama's "war on coal," stating that he would end restrictions on America's "untapped energy—some $50 trillion... in shale energy, oil reserves and natural gas on federal lands, in addition to hundreds of years of coal energy reserves."[51] Rarely has a presidential candidate called for such total repudiation of his predecessor's legacy.

Trump's appointments reflected these priorities. Scott Pruitt, the new EPA administrator, had filed 14 major lawsuits against the agency as attorney general of Oklahoma. With the support of fossil fuel interests, he had played a leading role in the multistate legal challenge to the Clean Power Plan, the suit that had resulted in a Supreme Court stay of the plan in early 2016.[52] He denied any scientific consensus on climate change and indicated that he would seek to halt all EPA and other agency actions directed toward limiting greenhouse gas emissions. Pruitt also hired several high-level aides from the staff of Sen. James Inhofe (R-Oklahoma), the leading critic of climate science in Congress, and called for the EPA to return to its "core functions" of protecting clean air and water. He also suggested that many of these responsibilities could be transferred to the states.[53]

Trump's other environmental and energy appointments were somewhat less controversial but no less business oriented. Ryan Zinke, a little-known congressman and outdoorsman from Montana, was named secretary of the interior; former Texas governor Rick Perry became secretary of energy; and ex-Georgia governor Sonny Perdue was appointed agriculture secretary. Perry had long-standing ties to the oil and gas industry, while Perdue had supported corporate agricultural and timber interests throughout his career. Indeed, nearly all of Trump's high-level appointees came from business and industry, red state governments, or conservative lobbying firms or foundations that had opposed environmental regulations.[54] Many lacked scientific qualifications, and nearly all were climate change deniers or skeptics.[55]

Many of Trump's initial appointees had little or no experience in the federal government and were soon enmeshed in ethical, legal, and personal scandals. Like Anne Gorsuch and James Watt in the Reagan administration and several top officials in the George W. Bush administration, Pruitt and Zinke were forced to resign in 2018 amidst multiple investigations into their conduct.[56] Perry also resigned at the end of 2019 after being implicated in the Ukrainian scandal that led to Trump's impeachment trial. These and other officials were replaced by more veteran operatives: Andrew Wheeler, a former coal lobbyist, replaced Pruitt at the EPA; former oil lobbyist David Bernhard became interior secretary; and deputy energy secretary Dan Brouillette succeeded Perry.[57] All of these replacements continued to pursue the president's deregulatory agenda, but with less publicity and greater legal expertise and effectiveness.[58]

Unlike previous presidents, Trump appointed no prominent science advisers to the White House early in his term, and removed or marginalized many other scientists throughout his administration (see Chapters 7 and 9).[59] He also disbanded numerous scientific advisory panels in the EPA and other agencies and removed information on climate change from government websites.[60] According to 2018 a survey of more than 63,000 scientists in 16 federal agencies conducted by the Union of Concerned Scientists, scientific advice was frequently ignored, distorted, or blocked by political appointees.[61] Trump himself ignored or summarily dismissed major scientific reports such as the Fourth National Climate Assessment, completed in 2017 by thirteen federal agencies, the Fifth Report of the UN Intergovernmental Panel on Climate Change in 2018, and threat assessments by the Defense Department and intelligence agencies regarding climate change.[62] The president eventually appointed Kelvin Droegemeier, a meteorologist from the University of Oklahoma, as his science advisor and as director of the Office of Science and Technology Policy, and brought retired Princeton physicist

William Happer, a well-known climate science opponent, into the National Security Council, but it was not clear that their advice was taken seriously.[63]

Trump's initial budget proposal in 2017 called for massive cuts in virtually all environmental and science programs. The EPA was singled out for the steepest funding cuts of all: its budget was to be reduced from about $8.3 billion to $5.7 billion, or about 31 percent, with 20 percent of its staff and dozens of programs eliminated.[64] Other agencies supporting research related to climate change were also to be reduced sharply: spending by the National Oceanic and Atmospheric Administration (NOAA) was to be cut 18 percent, with the largest reductions in the agency's environmental satellite programs; and the Energy Department's nonnuclear programs were to be reduced by about the same amount, including its national laboratories and Office of Science. "As to climate change," Trump's first budget director and later chief of staff, Mick Mulvaney, stated bluntly, "I think the president was fairly straightforward: We're not spending money on that anymore."[65]

Congress refused to accept these proposed cuts and has since passed a series of continuing resolutions that have roughly maintained previous levels of spending (see Appendix 2). Trump has nevertheless proposed deep cuts in environmental programs in each of his subsequent budgets. His most recent budget, for Fiscal Year 2021, requested a 26 percent cut for the EPA and called for elimination of nearly 50 programs; cut the Interior Department by 16 percent; and sliced the nonnuclear programs of the Energy Department by 28 percent. Virtually all science and research programs were to be cut drastically.[66] Although such draconian cutbacks were not likely to be acceptable to Congress, they reflected Trump's priorities. Moreover, the administration's attitude toward environmental programs has contributed to demoralization and loss of personnel at agencies like the EPA, whose staffing is now at the lowest levels since the Reagan administration (Appendix 3).[67]

Like previous presidents, Trump quickly utilized his authority to revoke or suspend many of his predecessor's executive orders and to issue new directives of his own.[68] All new regulations not yet in effect were frozen, as were all pending EPA contracts and awards. The president then issued a series of executive orders on regulatory reform: Executive Order (EO) 13771 required that any new regulation be offset by repeal of at least two other regulations that would reduce compliance costs by at least the same amount; and EO 13777 required the appointment of "regulatory reform officers" and "regulatory reform task forces" in all agencies to "evaluate existing regulations... and make recommendations to the agency head regarding their repeal, replacement, or modification, consistent with applicable law."[69] A further memorandum instructed the Commerce Department to solicit comments from manufacturing industries on which regulations to target. These mandates were similar to those imposed by Ronald Reagan, but they appeared to go even further in focusing on regulatory costs regardless of benefits.[70]

EO 13783 on "Promoting Energy Independence and Economic Growth," issued on March 28, 2017, took direct aim at President Obama's climate change policies. It required federal departments and agencies to "immediately review all existing regulations that potentially burden the development or use of domestically produced energy resources and appropriately suspend, revise, or rescind" any that are unnecessary; instructed the EPA administrator to begin the process of withdrawing the Clean Power

Plan; revoked all other presidential directives related to the consideration of greenhouse gas emissions, including environmental reviews under the National Environmental Policy Act and calculation of the "social costs of carbon"; ended the moratorium on coal leasing on federal lands; and ordered the review of Obama-era rules regarding the control of methane emissions and of hydraulic fracturing (fracking) on public lands.[71]

Trump delivered on all of these issues. Indeed, he rolled back, or was in the process of modifying or eliminating, at least 100 environmental rules and regulations by mid-2020.[72] Space obviously does not allow listing them all here, but many are discussed in detail in other chapters of this volume.

The Trump administration reversed Obama's climate change policies in six fundamental ways to promote the president's "energy dominance" agenda:

- withdrawal from the Paris Climate Agreement, once again isolating the United States in international environmental diplomacy;[73]

- revocation of the "Clean Power Plan" to reduce carbon emissions from power plants and replacement with a much weaker "Affordable Clean Energy" rule that would largely leave regulation of individual plants to the states;[74]

- replacement of the vehicle fuel economy standards negotiated by Obama for the years 2022–2025 and revocation of California's waiver under the Clean Air Act to set higher mileage standards;[75]

- opening virtually all federal public lands and offshore waters to oil and gas leasing, including parts of the Atlantic, Pacific, and Arctic coasts, the Arctic National Wildlife Refuge, and other areas protected for endangered species;[76]

- elimination or weakening of Obama restrictions on fossil fuel production and use, including stream protection, methane releases, fracking practices, offshore drilling safety, oil pipeline construction, power plant emissions of mercury and other toxic pollutants, and disposal of toxic wastes;[77]

- radical revision of cost–benefit analysis to discount future benefits of carbon emission reduction and to exclude consideration of climate change in federal rulemaking and environmental impact assessment.[78]

In addition to these reversals, the Trump administration altered the implementation of numerous other environmental laws dating back to the 1970s. For example, environmental impact reviews of "major federal actions" under the National Environmental Policy Act were to be curtailed in duration, scope, and length to accelerate projects, thus reducing public input and scientific evidence.[79] The EPA also proposed limiting the kinds of public health evidence that can be used in setting environmental standards under the Clean Air and Clean Water Acts and in making decisions on exposure to hazardous chemicals such as pesticides.[80] Trump rescinded the "Waters of the United States" rule issued by Obama in 2015 after years of negotiation to clarify which bodies of water are covered by the Clean Water Act. As a result, most smaller streams, seasonal ponds, and agricultural wetlands may no longer be protected.[81] Criteria for listing of threatened and endangered species under the Endangered Species

Act were also tightened and large areas of protected habitat were opened for development.[82] Cleanups of numerous Superfund sites were delayed, resulting in a growing backlog of untreated sites.[83] And in one of his most dramatic moves, Trump asserted unprecedented authority to drastically alter the size and shape of national monuments established by his predecessors: he reduced the Grand Staircase Escalante National Monument established by President Clinton in the red rock canyons of Utah by nearly half, and shrank the Bears Ears National Monument created by Obama on ancient tribal lands by more than 80 percent (Chapter 9).[84] It remains unclear whether these interventions will be ruled legal under the Antiquities Act of 1906, but at least some of the excluded areas have been opened to oil, gas, and mining leasing.

In addition to rolling back regulations, enforcement of environmental law declined sharply under Trump. For example, the amount of civil penalties for violation of pollution standards by EPA dropped by 85 percent during the first 2 years of the administration.[85] During the COVID-19 pandemic in early 2020, the EPA further relaxed requirements for detecting and reporting violations. The agency also refused to tighten national air quality standards for particulate matter, even though studies indicated that persons suffering from chronic illnesses linked to pollution such as asthma were more vulnerable to the virus.[86]

Overall, the pace of deregulation under President Trump was unprecedented—surpassing even that of the Reagan era—and was indicative of what former White House strategist Stephen Bannon called the "deconstruction of the administrative state."[87] However, undoing rules and regulations that have been adopted by formal administrative procedures is difficult and time consuming, often taking years. In many cases Trump appointees violated procedures required by the Administrative Procedure Act or simply ignored the advice of agency experts, inviting legal challenges on both legal and scientific grounds.[88] These and other major decisions are subjected to ongoing court challenges and resistance from cities and states, environmental and other interest groups, businesses, and citizens (Chapter 6).

CONCLUSION

The Obama and Trump presidencies make it clear that the president has become the most significant actor in national environmental policymaking. The powers of the presidency have continued to expand in the current era of ideological polarization and congressional gridlock. The result is violent pendulum swings between liberal and conservative policies, which reflect fundamentally different views on the role of government, the nature and place of science in designing environmental policies, and global economic forces and trends. Obama's reliance on executive powers alone to impose climate policies was clearly a weakness, albeit a necessary one given the opposition he faced in Congress. Donald Trump then utilized his executive powers to an unprecedented degree to erase his predecessor's environmental legacy. In doing so, he invoked nearly unlimited authority to impose his environmental and energy policies throughout a "unitary executive." Future presidents may do the same, threatening ongoing legal and constitutional crises. The uncertainty and division that are likely to result could permanently weaken the nation's commitment to environmental goals.

SUGGESTED WEBSITES

Center for Climate and Energy Solutions (www.C2ES.org) This site provides research from a leading think tank on climate change policies.

Department of the Interior (www.interior.gov) The official website for the department and bureaus within it.

Environmental and Energy Law Program, Harvard University (www.eelp.law.harvard.edu/regulatory-rollback-tracker/) This site tracks deregulation actions and their current status.

Environmental Protection Agency (www.epa.gov) The official website for the EPA offers information on environmental topics and US environmental policy, laws, and regulations.

Heritage Foundation (www.heritage.org) The website of this nonprofit think tank offers research and analysis on energy and environmental issues from a conservative perspective.

Sabin Center for Climate Change Law, Columbia University (https://climate.law. columbia.edu/climate-deregulation-tracker/) This site documents changes in laws and regulations affecting climate change.

White House (www.whitehouse.gov) The president's official website provides information on presidential actions, including executive orders, and pending, signed, and vetoed legislation.

NOTES

1. Brady Dennis and Dino Grandoni, "In Reversal, Trump to Ban Oil Drilling Off Coasts of Florida, Georgia and South Carolina," *Washington Post*, September 8, 2020; Annie Karni and Lisa Friedman, "Trump, Calling Himself 'The No. 1 Environmental President,' Green Washes His Record," *New York Times*, September 8, 2020.
2. Richard P. Nathan, *The Administrative Presidency* (New York: Wiley, 1983); Frank J. Thompson, Kenneth K. Wong, and Barry G. Rabe, *Trump, the Administrative Presidency, and Federalism* (Washington, DC: Brookings, 2020).
3. Nadja Popovich, "Climate Change Rises as a Public Priority. But It's More Partisan Than Ever," *New York Times*, February 20, 2020.
4. Charlie Savage, "Trump and His Lawyers Embrace a Vision of Vast Executive Power," *New York Times*, June 4, 2018; Michael Brice-Saddler, "While Bemoaning Mueller Probe, Trump Falsely Says the Constitution Gives Him 'the Right to Do Whatever I Want,'" *Washington Post*, July 23, 2019.
5. Dennis L. Soden, ed., *The Environmental Presidency* (Albany: State University of New York Press, 1999).
6. For references, see "National Defense Strategy: Climate Change in the Age of Great Power Competition," www.americansecurityproject.org/climate-change-in-the-age-of-great-power-competition/, and Bruce Lieberman, "A Brief Introduction to Climate Change and National Security," July 23, 2019, www.yaleclimateconnections.org/2019/07/a-brief-introduction-to-climate-change-and-national-security/.

7. Michael D. Shear, Maggie Haberman, Nicholas Confessore, Karen Yourish, Larry Buchanan and Keith Collins, "Office of Presidency Transformed by Force of Thousands of Tweets," *New York Times*, November 3, 2019.

8. Kenneth R. Mayer, "Going Alone: The Presidential Power of Unilateral Action," in *The Oxford Handbook of the American Presidency*, ed. George C. Edwards III and William G. Howell (Oxford: Oxford University Press, 2009), 427–54.

9. David E. Lewis and Terry M. Moe, "The Presidency and the Bureaucracy: The Levers of Presidential Control," in *The Presidency and the Political System*, 9th ed., ed. Michael Nelson (Washington, DC: CQ Press, 2010), 367–400.

10. Charles O. Jones, *The Presidency in a Separated System* (Washington, DC: Brookings Institution Press, 1994).

11. Robert F. Durant and William G. Resh, "Presidential Agendas, Administrative Strategies, and the Bureaucracy," in *The Oxford Handbook of the American Presidency*, ed. George C. Edwards III and William G. Howell (Oxford: Oxford University Press, 2009), 577–600.

12. Byron W. Daynes and Glen Sussman, *White House Politics and the Environment: Franklin D. Roosevelt to George W. Bush* (College City: Texas A&M University Press, 2010).

13. Daynes and Sussman, *White House Politics*, 210–15.

14. See Douglas Brinkley, *Rightful Heritage: Franklin D. Roosevelt and the Land of America* (New York: HarperCollins, 2016); Russell E. Train, "The Environmental Record of the Nixon Administration," *Presidential Studies Quarterly* 26, no. 1 (Winter 1996): 185–96.

15. For a more detailed analysis of Reagan's environmental record, see Norman J. Vig and Michael E. Kraft, eds., *Environmental Policy in the 1980s: Reagan's New Agenda* (Washington, DC: CQ Press, 1984).

16. Nathan, *The Administrative Presidency*.

17. Lewis and Moe, "The Presidency and the Bureaucracy," 390; Marc Allen Eisner, *Governing the Environment: The Transformation of Environmental Regulation* (Boulder, CO: Lynne Rienner, 2007), 80–5.

18. On the impact of the Reagan budget cuts, see especially two chapters in *Environmental Policy in the 1980s: Reagan's New Agenda*: Robert V. Bartlett, "The Budgetary Process and Environmental Policy," 121–42; J. Clarence Davies, "Environmental Institutions and the Reagan Administration," 143–60.

19. For a detailed summary of Watt's policies, see Paul J. Culhane, "Sagebrush Rebels in Office: Jim Watt's Land and Water Policies," in *Environmental Policy in the 1980s: Reagan's New Agenda*, 293–317.

20. See Gary C. Bryner, *Blue Skies, Green Politics: The Clean Air Act of 1990 and Its Implementation*, 2nd ed. (Washington, DC: CQ Press, 1995).

21. Bill Clinton and Al Gore, *Putting People First* (New York: Times Books, 1992), 89–99.

22. On the "reinvention" effort, see Daniel J. Fiorino, *The New Environmental Regulation* (Cambridge, MA: MIT Press, 2006), Chapter 5.

23. As an alternative way of implementing the Endangered Species Act, the Clinton administration supported completion of more than 250 habitat conservation plans protecting some 170 endangered plant and animal species while allowing controlled development on twenty million acres of private land. William Booth, "A Slow Start Built to an Environmental End-Run," *Washington Post*, January 13, 2001.

24. Bill Clinton, *My Life* (New York: Knopf, 2004), 948.

25. In particular, the Byrd–Hagel resolution (passed 95–0 on June 12, 1997) opposed any agreement that would harm the US economy or that did not require control of greenhouse gas emissions by developing countries.

26. Jonathan Weisman, "In 2003, It's Reagan Revolution Redux," *Washington Post*, February 4, 2003; Bill Keller, "Reagan's Son," *New York Times Magazine*, January 26, 2003.

27. "Bush's Budget: The Losers," *Washington Post*, April 10, 2001.

28. Budgets for the EPA and other agencies since 1976 can be found at www.whitehouse. gov/omb/budget/Historicals.

29. Joel Brinkley, "Out of Spotlight, Bush Overhauls U.S. Regulations," *New York Times*, August 16, 2004; Bruce Barcott, "Changing All the Rules," *New York Times Magazine*, April 4, 2004; and Christopher Klyza and David Sousa, *American Environmental Policy, 1990–2006* (Cambridge, MA: MIT Press, 2008), 135–52.

30. C. W. Copeland, "The Law: Executive Order 13422: An Expansion of Presidential Influence in the Rulemaking Process," *Presidential Studies Quarterly* 37 (2007): 531–44.

31. For details on these actions and their outcomes, see Klyza and Sousa, *American Environmental Policy: Beyond Gridlock, 2008* Chapters 4 and 9.

32. In his 2007 State of the Union address, Bush called for mandatory standards requiring that 35 million gallons of renewable and alternative fuels be produced by 2017, nearly a fivefold increase.

33. See John M. Broder, "Bush Signs Broad Energy Bill," *New York Times*, December 19, 2007.

34. "Barack Obama and Joe Biden: Promoting a Healthy Environment" (https://energy.gov/ sites/prod/files/edg/media/Obama_Cap_and_Trade_0512.pdf) and "Barack Obama and Joe Biden: New Energy for America" (https://energy.gov/sites/prod/files/edg/media/ Obama_New_Energy_0804.pdf).

35. Emmarie Huetteman, "Aides Say Obama Is Willing to Work with or without Congress to Meet Goals," *New York Times*, January 26, 2014; Charles M. Blow, "A Pen, a Phone and a Meme," *New York Times*, February 7, 2014.

36. Gardiner Harris, "4 Top Science Advisers Are Named by Obama," *New York Times*, December 21, 2008; "A New Respect for Science," editorial, *New York Times*, December 22, 2008.

37. See Note 28. These numbers refer to budget outlays rather than budget authority.

38. "Memorandum for the Heads of Executive Departments and Agencies—Subject: Scientific Integrity," March 9, 2009, https://obamawhitehouse.archives.gov/the-press-office/ memorandum-heads-executive-departments-and-agencies-3-9-09.

39. John M. Broder, "Obama Seeking a Steep Increase in Auto Mileage," *New York Times*, July 4, 2011. These rules were also converted to greenhouse gas emission limits; see Bill Vlasic, "U.S. Sets High Long-Term Fuel Efficiency Goals for Automakers," *New York Times*, August 29, 2012.

40. See "Remarks by the President on Climate Change," June 25, 2013, www.whitehouse. gov/the-press-office/2013/06/25/remarks-president-climate-change; "At Last, an Action Plan on Climate," editorial, *New York Times*, June 26, 2013.

41. See Center for Climate and Energy Solutions, "President Obama's Climate Action Plan: One Year Later," June 2014, https://www.c2es.org/publications/president-obamas-climate-action-plan-one-year-later, for a list of proposals and actions taken.

42. Julie Hirschfeld Davis, "Obama Unveils Plan to Sharply Limit Greenhouse Gas Emissions," *New York Times*, August 3, 2015.

43. Coral Davenport, "Deal on Carbon Emissions by Obama and Xi Jinping Raises Hopes for Upcoming Paris Climate Talks," *New York Times*, November 12, 2014; Davenport, "Nations Approve Landmark Climate Accord in Paris," *New York Times*, December 12, 2015.

44. Alexander Ovodenka, "140 Countries Will Phase Out HFCs. What Are These and Why Do They Matter?" *Washington Post*, November 3, 2016.

45. Coral Davenport, "McConnell Urges States to Defy U.S. Plan to Cut Greenhouse Gas," *New York Times*, March 5, 2015; Davenport, "Senate Rejects Obama Plan to Cut Emissions at Coal-Burning Plants," *New York Times*, November 18, 2015; and David M. Herszenhorn, "As Obama Pushes Climate Deal, Republicans Move to Block Emissions Rules," *New York Times*, December 1, 2015.

46. Coral Davenport, "Numerous States Will Sue to Stop New Climate Rules," *New York Times*, October 22, 2015; Adam Liptak and Coral Davenport, "Supreme Court Deals Blow to Obama's Efforts to Regulate Coal Emissions," *New York Times*, February 9, 2016. The 5–4 decision was unusual, as the Supreme Court rarely blocks a regulation prior to review by a federal appeals court.

47. Coral Davenport, "President Rejects Keystone Pipeline to Transport Oil," *New York Times*, November 7, 2015.

48. Brady Dennis, "Obama Administration Will Keep Tough Fuel Standards in Place," *Washington Post*, November 30, 2016; Coral Davenport, "Obama Leans On a 1953 Law to Ban Drilling," *New York Times*, December 21, 2016; and Juliet Eilperin, "In a Race to the Finish, Obama Administration Presses Ahead with Ambitious Rules," *Washington Post*, December 1, 2016.

49. Juliet Eilperin and Brady Dennis, "With New Monuments in Nevada, Utah, Obama Adds to His Environmental Legacy," *Washington Post*, December 28, 2016.

50. Coral Davenport, "Climate Policy Faces Reversal by New Leader," *New York Times*, November 11, 2016.

51. John W. Miller, "Donald Trump Promises Deregulation of Energy Production," *Wall Street Journal*, September 22, 2016.

52. Chris Mooney, Brady Dennis and Steven Mufson, "Trump to Name Scott Pruitt, Oklahoma Attorney General Suing EPA on Climate Change, to Head the EPA," *Washington Post*, December 7, 2016; Coral Davenport and Eric Lipton, "Choice for E.P.A. has Led Battles to Constrain It," *New York Times*, December 8, 2016.

53. Coral Davenport, "E.P.A. Nominee Criticizes Rules to Protect Climate," *New York Times*, January 19, 2017; Davenport, "Scott Pruitt is Seen Cutting the E.P.A. with a Scalpel, Not a Cleaver," *New York Times*, February 5, 2017; and Davenport, "E.P.A. Chief Doubts Consensus View of Climate Change," *New York Times*, March 9, 2017.

54. Eric Lipton, Ben Protess and Andrew W. Lehren, "Raft of Potential Conflicts in President's Appointments," *New York Times*, April 16, 2017; Lisa Friedman and Claire O'Neill, "Who Controls Trump's Environmental Policy?" *New York Times*, January 14, 2020. According to the latter report, "Of 20 key officials across several agencies, 15 came from careers in the oil, gas, coal, chemical or agricultural industries, while another three hail from state governments that have spent years resisting environmental regulations. At least four have direct ties to organizations led by Charles G. Koch and the late David H. Koch, who have spent millions of dollars to defeat climate change and clean energy measures."

55. Coral Davenport, "New Administrator Stacks the E.P.A. with Climate Change Skeptics," *New York Times*, March 8, 2017; Emily Holden, "Climate Change Skeptics Run the Trump Administration," *Politico*, March 7, 2018.

56. Scott Pruitt, "Man of Little Shame," editorial, *New York Times*, April 18, 2018; Lisa Friedman, "13 Reasons Scott Pruitt Lost His Job as E.P.A. Chief," *New York Times*, July 5, 2018; and Juliet Eilperin, Josh Dawsey and Darryl Fears, "Interior Secretary Zinke Resigns Amid Investigations," *Washington Post*, December 15, 2018.

57. Lisa Friedman, "Andrew Wheeler, Who Continued Environmental Rollbacks, Is Confirmed to Lead E.P.A.," *New York Times*, February 28, 2019; Juliet Eilperin, Josh Dawsey and Darryl Fears, "Trump to Nominate David Bernhardt, a Former Lobbyist, as the Next Interior Secretary," *Washington Post*, February 4, 2019; Coral Davenport, "Senate Confirms Bernhardt as Interior Secretary Amid Calls for Investigations Into His Conduct," *New York Times*, April 11, 2019; and Friedman, "Senate Confirms Dan Brouillette to Lead Energy Department," *New York Times*, December 2, 2019.

58. Coral Davenport and Lisa Friedman, "In Rush to Kill Obama-Era Rules, Pruitt Puts Own Agenda at Risk" *New York Times*, April 8, 2018; Juliet Eilperin, Josh Dawsey and

Brady Dennis, "Shift at EPA shows Technocrats are Replacing Big-Personality Cabinet Members," *Washington Post*, July 6, 2018.

59. Chris Mooney, "Trump Has Filled Just 15 Percent of the Government's Top Science Jobs," *Washington Post*, June 6, 2017; Mooney, "Trump Has Taken Longer to Name a Science Adviser than any Modern President," *Washington Post*, October 16, 2017; and Annie Gowen, Juliet Eilperin, Ben Guarino and Andrew Ba Tran, "Science Ranks Grow Thin in Trump Administration," *Washington Post*, January 23, 2020.

60. Toly Rinberg and Andrew Bergman, "Censoring Climate Change," *New York Times*, November 22, 2017.

61. Pamela Worth, "Federal Scientists Speak on the State of Science Under President Trump," Union of Concerned Scientists, *Catalyst*, Fall 2018. See also Coral Davenport, "In the Trump Administration, Science is Unwelcome. So Is Advice," *New York Times*, June 9, 2018; Dino Grandoni and Juliet Eilperin, "EPA Scraps Pair of Air Pollution Science Panels," *Washington Post*, October 13, 2018; Jacob Carter, Emily Berman, Anita Desikan, Charise Johnson and Gretchen Goldman, *The State of Science in the Trump Era*, Center for Science and Democracy, Union of Concerned Scientists, January 2019; and for a good short summary, Brad Plumer and Coral Davenport, "Science Under Attack: How Trump Is Sidelining Researchers and Their Work," *New York Times*, December 28, 2019.

62. Coral Davenport, "Trump Administration's Strategy on Climate: Try to Bury Its Own Scientific Report," *New York Times*, November 25, 2018; Juliet Eilperin, Brady Dennis and Missy Ryan, "As White House Questions Climate Change, U.S. Military Is Planning for It," *Washington Post*, April 8, 2019; Eilperin, Josh Dawsey and Dennis, "White House Blocked Intelligence Agency's Written Testimony Calling Climate Change 'Possibly Catastrophic,'" *Washington Post*, June 8, 2019; and Hiroko Tabuchi, "A Trump Insider Embeds Climate Denial in Scientific Research," *New York Times*, March 2, 2020.

63. Tony Room, "Trump Intends to Nominate Extreme-Weather Expert for Top White House Science and Tech Role," *Washington Post*, August 1, 2018; Scott Waldman, "Why a High-Profile Climate Science Opponent Quit Trump's White House," www.sciencemag. org/news/2019/09/why-high-profile-climate-science-opponent-quit-trump-white-house.

64. Brady Dennis and Juliet Eilperin, "Trump's Budget Takes a Sledgehammer to the EPA," *Washington Post*, March 16, 2017; Juliet Eilperin, Chris Mooney, and Steven Mufson, "New EPA Documents Reveal Even Deeper Proposed Cuts to Staff and Programs," *Washington Post*, March 31, 2017.

65. Henry Fountain and John Schwartz, "Researchers Bristle at Extent of Cuts," *New York Times*, March 17, 2017.

66. Joel Achenbach, Laurie McGinley, Amy Goldstein and Ben Guarino, "Trump Budget Cuts Funding for Health, Science, Environment Agencies," *Washington Post*, February 10, 2020; David Malakoff and Jeffrey Mervis, "Trump's 2021 Budget Drowns Science Agencies in Red Ink, Again," February 10, 2020, https://www.sciencemag.org/news/2020/ 02/trump-s-2021-budget-drowns-science-agencies-red-ink-again.

67. Lisa Friedman, Marina Affo and Derek Kravitz, "E.P.A. Officials, Disheartened by Agency's Direction, Leave in Droves," *New York Times*, December 23, 2017; Timothy Cama, "EPA Staffing Falls to Reagan-Era Levels," *The Hill*, January 9, 2018.

68. Peter Baker, "Analysis: Trump Adopts Obama Approach While Seeking to Undo a Legacy," *New York Times*, October 13, 2017.

69. See White House, "Presidential Executive Order on Reducing Regulation and Controlling Regulatory Costs," January 30, 2017; "Presidential Executive Order on Enforcing the Regulatory Agenda," February 24, 2017.

70. Juliet Eilperin, "Trump Undertakes Most Ambitious Regulatory Rollback Since Reagan," *Washington Post*, February 12, 2017.

71. White House, "Presidential Executive Order on Promoting Energy Independence and Economic Growth," March 28, 2017. For further summary and commentary see Juliet Eilperin and Brady Dennis, "Trump Moves Decisively to Wipe Out Obama's Climate-Change Record," *Washington Post*, March 28, 2017; Coral Davenport and Alissa J. Rubin, "Trump Signs Rule to Block Efforts on Aiding Climate," *New York Times*, March 29, 2017.

72. Nadia Popovich, Livia Albeck-Ripka and Kendra Pierre-Louise, "The Trump Administration Is Reversing 100 Environmental Rules. Here's the Full List," *New York Times*, July 10, 2020. Various other organizations such as the Brookings Institution, the Environmental and Energy Law Program at Harvard University, and the Sabin Center for Climate Change Law at Columbia University also track deregulation by the Trump administration; see websites listed at the end of this chapter.

73. Michael D. Shear, "Trump Abandoning Global Climate Accord," *New York Times*, June 2, 2017; Lisa Friedman, "Trump Serves Notice to Quit Climate Accord, as Diplomats Plot to Save It," *New York Times*, November 4, 2019.

74. Juliet Eilperin, "Trump Administration Proposes Rule to Relax Carbon Limits on Power Plants," *Washington Post*, August 21, 2018; Lisa Friedman, "E.P.A.'s New Coal Pollution Rules Will Lead to More Deaths, Agency's Numbers Show," *New York Times*, August 21, 2018; Friedman, "E.P.A. Plans to Get Thousands of Deaths Off the Books by Changing Its Math," *New York Times*, May 20, 2019; and Friedman, "E.P.A. Finalizes Its Plan to Replace Obama-Era Climate Rules," *New York Times*, June 19, 2019.

75. Coral Davenport and Hiroko Tabuchi, "U.S. Set to Blunt Pollution Rules for Automakers," *New York Times*, March 30, 2018; Davenport, "Top Trump Officials Clash Over Plan to Let Cars Pollute More," *New York Times*, July 27, 2018; Davenport, "Trump Unveils His Plan to Weaken Car Pollution Rules," *New York Times*, August 2, 2018; Juliet Eilperin and Brady Dennis, "Trump Administration to Revoke California's Power to Set Stricter Auto Emission Standards," *Washington Post*, September 17, 2019; and Davenport, "Trump Administration, in Biggest Environmental Rollback, to Announce Auto Pollution Rules," *New York Times*, March 30, 2020.

76. Lisa Friedman, "Trump Moves to Open Nearly All Offshore Waters to Drilling," *New York Times*, January 4, 2018; Darryl Fears and Juliet Eilperin, "The Trump Administration Is Opening Millions of New Acres to Drilling—And That's Just the Start," *Washington Post*, March 15, 2018; Eilperin and Steven Mufson, "Trump Administration Wants Drilling on More Than Two-Thirds of the Largest Swath of U.S. Public Land," *Washington Post*, June 25, 2020; Mufson and Eilperin, "Trump Administration Opens Huge Reserve in Alaska to Drilling," *Washington Post*, September 13, 2019; and Eilperin, "Trump Finalizes Drilling Plan for Iconic Arctic National Wildlife Refuge," *Washington Post*, August 17, 2020.

77. Hiroko Tabuchi, "G.O.P. Reverses Obama-Era Rule to Protect Streams from Coal Mining," *New York Times*, February 3, 2017; Steven Mufson, "Trump Administration Scraps Limits on Methane Leaks at Oil and Gas Sites," *Washington Post*, August 13, 2020; Lisa Friedman and Tabuchi, "U.S. to Roll Back Safety Rules Created After Deepwater Horizon Spill," *New York Times*, December 28, 2017; Friedman, "E.P.A. Proposes Rule Change That Would Let Power Plants Release More Toxic Pollution," *New York Times*, December 28, 2018; Juliet Eilperin and Brady Dennis, "The EPA Is About to Change a Rule Cutting Mercury Pollution. The Industry Doesn't Want It," *Washington Post*, February 17, 2020; and Dennis and Eilperin, "EPA Moves to Overhaul Obama-Era Safeguards on Coal Ash Waste," *Washington Post*, March 1, 2019.

78. Brad Plumer, "Trump Put a Low Cost on Carbon Emissions. Here's Why It Matters," *New York Times*, August 23, 2018; Lisa Friedman, "Trump Rule Would Exclude Climate Change in Infrastructure Planning," *New York Times*, January 2, 2020; Friedman,

"G.A.O.: Trump Boosts Deregulation by Undervaluing Cost of Climate Change," *New York Times*, July 14, 2020; and Jean Chemnick, "Trump Slashed the Social Cost of Carbon. A Judge Noticed," *E&E News*, July 28, 2020.

79. Juliet Eilperin and Brady Dennis, "Nixon Signed This Key Environmental Law. Trump Plans to Change It to Speed Up Pipelines, Highway Projects and More," *Washington Post*, July 14, 2020; Lisa Friedman, "Trump Weakens Major Conservation Law to Speed Construction Permits," *New York Times*, July 15, 2020.

80. Danny Hakim and Eric Lipton, "Once Trusted Studies Are Scorned by Trump's E.P.A.," *New York Times*, August 26, 2018; Steven Mufson and Chris Mooney, "EPA Excluded Its Own Top Science Officials When It Rewrote Rules on Using Scientific Studies," *Washington Post*, October 3, 2018; Lisa Friedman, "E.P.A. to Limit Science Used to Write Public Health Rules," *New York Times*, November 11, 2019; Friedman, "E.P.A. Updates Plan to Limit Science Used in Environmental Rules," *New York Times*, March 4, 2020; and Rebecca Beitsch, "EPA Looks to Other Statutes to Expand Scope of Coming 'Secret Science' Rule," *The Hill*, July 29, 2020.

81. Coral Davenport, "Trump Prepares to Unveil a Vast Reworking of Clean Water Protections," *New York Times*, December 10, 2018; Juliet Eilperin and Brad Dennis, "Administration Finalizes Repeal of 2015 Water Rule Trump Called 'Destructive and Horrible,'" *Washington Post*, September 11, 2019; and Davenport, "Trump Removes Pollution Controls of Streams and Wetlands," *New York Times*, January 22, 2020.

82. Elizabeth Kolbert, "The Trump Administration Takes On the Endangered Species Act," *New Yorker*, July 26, 2018; Coral Davenport and Lisa Friedman, "Lawmakers, Lobbyists and the Administration Join Forces to Overhaul the Endangered Species Act," July 22, 2018; and Friedman, "Trump Administration Weakens Protection for Endangered Species," *New York Times*, August 12, 2019.

83. Ellen Knickmeyer, Matthew Brown and Ed White, "Backlog of Toxic Superfund Cleanups Grows Under Trump," Associated Press, January 2, 2020.

84. Julie Turkewitz, "Trump Reverses U.S. Protections for 2 Utah Sites," *New York Times*, December 5, 2017; Nadia Popovich, "How Trump's Order Shrinks Bears Ears," *New York Times*, December 9, 2017; and Eric Lipton and Lisa Friedman, "Oil Was Central in Decision to Shrink Protected Utah Site, Emails Show," *New York Times*, March 5, 2018.

85. Eric Lipton and Danielle Ivory, "Under Trump, E.P.A. Has Slowed Actions Against Polluters, and Put Limits on Enforcement Officers," *New York Times*, December 10, 2017; Juliet Eilperin and Brady Dennis, "Civil Penalties for Polluters Dropped Dramatically in Trump's First Two Years, Analysis Shows," *Washington Post*, January 24, 2018; and Lindsay Dillon, Chris Sellers, and Phil Brown, "EPA Staff Say the Trump Administration Is Changing Their Mission from Protecting Human Health and the Environment to Protecting Industry," *Salon*, June 10, 2018, https://www.salon.com/2018/06/10/epa-staff-claims-trump-administration-is-changing-their-mission_partner/.

86. Lisa Friedman, "E.P.A., Citing Coronavirus, Drastically Relaxes Rules for Polluters," *New York Times*, March 26, 2020; Juliet Eilperin, Dino Grandoni, and Brady Dennis, "Trump Officials Reject Stricter Air Quality Standards, Despite Link Between Air Pollution, Coronavirus Risks," *Washington Post*, April 14, 2020; and Siddhi Doshi and Cayll Baker, "Six COVID-Related Deregulations to Watch," *Brookings*, April 30, 2020.

87. Philip Rucker, "Bannon: Trump Administration Is in Unending Battle for 'Deconstruction of the Administrative State,'" *Washington Post*, February 23, 2017.

88. See, e.g., Eric Lipton, "Resolute Rush to Deregulate Hits Obstacles," *New York Times*, October 7, 2017; Juliet Eilperin, "Judge Voids Nearly 1 Million Acres of Oil and Gas Leases, Saying Trump Policy Undercut Public Input," *Washington Post*, February 27, 2020; and Juliet Eilperin and Brady Dennis, "EPA Staff Warned That Mileage Rollbacks Had Flaws. Trump Officials Ignored Them," *Washington Post*, May 19, 2020.

5 ENVIRONMENTAL POLICY IN CONGRESS

Michael E. Kraft

The sun is setting on the dirty energy of the past. Today marks the dawn of a new era of climate action.

Senator Edward Markey, D-Massachusetts, co-chair of the Senate Climate Task Force, upon introducing the Green New Deal resolution, February 2019[1]

On February 7, 2019, Rep. Alexandria Ocasio-Cortez (D-New York) and Sen. Edward Markey introduced resolutions in both the US House and Senate to create a Green New Deal (GND). They called for a "10-year national mobilization" that would speed the development of clean energy and shift the nation away from fossil fuels as well as "provide unprecedented levels of prosperity and economic security for Americans, and counteract systemic injustices" in the nation. These goals could be achieved, they said, as part of an ambitious effort to respond to "the existential challenge of climate change."[2]

The nonbinding resolution was brief at only six pages, yet it reflected the convictions of many progressives in the Democratic Party, and later in the presidential campaign of Sen. Bernie Sanders, that business as usual was no longer acceptable in light of recent scientific reports on climate change. They argued that dire forecasts of the damage that climate change was likely to inflict on the nation made it essential to act boldly and quickly to address the challenge. They hoped that the congressional resolution would stimulate national debate and spur Congress to act.

In the House, the resolution, H. Res. 109, with 91 cosponsors (all Democrats), was referred to multiple committees and subcommittees for consideration. The Senate measure, S. Res 59, with 12 cosponsors (11 Democrats and 1 Independent, Sanders), was referred to the Senate Committee on Environment and Public Works. The 116th Congress (2019–2021) saw many other resolutions and bills on energy use and climate change, some 17 of them by early May of 2019, and a number of those bills attracted some bipartisan support, unlike the GND.[3]

Although unsuccessful legislatively, the GND spurred widespread responses by the nation's media, think tanks, and partisan organizations, with some comparing it to long-standing calls for sustainable development or green growth (see Chapter 14). For example, by July 2019, the Center for American Progress and more than 70 environmental and other progressive groups released a major climate change proposal for the next Democratic administration called the Equitable and Just National Climate Platform that reflected views similar to the GND.[4] As might be expected, Republican and conservative groups that place a higher priority on conventional economic development were sharply critical of the resolution. They also indicated they would use it aggressively against Democrats in the 2020 elections.[5] Even with that risk, introducing the GND resolution illustrated well the role that members of Congress can play in framing issues and setting the political and policy agendas even when there is insufficient support to enact legislation.

ENVIRONMENTAL CHALLENGES AND POLITICAL CONSTRAINTS

The GND resolution and its short-term failure to move forward tells us much about the way Congress deals with environmental, energy, and natural resource issues today. It also speaks to the many obstacles Congress will face in the future in trying to chart new policy directions to better address twenty-first-century challenges such as climate change and how to foster sustainable development that were discussed in Chapter 1.[6] The capacity of the 116th Congress to act on these problems, like that of many Congresses before it, was understandably affected by what political scientists have long called an era of "partisan warfare" on Capitol Hill. Compromise between the parties became exceptionally difficult as each sought to deny the other any semblance of victory, even at the cost of stalemate in dealing with pressing national environmental issues.[7]

Whichever party dominates Congress in the years ahead, we are unlikely to see the kind of broad bipartisan support for environmental policies that prevailed during the 1970s and even through the 1980s. This is particularly the case for major policy actions, such as rewriting the basic environmental laws, most of which were adopted close to 50 years ago and ideally would be modernized and integrated to better address today's needs. It was not always so. For nearly three decades, from the late 1960s to the mid-1990s, Congress enacted—and over time strengthened—an extraordinary range of environmental policies, typified by the 1970 Clean Air Act, the 1972 Clean Water Act, the 1973 Endangered Species Act, and the 1976 Resource Conservation and Recovery Act (see Chapter 1 and Appendix 1). In doing so, members within both political parties recognized and responded to rising public concern about environmental degradation and its impacts on public health. For the same reasons, they stoutly defended and even expanded those policies during the 1980s when environmental legislation was assailed by Ronald Reagan's White House.[8]

This pattern of bipartisan cooperation and compromise changed dramatically with the election of the 104th Congress in 1994, as the new Republican majority brought to the Hill a very different position on the environment. It was far more critical of regulatory bureaucracies, such as the EPA, and of the environmental and public health policies they are charged with implementing.[9] On energy and natural resource issues—from drilling for oil in the Arctic National Wildlife Refuge (ANWR) or on offshore public lands to, more recently, the Keystone XL pipeline controversy, oil and gas drilling on public land, and especially actions to mitigate climate change—Republicans have tended to lean heavily toward increasing resource use and economic development rather than conservation or environmental protection.

As Republican Party leaders pursued these goals from 1995 through 2020, they invariably faced intense opposition from Democrats who were just as determined to block what they characterized as ill-advised attempts to roll back years of progress in protecting public health and the environment.[10] The result has been that Congress has been unable to approve either the sweeping changes sought by Republicans or the moderate policy reforms preferred by most Democrats.

We saw much the same pattern in the 116th Congress. Republicans retained control of the Senate while Democrats won a solid majority in the House. The result was frequent oversight and criticism of the Trump administration by the House, but little comparable action in the Senate, which was strongly supportive of the administration. Moreover, legislative gridlock on public policy continued with only a few exceptions, as noted later in the chapter. But misinterpretation of this failure to act legislatively is common. For example, President Trump frequently criticized what he called the "do-nothing Democrats" in Congress, implying that Democrats were accomplishing little beyond criticism of him and his administration. In reality, the Democratic House under Speaker Nancy Pelosi, D-California, passed hundreds of major bills, many with bipartisan support, that the Senate majority leader, Mitch McConnell (R-Kentucky), chose not to consider.[11] Many of those bills focused on environmental issues.

Unfortunately, the ongoing partisan fights in Congress mean that existing policies—with their many acknowledged flaws—have largely continued in force. State and local governments have fostered innovative policy actions that to some extent help to fill this void (see Chapter 2). At the federal level, the Trump administration chose to rely on administrative actions rather than seek congressional approval for major policy changes (see Chapters 4 and 7).[12] This pattern is problematic for several reasons. Perhaps the most important is that only Congress can redesign federal environmental policy in a comprehensive and coherent way so that it might successfully address complex problems such as climate change and sustainability. Thus, it is important to understand how Congress makes decisions on environmental issues and why members adopt the positions and take the actions they do. The sections that follow examine attempted policy changes on Capitol Hill and compare them with the way Congress dealt previously with similar issues. I give special consideration to the phenomenon of policy stalemate or gridlock, which at times, including most of the 2010s, has been a defining characteristic of congressional involvement with environmental policy, and may well reappear during the 2020s.

CONGRESSIONAL AUTHORITY AND ENVIRONMENTAL POLICY

Under the Constitution, Congress shares authority with the president for federal policymaking on the environment. In most years, members of Congress make critical decisions on hundreds of measures that affect environmental policy broadly defined. These range from funding the operations of the EPA, the Department of the Interior, and other agencies to supporting highways, mass transit, forestry, farming, oil and gas exploration, energy research and development, the protection of wilderness areas, and international population and development assistance programs. These actions are rarely front-page news and are not often the focus of television news programs or social media commentary. Hence, the public may hear little about them.[13]

As discussed in Chapter 1, we can distinguish congressional actions in several different stages of the policy process: agenda setting, formulation and adoption of policies, and implementation of them in executive agencies. Largely because of extensive media coverage, presidents have far greater opportunities than does Congress to set the political agenda, that is, to call attention to specific problems and define or frame the terms of debate. Yet members of Congress can have a major impact on the agenda through legislative and oversight hearings as well as through the abundant opportunities they have for introducing legislation, requesting and publicizing studies and reports, making speeches, taking positions, voting, and campaigning for reelection.

All of these actions can assist them in framing issues in a way that promotes their preferred solutions, as was evident in both House and Senate resolutions on the GND, and on climate change, energy use, and environmental issues broadly. One illustration is the introduction of bills to put a price on carbon, such as the Energy Innovation and Carbon Dividend Act of 2019, H.R. 763, introduced in both the House and Senate with 80 cosponsors drawn from both parties. Even when such legislation has little chance of approval, its introduction and possible hearings, and thus publicity, can help to shape national debate, which can be just as important. House Speaker, Nancy Pelosi, for example, created a special House committee on climate change to address what she called "the existential threat of the climate crisis." Yet, the committee's purpose was more to raise public awareness rather than to consider legislation in the usual manner. It was to recommend policy actions for other committees to consider.[14]

Because of their extensive executive powers, presidents also can dominate the process of policy implementation in the agencies (see Chapters 4 and 7). Here too, however, Congress can significantly affect agency actions, partly through oversight hearings, but also through its critical budgetary decisions. For example, shortly after he took office, and in succeeding years, President Trump proposed exceptionally deep budgetary and staffing cuts for the EPA and for environmental science and energy technology research at the EPA, Department of Energy, National Aeronautics and Space Administration (NASA), and National Oceanic and Atmospheric Administration (NOAA). However, Congress chose not to go along for the most part, and agency budgets remained much the same. Even so, as noted in Chapter 1, those budgets have rarely been sufficient to support full implementation of existing laws and programs, and many are now lower than in previous years.[15]

These congressional powers translate into an influential and continuing role of overseeing, and often criticizing, actions in executive agencies as varied as the EPA, Department of Energy, US Geological Survey, Fish and Wildlife Service, Bureau of Land Management, and Forest Service.[16] As one example of such congressional oversight, the former Republican chair of the House Science Committee, Rep. Lamar Smith (R-Texas), frequently used his influential position to criticize government scientists and their research. He also held hearings to try to rein in the use of what he termed "secret science" by the EPA, a controversial action to ignore certain kinds of established scientific findings that the Trump EPA also pursued through 2020 (see Chapter 7).[17] For their part, Democrats in the 116th Congress held some 15 hearings on climate change just through early March 2020. They were not intended to promote legislation, but as the press reported, to "gather testimony from climate experts—and to stump about the urgency needed to address global warming," that is, to speak out about it.[18]

In addition to committee oversight authority, the Senate plays a major role in choosing who will fill critical positions through its constitutional power to advise and consent on presidential nominations to the agencies and the courts. The Senate almost always approves presidential nominees when the same party controls both institutions, although when a president faces a Congress controlled by the other party, approval is far less certain. For example, President Trump's nomination of Scott Pruitt to head the EPA was among the most controversial of his appointments in 2017, and it drew widespread media coverage and condemnation from Senate Democrats and environmental, scientific, and public health organizations. Hundreds of current and former EPA employees also urged the Senate to reject his nomination; the Senate confirmed Pruitt, although on a relatively close party-line vote of 52 to 46. In early 2019, Pruitt's replacement at the EPA, Andrew Wheeler, also faced tough questions from Democrats on the Senate Committee on Environment and Public Works, as they pressed him on the agency's regulatory rollbacks and the president's stance on climate change.[19]

Even if it cannot compete on an equal footing with the president in some of these policymaking activities and personnel decisions, in many ways, Congress has been more influential than the White House in the formulation and adoption of environmental policies. It also brings exceptional institutional capacity to its many roles in policymaking and oversight of implementation, including the work of its support agencies and staffs, such as the Congressional Budget Office, Government Accountability Office, and the Congressional Research Service.[20] Yet the way in which Congress exercises its formidable powers at any given time is shaped by several key variables, such as its institutional structure and norms, public opinion and interest group pressure, whether the president's party also controls Congress—and by what margins—and members' willingness to defer to the president's recommendations.

Congress's actions on the environment also invariably reflect its dual mission to be deeply engaged with both lawmaking and representation of constituents. In addition to serving as a national legislative body, the House and Senate are assemblies of elected officials who represent politically disparate districts and states. It is hardly surprising that members are politically motivated to consider local, state, and regional concerns and interests, such as coal mining, oil and gas development, or vehicle manufacturing, which can put them at odds with the president or their own party leaders. Indeed, electoral incentives induce members of Congress to think at least as much about local and regional impacts of environmental policies as they do about the larger national interest.[21] Such political pressures led members in the early 2000s, for example, to drive up the cost of President Bush's energy proposals with what one journalist called an "abundance of pet projects, subsidies and tax breaks" to specific industries in their districts and states. Much the pattern is evident today, for example, in the laws that Congress enacted in spring 2020 to aid economic recovery in the midst of the coronavirus pandemic.[22]

Another distinctive institutional characteristic is the system of House and Senate standing committees, where most significant policy decisions take place. Dozens of committees and subcommittees have jurisdiction over environmental policy (see Table 1.1 in Chapter 1), which tends to fragment decisions and erect barriers to integrated or holistic approaches to energy, the environment, and sustainable development. Such a committee structure also means that the outcomes of specific legislative battles often turn on which members sit on and control those committees. For example, in the Senate, James Inhofe often used his positions on the Senate Environment and Public Works Committee (both as chair and ranking Republican member) to question the science of climate change. Many of Inhofe's former aides on the committee were later hired for high-level positions at the EPA or at President Trump's White House.[23]

Taken together, these congressional characteristics have important implications for environmental policy. First, building policy consensus in Congress is rarely easy because of the diversity of interests and of members whose concerns need to be met and because of the conflicts that can arise among committees and leaders. Second, policy compromises invariably reflect members' preoccupation with the local and regional impacts of environmental decisions, for example, how climate change policy will affect industries and homeowners in coal-producing states and those heavily dependent on coal-fired power plants, such as West Virginia, Kentucky, Ohio, and Indiana. Third, the White House matters a great deal in how the issues are defined and whether policy decisions can be made acceptable to all concerned, but the president's influence is nevertheless limited by the independent political calculations made on Capitol Hill.

Given these constraints, Congress frequently finds itself unable to make crucial decisions on environmental policy. The US public may see a "do-nothing Congress," yet the reality is that all too often members can find no way to reconcile the conflicting views of multiple interests and constituencies. It remains to be seen if this pattern will change in 2021 and beyond.

However, there are some important exceptions to this common pattern of policy stalemate. A brief examination of the way Congress has dealt with environmental issues since the early 1970s helps to explain this seeming anomaly. Such a review also provides a useful context in which to examine and assess the actions of recent Congresses and the outlook for environmental policymaking for the early twenty-first century.

CAUSES AND CONSEQUENCES OF ENVIRONMENTAL GRIDLOCK

Policy gridlock refers to an inability to resolve conflicts in a policymaking body such as Congress, which results in government inaction in the face of important public problems. There is no consensus on *what* to do, and therefore no movement occurs in any direction. Present policies, or slight revisions of them, continue until agreement is reached on the direction and magnitude of change. Sometimes environmental or other programs officially expire but continue to be funded by Congress through a waiver of the rules governing the annual appropriations process. The failure to renew the programs, however, contributes to administrative and public policy drift, ineffectual congressional oversight, and a propensity for members to use the appropriations process to achieve what cannot be gained through statutory change.[24]

While usually seen as a negative characteristic, policy gridlock in Congress also has had what environmentalists would see as positive effects. For example, it has stimulated environmental policy change at the state and local levels, which in turn may illustrate the political feasibility or effectiveness of new policy directions (see Chapter 2). It has pushed executive agencies to alter administrative decision-making in creative ways to compensate for the lack of new congressional directives (see Chapter 7). It also means that multiple parties in policy disputes are more likely to turn to the courts to resolve disagreements, which may lead to decisions that facilitate environmental policy reforms even in the face of intense partisan conflicts (see Chapter 6).[25]

Why does policy gridlock occur so frequently in Congress? There is no single answer that fits every situation, but among the major reasons are the sharply divergent policy views of Democrats and Republicans, the influence of organized interest groups in both elections and policymaking, and the inherent complexity of environmental problems. The lack of public consensus on the issues or unclear public preferences, the constitutionally mandated structure of US government (especially the separation of powers between the presidency and Congress), and weak or ineffectual political leadership also make a difference.[26]

Most of these factors are easy to understand. For example, Republicans and Democrats bring different political philosophies and beliefs to the table, and those views are reinforced by the nature of congressional elections today. Most members come from relatively safe districts or states that lean strongly toward one party or the other. For the House, district safety often is a consequence of strong partisan gerrymandering that is intended to expand the party's political power. In recent elections, only about two to three dozen House seats out of 435 have been truly competitive between the major parties. One result is that the electoral system gives exceptional clout to the majority party's base or core voters, particularly in low-turnout, off-year elections. Thus members are forced to appeal more to these voters (such as the Tea Party forces within the Republican Party in the early 2010s or to Trump's base of voters in the late 2010s) than to those in the political center or in the other party. That political imperative tends to promote ideological rigidity among members and also to discourage the kind of compromise and consensus building in policymaking that long prevailed in Congress until recent years.[27]

It is also understandable that when a given issue sparks involvement by diverse and opposing interests (such as oil and gas companies, renewable energy companies, electric utilities, the coal industry, the automobile industry, labor unions, and environmentalists), finding politically acceptable solutions can be exceptionally difficult. Similarly, public opinion polls may show a strong public preference for certain actions on the environment, energy, or climate change, but the public tends not to be well informed on the issues, and these issues also typically are low in salience for most people (see Chapter 3). Both factors limit the extent to which members of Congress are likely to give public preferences much weight in their voting decisions.[28]

Unfortunately, these conditions also significantly diminish the public's political influence. Absent a clear and forceful public voice, members of Congress look elsewhere when deciding how to vote on measures before them, and most people are not likely to notice how their representatives are voting.[29] It is for that reason that some groups, such as the Citizens' Climate Lobby, have tried to change the political calculus. They organize grassroots campaigns on climate change in congressional districts nationwide that are designed to convince members of Congress from both parties that their constituents do desire strong policies and that they are watching the votes.[30]

The notable differences in policy preferences between Republicans and Democrats on environmental and energy issues today reflect a striking trend toward ideological polarization that has developed over the past several decades. The shift has been well documented by scholars, and it can be seen in public opinion surveys as well as rankings of members' voting on environmental issues.[31] For example, based on rankings by the League of Conservation Voters (LCV), the parties showed increasing divergence from the early 1970s through the early 2000s. On average, they differed by nearly 25 points on a 100-point scale, and those differences grew much wider over time.[32] The gap is exceptionally large today, with Republicans in both chambers in 2019 averaging about 13 percent in comparison to an average for Democrats of 95 percent.[33] Smaller but still significant differences appear in surveys of the general public, with Republicans generally less supportive of environmental policy action than Democrats. Environmental policies have become at least as polarizing as any other issue in the late 2010s, a striking shift from the public consensus and bipartisanship that prevailed in the 1970s (see Chapter 3).[34]

Despite these disagreements on the issues, members of Congress come together on occasion to form bipartisan caucuses or congressional member organizations that seek practical solutions with broad public appeal. One illustration is the Climate Solutions Caucuses. They were originally designed to have equal Republican and Democratic membership, but today membership is equal only in the Senate; in the House Democrats outnumber Republicans.[35]

Some of the caucuses and coalitions, however, attract most or all of their members from one party. This has long been the case, for example, for the House Sustainable Energy and Environment Coalition; in 2020, all of its members were Democrats. In 2019, Republicans in the House and Senate formed a bicameral conservation caucus that they said was aimed at overcoming the perception that their party lacked concern for climate change and other environmental issues. They called it the Roosevelt Conservation Caucus in recognition of the historic role of President Theodore Roosevelt in public lands conservation. In a somewhat parallel move, the Senate's Minority Leader, Chuck Schumer (D-New York), established a new Senate Democrats' Special Committee on the Climate Crisis when the Republican majority leader, Mitch McConnell, refused to consider a resolution to create a bipartisan select committee on climate change.[36]

FROM CONSENSUS IN THE ENVIRONMENTAL DECADE TO DEADLOCK IN THE 1990S

As Chapter 1 recounts, the legislative record for the environmental decade of the 1970s is truly remarkable. The National Environmental Policy Act, Clean Air Act, Clean Water Act, Endangered Species Act, and Resource Conservation and Recovery Act, among others, were all signed into law in that decade, and most of them were enacted in the 6-year period from 1970 to 1976. We can debate the merits of these early statutes with the clarity of hindsight and in light of contemporary criticism of them. Yet their adoption demonstrates vividly that the US political system is capable of developing major environmental policies in fairly short order under the right conditions.

Consensus on environmental policy prevailed in the 1970s, in part, because the issues were new and politically popular, and attention was focused on broadly supported goals such as cleaning up the nation's air and water rather than on the means to be used (command-and-control regulation) or program costs. At that time, there was also little overt and sustained opposition to these measures.

Environmental Gridlock Emerges

These patterns of the 1970s did not last long. Congress's enthusiasm for environmental policy gradually gave way to apprehension about its impacts on the economy, and policy stalemate and retrenchment became the norm in the early 1980s. Ronald Reagan's election as president in 1980 also altered the political climate and threw Congress into a defensive posture. It was forced to react to the Reagan administration's aggressive policy actions to turn back the clock. Rather than proposing new programs or expanding old ones, Congress focused its resources on oversight and criticism of the administration's policies. Members were increasingly cross-pressured by environmental and industry groups, partisanship on these issues increased, and Congress and President Reagan battled repeatedly over budget and program priorities.[37] The cumulative effect in the early 1980s was that Congress was unable to agree on new environmental policy directions.

Gridlock Eases: 1984–1990

The legislative logjam began breaking up in late 1983, as the US public and Congress repudiated Reagan's antienvironmental agenda (see Chapter 4). The new pattern was evident by 1984 when, after several years of deliberation, Congress approved major amendments to the 1976 Resource Conservation and Recovery Act; these strengthened the program and set tight new deadlines for EPA rulemaking on control of hazardous chemical wastes. The 99th Congress (1985–1987) compiled a record very much at odds with the deferral politics of the 97th and 98th Congresses (1981–1985). In 1986, the Safe Drinking Water Act was strengthened and expanded, and Congress approved the Superfund Amendments and Reauthorization Act, adding a separate Title III, the Emergency Planning and Community Right-to-Know Act (EPCRA), which created the Toxics Release Inventory. Democrats regained control of the Senate following the 1986 election, and Congress reauthorized the Clean Water Act over a presidential veto. Still, Congress could renew neither the Clean Air Act nor the Federal Insecticide, Fungicide, and Rodenticide Act—the nation's key pesticide control act—nor could it pass new legislation to control acid rain.

However, with the election of George H. W. Bush in 1988, Congress and the White House were able to agree on enactment of the innovative and stringent Clean Air Act Amendments of 1990 and the Energy Policy Act of 1992. The latter was an important, if modest, advancement in promoting energy conservation, and it restructured the electric utility industry to promote greater competition and efficiency. Success on the Clean Air Act, a truly landmark and far-reaching piece of legislation, was particularly important because for years it was a stark symbol of Congress's inability to reauthorize controversial environmental programs. Passage was possible in 1990 because of improved scientific research that clarified the risks of dirty air, reports of worsening ozone in urban areas, and President Bush's leadership. He had vowed to "break the gridlock" and support renewal of the Clean Air Act, and Sen. George Mitchell, D-Maine, newly elected as Senate majority leader, was equally determined to enact a bill.[38] Unsurprisingly, the Senate approved the act by a vote of 89 to 10 and the House by 401 to 25.

Policy Stalemate Returns in the 1990s

Unfortunately, approval of the 1990 Clean Air Act Amendments was no signal that a new era of cooperative and bipartisan policymaking on the environment was about to begin. Nor was the election of Bill Clinton and Al Gore in 1992, even as Democrats regained control of both houses of Congress. Most of the major environmental laws were once again up for renewal. Yet despite an emerging consensus on many of the laws, in the end, the 103rd Congress (1993–1995) remained far too divided to act. Coalitions of environmental groups and business interests clashed regularly on all of these initiatives, and congressional leaders and the Clinton White House were unsuccessful in resolving the disputes.

Partisan bickering escalated after the 1994 midterm elections. After one of the most expensive, negative, and anti-Washington campaigns in modern times, Republicans captured both houses of Congress, picking up an additional fifty-two seats in the House and eight in the Senate. They also did well in other elections throughout the country, contributing to their belief that voters had endorsed the Contract with America, which symbolized the new Republican agenda.[39] The contract had promised a general rolling back of government regulations and a shrinking of the federal government's role, but with no specific mention of environmental policy. It drew heavily from the work of conservative and probusiness think tanks, which for years had waged a multifaceted campaign to discredit environmentalist thinking and policies.[40] In recent years, much the same kind of effort can be seen in the reports of the Heritage Foundation, the Competitive Enterprise Institute, the CATO Institute, and the Heartland Institute, all of which also were influential in shaping the policy agenda of the Trump administration.[41]

Despite evidence showing no public mandate to roll back environmental protection, the political result was clear. It put Republicans in charge of the House for the first time in four decades and initiated an extraordinary period of legislative action on environmental policy characterized by bitter relations between the two parties, setting the stage for a similar confrontation that emerged following the 2010 elections and again after the 2016 elections.

The environmental policy deadlock in the mid-1990s should have come as no surprise. With several notable exceptions, consensus on the issues simply could not be built, and the Republican revolution under Speaker Newt Gingrich failed for the most part. The lesson seemed to be that a direct and well-publicized attack on popular environmental programs could not work because it would provoke a political backlash. Those who supported a new conservative policy agenda turned instead to a strategy of evolutionary or incremental environmental policy change through a subtler and far less visible exercise of Congress's appropriations and oversight powers. This strategy included what is usually termed "regulatory reform." Here they were more successful.[42]

The George W. Bush administration relied on a similar strategy from 2001 to early 2009, the quiet pursuit of a deregulatory agenda, and there is evidence of a similar strategy in the 115th and 116th Congresses in the late 2010s as well as in the Trump administration's reliance on environmental deregulation and defunding instead of statutory reforms (see Chapter 4).

ENVIRONMENTAL POLICY ACTIONS IN RECENT CONGRESSES

As discussed earlier, Congress influences nearly every environmental and resource policy through exercise of its powers to legislate, oversee executive agencies, advise and consent on nominations, and appropriate funds. Sometimes these activities take place largely within the specialized committees and subcommittees, and sometimes they reach the floor of the House and Senate, where they may attract greater media attention. Some of the decisions are made routinely and are relatively free of controversy (for example, appropriations for the national parks), whereas others stimulate often intense political conflict.

The capacity of members of Congress to cooperate in the search for environmental policy reforms became even more problematic during the Trump administration as political polarization in both the media and the nation increased sharply. Adding to the challenge was the adoption by the administration of a largely uncompromising posture on major environmental, energy, and natural resource issues, most notably EPA regulation, oil and gas development on public lands, and climate change (see Chapters 4, 7, 9, and 12). To put the recent policy debates in context, a brief look backward at some of the most notable congressional actions of the past 25 years is instructive. These are presented within three broad categories: regulatory reform initiatives (directed at the way agencies make decisions), appropriations (funding levels and use of budgetary riders), and proposals for changing the substance of environmental policy, with emphasis on the last.

Regulatory Reform: Changing Agency Procedures

Regulatory reform has long been of concern in US environmental policy (see Chapters 1, 4, and 7). There is no real dispute about the need to reform agency rulemaking, which has been long faulted for being too inflexible, intrusive, cumbersome, and adversarial, and sometimes based on insufficient consideration of science and economics.[43] However, much disagreement exists over precisely what elements of the regulatory process need to be reformed and how best to ensure that the changes are both fair and effective.

Beginning in 1995 and continuing for several Congresses, the Republican Party and conservative Democrats favored omnibus regulatory reform legislation that would affect all environmental policies by imposing broad and stringent mandates on executive agencies, particularly the EPA. Those mandates were especially directed at the use of benefit–cost analysis and risk assessment in proposing new regulations. Proponents of such legislation also sought to open agency technical studies and rulemaking to additional legal challenges to help protect the business community against what they viewed as unjustifiable regulatory action. Opponents of both kinds of measures argued that such impositions and opportunities for lawsuits were not reform in any meaningful sense and would wreak havoc within agencies that already faced daunting procedural hurdles and frequent legal disputes as they developed regulations (see Chapters 6 and 7).[44]

Ultimately, Congress did approve several bills that Republicans characterized as regulatory reform, including the Unfunded Mandates Reform Act (1995) and the Small Business Regulatory Enforcement Fairness Act (1996). As part of the latter, the separately named Congressional Review Act gives Congress the ability to reject a newly approved agency rule if within sixty legislative working days a simple majority in each house approves a "resolution of disapproval" that is also signed by the president.

With the presidency of George W. Bush, the regulatory reform agenda shifted from imposing these kinds of congressional mandates on Clinton administration agencies to direct intervention by the White House. Bush appointed conservative and probusiness officials to nearly all environmental and natural resource agencies, and rulemaking shifted decisively toward the interests of the business community (see Chapter 4).[45] Bush and his supporters saw less need for congressional involvement in regulatory reform.

Barack Obama's election coincided with another shift in regulatory philosophy. In light of the financial meltdown on Wall Street in 2008 and reports of ineffective federal regulation of banking institutions and of food, drugs, consumer products, and the environment, public sentiment at least temporarily shifted back in favor of strong, or at least "smart," regulation that achieves its purposes without imposing unreasonable burdens.[46] However, following the 2010 elections, Republicans once again controlled the House, and antiregulatory sentiment returned as members sought to reduce perceived burdens on the business community in a slowly recovering economy and to limit or repeal regulations that they believed were hindering job creation. Given opposition by the Obama White House, these efforts had limited impact beyond helping to set a regulatory reform agenda for the future.

By 2017, however, the political alignment changed radically with the election of President Trump. This shift was evident in repeated efforts in both Congress and in the administration to block the EPA from using scientific studies in regulatory processes if those studies were not fully public or transparent (see Chapter 7). It was equally apparent in a broad Regulatory Accountability Act Congress considered in 2017 and in a measure that the House passed on a largely party-line vote in May of 2017: the Reducing Regulatory Burdens Act of 2017, a measure intended to weaken regulation of pesticides under the Clean Water Act.[47] It was even more palpable in the unprecedented use of the Congressional Review Act to overturn recently approved regulations and in the appointments of Scott Pruitt and Andrew Wheeler as administrators of the EPA, as well as in their recruitment of top-level political staff to the agency who were dedicated to the same deregulatory agenda.

Until 2017, the Congressional Review Act was not a useful tool for regulatory reform, and it was employed only once by Congress, in March 2001, to overturn a Clinton-era workplace ergonomics rule. With the election of President Trump and Republican majorities in both the House and Senate, the act became a convenient way to roll back Obama-era environmental rules, among others, and to do so quickly. Less than 1 month into the new administration, the House had used the review act to nullify eight rules, as it considered dozens more. Some quickly cleared Congress and were sent to the president for approval. These included an EPA rule intended to prevent coal mining waste from contaminating streams (the stream protection rule) and a Fish and Wildlife rule that forbade the baiting, trapping, and killing of bears and wolves in their winter dens in Alaska's national wildlife refuges.[48] By May 1, the president had signed thirteen such bills to erase new rules, leading one journalist to describe his use of the act as a "regulatory wrecking ball."[49] The Senate narrowly rejected one of the proposed rule changes, aimed at overturning a Bureau of Land Management regulation that set standards for limiting release of methane (a potent greenhouse gas) by oil and gas drillers on federal land; three Republicans joined all Democrats in opposing the measure.

Because political appointees at the EPA under Trump were committed to an aggressive regulatory reform agenda, there was little further need for congressional action on such reform. Conservative members were confident that their concerns would be addressed in the agency. As discussed in Chapters 6 and 7, these regulatory rollbacks, however, faced strong opposition by scientific, environmental, and public health groups, and eventually will be addressed in multiple federal court challenges.

Appropriation Politics: Budgets and Riders

The implementation of environmental policies depends heavily on the funds that Congress appropriates each year. Thus, if certain policy goals cannot be achieved through changing the governing statutes or altering the rulemaking process through regulatory reform, attention may turn instead to the appropriations process. This was the case during the Reagan administration in the 1980s, which severely cut environmental budgets. Limiting such budgets became a major element of the Republican strategy in Congress from 1995 to 2006, as well as in the George W. Bush administration. The importance of budgetary politics depends in part on which party controls Congress and the White House. Democrats tend to favor increased spending on the environment, and Republicans generally favor decreased spending. Regardless of which party controls Congress, however, the appropriations process has been used in two distinct ways to achieve policy change. One is through reliance on riders, loosely related legislative stipulations attached to appropriations bills; they ride along with the bill, hence the name. The other is through changes in the level of funding, either a cut in spending for programs that are not favored or an increase for those that are endorsed.

Appropriation Riders

The use of riders became a common strategy following the 1994 election. For example, in the 104th Congress, more than fifty antienvironmental riders were included in seven different appropriations bills, largely with the purpose of slowing or halting enforcement of laws by the EPA, the Interior Department, and other agencies until Congress could revise them.[50]

The use of riders continued in subsequent years, as has opposition to the strategy by environmental groups. In 2011, a rider attached to the fiscal year 2012 spending bill would have kept the Obama administration EPA from issuing any proposed regulations on emissions of greenhouse gases from power plants or industrial facilities. Another would have prevented the Interior Department from using any federal funds to limit oil, gas, or other commercial development on public lands that might qualify in the future for wilderness designation.[51]

Why use budgetary riders to achieve policy change rather than introducing free-standing legislation to pursue the same goals? Such a strategy is attractive to its proponents because appropriations bills, unlike authorizing legislation, typically move quickly, and Congress must enact them each year to keep the government operating. Many Republicans and business lobbyists also argue that the use of riders is one of the few ways they have to rope in a bureaucracy that they believe needs additional constraints. They argue that they are unable to address their concerns through changing the authorizing statutes themselves, a far more controversial and uncertain path to follow.[52] Even if members fail, their efforts on such riders help to assure critical constituency groups that their representatives are determined to meet their needs.

Critics of the process, however, say that relying on riders is an inappropriate way to institute policy change because the process provides little opportunity to debate the issues openly, and there are no public hearings or public votes. One example is the Data Quality Act of 2000, a rider seemingly designed to ensure the accuracy of data on which agencies base their rulemaking to which few would object. Yet the act's provisions were written largely by an industry lobbyist with less benign motives—to make it easier for business groups to challenge agency rulemaking. To escape public notice, the rider's twenty-seven lines of text were buried in a massive budget bill that President Clinton had to sign.[53] In a retrospective review in 2001, the Natural Resources Defense Council (NRDC) counted hundreds of antienvironmental riders attached to appropriations bills since 1995. Clinton blocked more than seventy-five of them, but many became law, including the Data Quality Act. The practice has continued since then.[54]

Cutting Environmental Budgets

The history of congressional funding for environmental programs was discussed in Chapter 1, and it is set out in Appendix 2 for selected agencies and in Appendix 4 for overall federal spending on natural resources and the environment. These budgets have been the focus of continuing conflict within Congress since the 1980s. For example, in the 104th Congress in the mid-1990s, GOP leaders enacted deep cuts in environmental spending only to face President Clinton's veto of the budget bill. Those conflicts led eventually to a temporary shutdown of the federal government, with the Republicans receiving the brunt of the public's wrath for the budget wars. Most of the environmental cuts were reversed, but disagreements over program priorities have continued since that time.[55]

George W. Bush regularly sought to cut the EPA's budget but was rebuffed by Congress until 2004, after which it tended to go along with the president. Since then, overall appropriations for the environment and natural resources have increased, although only slightly in real terms, while spending on pollution control (by the EPA, for example) has declined markedly after adjusting for inflation (see Chapter 1 and Appendix 4).

As reviewed in Chapter 1, Congress generally opposed deep cuts to environmental spending proposed by the Trump White House, and that pattern seems likely to continue, in part because both Republican and Democratic members' states and districts benefit directly from that spending, and they recognize public support for many of the programs that are affected by such budget cuts.[56]

Legislating Policy Change

As noted earlier, in most years, Congress makes decisions that affect nearly all environmental or resource programs. In this section, I highlight selective actions in recent Congresses that demonstrate both the ability of members to reach across party lines to find common ground and the continuing ideological and partisan fights that often prevent legislative action.

Pesticides and Drinking Water

Among the most notable achievements of the otherwise antienvironmental 104th and 105th Congresses are two conspicuous success stories involving control of pesticides and agricultural chemicals and drinking water. Years of legislative gridlock were overcome as Republicans and Democrats uncharacteristically reached agreement on new policy directions.

The Food Quality Protection Act of 1996 was a major revision of the nation's pesticide law, long a prime example of policy gridlock as environmentalists battled with the agricultural, chemical, and food industries. The act required the EPA to develop a new, uniform, reasonable-risk approach to regulating pesticides used on food, fiber, and other crops, and it required that special attention be given to the diverse ways in which both children and adults are exposed to such chemicals. The act sped through Congress in record time without a single dissenting vote because the food industry was desperate to get the new law enacted after court rulings that would have adversely affected it without the legislation. In addition, after the bruising battles of 1995, GOP lawmakers were eager to adopt an election year environmental measure.[57]

The 1996 rewrite and reauthorization of the Safe Drinking Water Act sought to address many long-standing problems with the nation's drinking water program. It dealt more realistically with regulating contaminants based on their risk to public health and authorized $7 billion for state-administered loan and grant funds to help localities with compliance costs. It also created a new right-to-know provision that requires large water systems to provide their customers with annual reports on the safety of local water supplies. Bipartisan cooperation on the bill was made easier because it aided financially pressed state and local governments and, like the pesticide bill, allowed Republicans to score some election-year points with environmentalists. In late 2018 Congress also enacted the America's Water Infrastructure Act, one section of which added new requirements for community drinking water systems to develop or modify chemical risk assessments and emergency response plans.[58]

Brownfields, Healthy Forests, and Wilderness

Congress also completed action on a number of somewhat less visible issues that demonstrated its potential to fashion bipartisan compromises. In 2001, President Bush gained congressional approval of important legislation to reclaim so-called urban brownfields. House Republicans sought to reduce liability for small businesses under the Superfund program, and Democrats wanted to see contaminated and abandoned industrial sites in urban areas cleaned up.[59]

In a somewhat similar action in 2003, the 108th Congress approved the Healthy Forests Initiative over the objections of environmental groups. The measure was designed to permit increased logging in national forests, allegedly to lessen the risk of wildfires. Bipartisan concern over communities at risk from wildfires was sufficient for enactment. Wildfires struck Southern California only days before the Senate voted 80 to 14 to approve the bill.[60]

Finally, throughout 2007 and 2008, Congress considered a dozen proposals for setting aside large parcels of federal land for wilderness protection, totaling about two million acres in eight states, largely without significant media coverage. The measures were broadly supported within both parties, in part because environmentalists helped to build public support by working with opposing interests at the local level. Progress like this was also possible because of Democratic victories in the 2006 election, which switched control of the House Natural Resources Committee from Republican Richard Pombo of California to Democrat Nick J. Rahall of West Virginia. Pombo was a fierce opponent of such wilderness protection, and Rahall strongly favored it. Congress could not approve the wilderness bills in 2008, but by March 2009, in a more favorable political climate, they were approved as part of an Omnibus Public Land Management Act.[61]

In a similar display of unusual bipartisanship, in early 2019 Congress approved what the press called "the most sweeping conservation legislation in a decade." The John D. Dingell, Jr. Conservation, Management, and Recreation Act protected millions of acres of land as well as hundreds of miles of wild rivers, and also created four more national monuments, as well as reauthorizing the broadly supported Land and Water Conservation Fund (see Chapter 9). The expansive bill (662 pages long) passed the Senate by a vote of 92 to 8 and the House by a similarly large margin of 363 to 62. Its success owed much to the way the bill was crafted. As one journalist put it in describing Senate action, "the package is crammed full of provisions for nearly every senator who cast a vote" for it, quoting one of the bill's Senate sponsors, Lisa Murkowski, R-Alaska, on the strategy employed: "We have also worked for months on a bipartisan, bicameral basis to truly negotiate every single word in this bill." Her House counterpart, Raul Grijalva, D-Arizona, called it "an old-school green deal," and noted that he and the top Republican on the House Natural Resources Committee "are happy to work together to get this across the finish line." Much the same could be said for yet another major land conservation measure enacted in mid-2020, the Great American Outdoors Act. It provided more than $9 billion over the next five years for long-delayed repairs and maintenance of national park lands and facilities, and also assured $900 million in annual support for the popular Land and Water Conservation Fund. The measure easily passed both the Senate and House, with President Trump endorsing its passage.[62]

National Energy Policy and Chemical Safety

In 2005, Congress finally enacted one of the Bush administration's priorities that the president had sought since 2001, the Energy Policy Act of 2005. It was the first major overhaul of US energy policy since 1992. The original Bush energy plan, formulated in 2001 by a task force headed by Vice President Dick Cheney, called for an increase in the production and use of fossil fuels and nuclear energy, gave modest attention to the role of energy conservation, and sparked intense debate on Capitol Hill because of its emphasis on oil and gas drilling in ANWR. The Republican House quickly approved the measure in 2001, after what the press called "aggressive lobbying by the Bush administration, labor unions and the oil, gas, and coal industries."[63] The vote largely followed party lines.

Competing energy bills were debated on the Hill through mid-2005 without resolution and served as another prominent example of legislative gridlock. Neither side was prepared to compromise as lobbying by car manufacturers, labor unions, the oil and gas industry, and environmentalists continued. As one writer put it in 2002, the "debate between energy and the environment is important to core constituencies of both parties, the kind of loyal followers vital in a congressional election year."[64] Finally, the House and Senate reached agreement on an energy package, and the president signed the 1,700-page bill on August 8, 2005.[65] The thrust of the legislation remained largely what Bush and Cheney had sought in 2001, although the final measure included significant funding for energy research and development and other measures that Democrats had supported, such as new energy efficiency standards for federal office buildings and short-term tax credits for the purchase of hybrid vehicles.[66]

In 2007, Congress also approved the Energy Independence and Security Act of 2007, which set a national automobile fuel economy standard of 35 miles per gallon by 2020. In one of his most consequential acts, President Obama later negotiated with automakers to raise the fuel economy standard to 54.5 miles per gallon by 2025. However, after several years of contentious negotiations with auto makers, many of whom opposed the change, and states that had actively sought them, the Trump administration significantly weakened those standards, a decision certain to be reviewed by federal courts (see Chapter 6).[67]

President Obama's economic stimulus measure, which Congress approved in February 2009, contained about $80 billion in spending, tax incentives, and loan guarantees, including funds for energy efficiency, renewable energy sources, mass transit, and technologies for the capture and storage of greenhouse gases produced by coal-fired power plants. Had these energy components been a stand-alone measure, the *New York Times* observed, they would have amounted to "the biggest energy bill in history." In this case, however, bipartisan cooperation was largely absent. In the end, only three Republicans in Congress, all in the Senate, voted for the bill.[68]

Similarly, after several years of partisan gridlock, in April 2016, the Senate approved a modest but bipartisan energy efficiency and electrical grid infrastructure measure, the Energy Policy Modernization Act, by a vote of 85 to 12. Republican Lisa Murkowski of Alaska and Democrat Maria Cantwell of Washington State had worked together for several years on the legislation. Nearly 5 months earlier, in December of 2015, the House had approved a companion bill. However, House Republicans attached a number of unrelated and controversial environmental policy provisions that eroded almost all Democratic support for the measure. As a result, the final House vote was 249 to 174, with only nine Democrats voting in favor. The Senate and House measures were so at odds that no compromise legislation could clear the 114th Congress.

Murkowski vowed to reintroduce the bipartisan Senate bill in a later Congress.[69] By 2020, she had assembled a massive 555-page bill to shift the nation toward cleaner sources of energy, this time in cooperation with Sen. Joe Manchin III (D-West Virginia). The American Energy Innovation Act combined about a dozen bills from the Senate Committee on Energy and Natural Resources that had bipartisan support, but it failed on a procedural vote on the Senate floor in early March 2020. In the House, wide-ranging energy and climate change packages considered in 2020 were supported by most Democrats, but not Republicans, and stood little chance of passage in the Senate.[70]

Successful bipartisanship was more in evidence during a significant reform of the widely criticized Toxic Substances Control Act (TSCA) of 1976. After decades of unsuccessful efforts to modernize and strengthen the act, in 2016, Congress approved the Frank R. Lautenberg Chemical Safety for the 21st Century Act, a major revision of TSCA. The new act mandated that the EPA evaluate existing chemicals with a new risk-based safety standard, that it do so with clear and enforceable deadlines, with increased transparency for chemical information, and with assurances that the agency would have the budgetary resources to carry out its responsibilities for chemical safety.[71]

Continuing Partisan Conflict and Stalemate

The examples discussed previously and many more that could be cited, such as approval of a historic Great Lakes Compact in 2008 to prevent water diversion from the lakes and a new Higher Education Sustainability Act in 2008, show that over the past decade Congress has been able to move ahead on a wide range of environmental and natural resource policies.[72] Yet continuing partisan conflict has blocked action on key federal laws such as the Superfund program, the Endangered Species Act, the Clean Water Act, and the Clean Air Act, as well as on legislation to address climate change.

The Superfund program, for example, has not been reauthorized for well over two decades, and except for the brownfields measure discussed earlier, congressional agreement has not been forthcoming. The Endangered Species Act presents a similar level of conflict and lack of resolution. In 2001, then House Resources Committee chair James V. Hansen, R-Utah, captured the dilemma well: "We haven't reauthorized it because no one could agree on how to reform and modernize the law. Everyone agrees there are problems with the Act, but no one can agree on how to fix them."[73]

Climate change policy, of course, has long been something of a poster child for legislative gridlock, whether it is a stand-alone climate measure or part of an integrated energy bill. In 2009, the House approved a cap-and-trade bill, the American Clean Energy and Security Act of 2009, after extensive and prolonged negotiations and major concessions for the various industries likely to be affected by it. These included automakers, steel companies, natural gas drillers, oil refiners, utilities, and farmers, among others. The final vote was 219 to 212, with all but eight Republicans voting in opposition, along with forty-four Democrats. However, the Senate failed to pass comparable legislation, and Republicans won control of Congress in the 2010 elections, ending any prospect of a legislative solution to climate change. President Obama then had little choice but to pursue administrative solutions, such as the auto fuel economy rules and the EPA's Clean Power Plan. As indicated throughout the chapter, partisan disputes over climate change continue even as a diverse assortment of climate and energy bills have been introduced in both the House and Senate. In the absence of congressional action, the states have become centers of innovation on climate change (see Chapters 2 and 12).

CONCLUSION

The political struggles on Capitol Hill over the last several decades reveal sharply contrasting visions for environmental policy. The revolutionary rhetoric and anti-environmental stance of the 104th Congress had dissipated by the 2000s, but it was replaced in the 2010s by similar views held by Tea Party Republicans, especially in the House Freedom Caucus, and then by a solidly conservative Republican Party in both houses following Donald Trump's election in 2016. The political alignment shifted once again when Democrats recaptured the House in the 2018 midterm elections. As reviewed in the chapter, from the mid-1990s through 2020, Congress did revise several major environmental statutes and it adopted some new laws in an uncommon display of bipartisan cooperation. Nonetheless, for many other environmental programs, policy gridlock continued to frustrate all participants, and deep partisan differences prevented emerging issues such as climate change from being fully addressed.

Neither the election of President Obama in 2008 and 2012 nor the success of Donald Trump in the 2016 election did much to alter legislative prospects on the Hill. Republican opposition thwarted most legislative proposals on the environment during the Obama years, and Democratic resistance limited what the Trump administration could hope to do legislatively. Instead, the Trump White House turned to use of executive authority to further its preferred policies. Congress did, however, successfully block the Trump administration's proposals for severe budgetary cuts for environmental and natural resource agencies, indicating its continuing support for long-established federal programs and agencies.

Under these political conditions, it is no surprise that, by 2020, public ratings of Congress remained near historic lows, with strong public disapproval of both parties, even as the institution received somewhat higher rating than previously following passage of a series of congressional bills related to the coronavirus pandemic. Moreover, Congress ranked near the bottom in the public's confidence in American institutions, well below big business, newspapers, and television news, and far below the presidency and the courts.[74] The constitutional divisions between the House and Senate, and between Congress and the White House, guarantee that newly emergent forces, whether on the left or the right of the political spectrum, cannot easily push a particular legislative agenda. The 2016 elections did not change this outlook, nor did the 2018 elections.

As of the fall of 2020, the outlook for the November elections and environmental policy choices remained uncertain, particularly as the nation and world struggled with the devastating effects of the COVID-19 pandemic. What impacts will the pandemic and governmental responses to it have on the American public's views and behavior? Will people now see other threats to their health and safety, including environmental risks, as more important? Will people be more likely in the future to turn to the federal government for leadership on national and global challenges such as climate change, or less likely given the halting federal response to the pandemic and its inability to prevent horrific loss of life and great damage to the economy? Will we see more or less bipartisanship, and how will that affect congressional action on environmental challenges? It is encouraging that the House and Senate acted swiftly and mostly cooperatively in the spring of 2020 to approve trillions of dollars of economic support in response to the pandemic.[75]

The environmental policy battles of the past several decades offer other lessons. They remind us that in the US political system, effective policymaking will always require cooperation between the two branches and leadership within both to advance sensible policies and secure public approval for them. The public has a role to play in these deliberations, and the history of congressional policymaking on the environment strongly suggests the power of public beliefs and action. Public disillusionment with government and politics both before and after the pandemic of 2020 creates significant barriers to policy change. Regrettably, it also strengthens the power of special interests to secure the changes they desire, which often differ significantly from the public's own preferences. Ultimately, the solution can be found only in a heightened awareness of the problems and active participation by the American public in the political process.

SUGGESTED WEBSITES

Congress.gov (www.congress.gov) This site is one of the most comprehensive public sites available for legislative searches. See also www.house.gov and www.senate.gov for portals to the House and Senate and to the committee and individual member websites.

Environmental Protection Agency (www.epa.gov/epahome/rule.html) The EPA site for laws, rules, and regulations includes the full text of the dozen key laws administered by the EPA. It also has a link to current legislation before Congress.

Heritage Foundation (www.heritage.org) One of the most prominent and influential conservative think tanks, Heritage has strong ties to Republican members of Congress and issues frequent reports on a range of environmental, energy, and natural resource issues.

League of Conservation Voters (www.lcv.org) The LCV compiles environmental voting records for all members of Congress.

National Association of Manufacturers (www.nam.org) This leading business organization offers policy news, studies, and position statements on environmental issues, as well as extensive resources for public action on the issues.

Natural Resources Defense Council (www.nrdc.org) The NRDC is perhaps the most active and influential of the national environmental groups that lobby Congress.

Sierra Club (www.sierraclub.org) The Sierra Club is one of the leading national environmental groups that track congressional legislative battles.

US Chamber of Commerce (www.uschamber.com) The US Chamber of Commerce is one of the nation's leading business organizations, and it frequently challenges legislative proposals that it believes may harm business interests.

NOTES

1. The quotation comes from Sen. Markey's press release on introduction of the Green New Deal on February 7, 2019, available at www.markey.senate.gov/news/press-releases/senator-markey-and-rep-ocasio-cortez-introduce-green-new-deal-resolution.

2. Zack Colman, "Green New Deal Resolution Calls for 'National Mobilization' on Climate, Economy," *Politico*, February 7, 2019. Quotations from the resolution are taken from Sen. Markey's news release, cited above.

3. See, for example, the compilation by the Environmental and Energy Study Institute, at www.eesi.org/articles/view/how-the-116th-congress-is-addressing-climate-change.

4. See "Want a Green New Deal? Here's a Better One," by the Editorial Board of the *Washington Post*, February 24, 2019; Steven Ratner, "Yes, We Need a Green New Deal. Just Not the One Alexandria Ocasio-Cortez Is Offering," *New York Times*, May 20, 2019; and Brad Plumer, "A "Green New Deal" Is Far from Reality, but Climate Action Is Picking Up in the States," *New York Times*, February 8, 2019. Support for the original proposal came well before the resolution itself was introduced. See, for example, Greg Carlock and Sean McElwee, "Why the Best New Deal Is a Green New Deal," *The Nation*, September 18, 2018. On the July proposal by the Center for American Progress and the other groups, see Dino Grandoni, "Broad Group of Green Organization Releases Climate Platform Ahead of 2020 Election," *Washington Post*, July 18, 2019.

5. Dino Grandoni, "The GOP Campaign Against the Green New Deal May Be Working," *Washington Post*, May 10, 2019.

6. See Daniel J. Fiorino, *The New Environmental Regulation* (Cambridge, MA: MIT Press, 2006); Marc Allen Eisner, *Governing the Environment: The Transformation of Environmental Regulation* (Boulder, CO: Lynne Rienner, 2007); and Daniel A. Mazmanian and Michael E. Kraft, eds., *Toward Sustainable Communities: Transition and Transformation in Environmental Policy*, 2nd ed. (Cambridge, MA: MIT Press, 2009).

7. See Eric Schickler and Kathryn Pearson, "The House Leadership in an Era of Partisan Warfare," in *Congress Reconsidered*, 8th ed., ed. Lawrence C. Dodd and Bruce I. Oppenheimer (Washington, DC: CQ Press, 2005), 207–26. See also Thomas E. Mann and Norman J. Ornstein, *The Broken Branch: How Congress Is Failing America and How to Get It Back on Track* (New York: Oxford University Press, 2006).

8. See Chapter 1 in this volume and Michael E. Kraft, "Congress and Environmental Policy," in *The Oxford Handbook of U.S. Environmental Policy*, ed. Sheldon Kamieniecki and Michael E. Kraft (New York: Oxford University Press, 2013), 280–305.

9. Ed Gillespie and Bob Schellhas, eds., *Contract with America* (New York: Times Books/Random House, 1994); Bob Benenson, "GOP Sets the 104th Congress on New Regulatory Course," *Congressional Quarterly Weekly Report*, June 17, 1995, 1693–705.

10. For a general review of much of this period, see Lawrence C. Dodd and Bruce I. Oppenheimer, "A Decade of Republican Control: The House of Representatives, 1995–2005," in *Congress Reconsidered*, 8th ed., ed. Lawrence C. Dodd and Bruce I. Oppenheimer (Washington, DC: CQ Press, 2005), 23–54.

11. By one count, the House passed over 400 bills by November of 2019, with 275 of them having bipartisan support, on which the Senate majority leader, Mitch McConnell, would not allow debates or votes. See Katrina vanden Heuvel, "Don't Fall for Trump's Lie. Democrats Have Been Very Productive," *Washington Post*, February 11, 2020. For an assessment of why legislative action is so difficult in the Senate today, see an editorial by 70 former senators, "70 Former U.S. Senators: The Senate Is Failing to Perform Its Constitutional Duties," *Washington Post*, February 25, 2020.

12. See Mazmanian and Kraft, *Toward Sustainable Communities*; Eisner, *Governing the Environment*; Christopher McGrory Klyza and David Sousa, *American Environmental Policy: Beyond Gridlock*, updated and expanded edition (Cambridge, MA: MIT Press, 2013); and Robert F. Durant, Daniel J. Fiorino, and Rosemary O'Leary, eds., *Environmental Governance Reconsidered: Challenges, Choices, and Opportunities*, 2nd ed. (Cambridge, MA: MIT Press, 2017).

13. For a general analysis of roles that Congress plays in the U.S. political system, see Roger H. Davidson, Walter J. Oleszek, Frances E. Lee, and Eric Schickler, *Congress and Its Members*, 17th ed. (Thousand Oaks, CA: CQ Press, 2020).

14. Quoted in "Leader of New Climate Panel Talks of Need for 'Bold Action,'" MSN.com, February 8, 2019.

15. See, for example, Juliet Eilperin and Brady Dennis, "White House Eyes Plan to Cut EPA Staff By One-Fifth, Eliminating Key Programs," *Washington Post*, March 1, 2017; Jeffrey Mervis, "Congress Again Rejects Trump Cuts, Smiles on Science Agencies," *Science* 367 (January 9, 2020): 13–4.

16. On the broad oversight powers that Congress can use, see Paul C. Light, "How the House Should Investigate the Trump Administration: Lessons from the Most Important House Probes Since WW II," Brookings Institution, March 8, 2019.

17. Rein, "Meet the House Science Chairman." See also Smith's press release, "Statement of Chairman Lamar Smith (R-Texas), H.R. 1430, Honest and Open New EPA Science Treatment Act of 2017, March 9, 2017." For an update, see David Malakoff, "'Secret Science' Plan Is Back, and Critics Say It's Worse," *Science* 366 (November 15, 2019): 783–4; Lisa Friedman, "E.P.A. to Limit Science Used to Write Public Health Rules," *New York Times*, November 11, 2019; and Friedman, "E.P.A. Updates Plan to Limit Science Used in Environmental Rules," *New York Times*, March 4, 2020.

18. Dino Grandoni, "Congress Has Held At Least 15 Climate Hearings Since Democrats Won the House," *Washington Post*, March 7, 2020.

19. See Coral Davenport, "Scott Pruitt, Trump's E.P.A. Pick, Is Approved by Senate Committee," *New York Times*, February 2, 2017; Brady Dennis and Juliet Eilperin, "Hundreds of Current, Former EPA Employees Urge Senate to Reject Trump's Nominee for the Agency," *Washington Post*, February 6, 2017. On Wheeler's appearance before the committee, see Lisa Friedman, "Senators Grill Andrew Wheeler, Former Coal Lobbyist and Trump's Choice to Lead the E.P.A.," *New York Times*, January 16, 2019.

20. Kraft, "Congress and Environmental Policy." For a recent review of the strengths and weakness of Congress's institutional capacity, and recommendations for improvement, see the "Report of the Task Force Project on Congressional Reform" (Washington, DC: American Political Science Association, 2019). In addition to the three support agencies noted, the report suggested revival of the former Office of Technology Assessment, which for years aided Congress on environmental policy, as described at length in Gary C. Bryner, ed., *Science, Technology, and Politics: Policy Analysis in Congress* (Boulder, Colo.: Westview Press, 1992).

21. Davidson, Oleszek, Lee, and Schickler, *Congress and Its Members*. See also Gary C. Jacobson and Jamie L. Carson, *The Politics of Congressional Elections*, 10th ed. (New York: Rowman and Littlefield, 2020).

22. Carl Hulse, "Consensus on Energy Bill Arose One Project at a Time," *New York Times*, November 19, 2003; Kenneth P. Vogel, Catie Edmondson, and Jesse Drucker "Coronavirus Stimulus Package Spurs a Lobbying Gold Rush," *New York Times*, March 20, 2020.

23. See John M. Broder, "At House E.P.A. Hearing, Both Sides Claim Science," *New York Times*, March 8, 2011; Juliet Eilperin and Brady Dennis, "How James Inhofe Is Upending the Nation's Energy and Environmental Policies," *Washington Post*, March 14, 2017.

24. On the general idea of policy drift and failure to reform key public policies, see Jacob S. Hacker and Paul Pierson, *Winner-Take-All Politics: How Washington Made the Rich Richer—and Turned Its Back on the Middle Class* (New York: Simon and Schuster, 2010).
25. On the role of administrative agencies and the courts in policy change, see Klyza and Sousa, *American Environmental Policy*.
26. One of the few scholarly analyses of the subject is Sarah A. Binder, *Stalemate: Causes and Consequences of Legislative Gridlock* (Washington, DC: Brookings Institution Press, 2003). Aside from what the chapter covers, other factors also affect legislative gridlock today. Among them are the constitutional specification that the Senate be composed of two senators for each state (thus giving small and often conservative states an oversized representation in that body), the effects of legislative redistricting or gerrymandering that can distort the public's partisan preferences, and the weak national laws on campaign financing in the wake of the Supreme Court's *Citizens United v. FEC* decision. Even the filibuster in the Senate plays a role; it prevents the majority party from taking action without the sixty votes needed to overcome frequent filibusters by the minority party.
27. See, for example, James A. Thurber and Antoine Yoshinaka, eds., *American Gridlock: The Sources, Character, and Impact of Political Polarization* (New York: Cambridge University Press, 2015); Jeffery A. Jenkins and Eric M. Patashnik, eds., *Congress and Policy Making in the 21st Century* (New York: Cambridge University Press, 2016); and Theda Skocpol and Caroline Tervo, eds., *Upending American Politics: Polarizing Parties, Ideological Elites, and Citizens Activists from the Tea Party to the Anti-Trump Resistance* (New York: Oxford University Press, 2020).
28. This was one of the more important conclusions about constituency influence on congressional voting John Kingdon reached in his influential *Congressmen's Voting Decisions*, 3rd ed. (Ann Arbor: University of Michigan Press, 1989). For one of the most thorough studies of public opinion on the environment, see David P. Daniels, Jon A. Krosnick, Michael P. Tichy, and Trevor Tompson, "Public Opinion on Environmental Policy in the United States," in *The Oxford Handbook of U.S. Environmental Policy*, ed. Sheldon Kamieniecki and Michael E. Kraft (New York: Oxford University Press, 2013), 461–86. See also a fascinating study of how constituents' preferences may be undercut by the way legislative staff perceives and filters them, Alexander Hertel-Fernandez, Matto Mildenberger, and Leah C. Stokes, "Legislative Staff and Representation in Congress," *American Political Science Review* 113, 1 (2019): 1–18.
29. Studies show that members of Congress generally do vote in a way that is consistent with their campaign promises on environmental issues but that Republicans are "far more likely to break their campaign promises" and that proenvironmental campaign promises are more likely to be broken than are others. See Evan J. Ringquist and Carl Dasse, "Lies, Damned Lies, and Campaign Promises? Environmental Legislation in the 105th Congress," *Social Science Quarterly* 85 (June 2004): 400–19. The quotation is from p. 417. See also Evan J. Ringquist, Milena I. Neshkova, and Joseph Aamidor, "Campaign Promises, Democratic Governance, and Environmental Policy in the U.S. Congress," *Policy Studies Journal* 41, no. 2 (2013): 365–87.
30. See David Bornstein, "Cracking Washington's Gridlock to Save the Planet," *New York Times*, May 19, 2017. Recent polls document the continued low saliency of environmental issues, including climate change, even though the public favors many policy actions on climate change and has for several years, according to both the Gallup poll and the Yale Project on Climate Change Communication (See Chapter 3).
31. See, for example, Nolan McCarty, Keith Poole, and Howard Rosenthal, *Polarized America: The Dance of Ideology and Unequal Riches* (Cambridge, MA: MIT Press, 2008).

32. See Charles R. Shipan and William R. Lowry, "Environmental Policy and Party Divergence in Congress," *Political Research Quarterly* 54 (June 2001): 245–63.

33. See the League of Conservation Voters (LCV), *National Environmental Scorecard* (Washington, DC: LCV, 2020). The scorecard is published annually at the league's website (www.lcv.org); the party averages are taken from Rachel Frazin, "Both Parties See Gains on Environmental Scorecard, *The Hill*, March 12, 2020. For a detailed examination of these patterns over time, see Riley E. Dunlap, Aaron M. McCright, and Jerrod H. Yarosh, "The Political Divide on Climate Change: Partisan Polarization Widens in the U.S." *Environment* 58, no. 5 (September/October 2016): 4–22; Nadja Popovich, "Climate Change Rises As a Public Priority. But It's More Partisan Than Ever," *New York Times*, February 20, 2020.

34. See Aaron M. McCright, Chenyang Xiao, and Riley E. Dunlap, "Political Polarization on Support for Government Spending on Environmental Protection in the USA, 1974–2012," *Social Science Research* 48 (2014): 251–60.

35. The membership of both Senate and House caucuses can be found at the Citizen Climate Lobby website: https://citizensclimatelobby.org/climate-solutions-caucus/.

36. Cecelia Smith-Schoenwalder, "Senate Democrats Form Special Committee on Climate Change," *U.S. News and World Report*, March 27, 2019.

37. Mary Etta Cook and Roger H. Davidson, "Deferral Politics: Congressional Decision Making on Environmental Issues in the 1980s," in *Public Policy and the Natural Environment*, ed. Helen M. Ingram and R. Kenneth Godwin (Greenwich, CT: JAI, 1985). See also Norman J. Vig and Michael E. Kraft, eds., *Environmental Policy in the 1980s: Reagan's New Agenda* (Washington, DC: CQ Press, 1984); Michael E. Kraft and Norman J. Vig, "Environmental Policy in the Reagan Presidency," *Political Science Quarterly* 99, no. 3 (Fall 1984): 415–39.

38. For a fuller discussion of the gridlock over clean air legislation, see Gary C. Bryner, *Blue Skies, Green Politics: The Clean Air Act of 1990 and Its Implementation* (Washington, DC: CQ Press, 1995).

39. Rhodes Cook, "Rare Combination of Forces May Make History of '94," *Congressional Quarterly Weekly Report*, April 15, 1995, 1076–81.

40. Katharine Q. Seelye, "Files Show How Gingrich Laid a Grand G.O.P. Plan," *New York Times*, December 3, 1995. On the broader history of conservative and business campaigns against environmental policy, see Judith A. Layzer, *Open for Business: Conservatives' Opposition to Environmental Regulation* (Cambridge, MA: MIT Press, 2012). For a somewhat parallel development in recent years to discredit climate change, see Riley E. Dunlap and Aaron M. McCright, "Challenging Climate Change: The Denial Countermovement," in *Climate Change and Society: Sociological Perspectives*, ed. R. E. Dunlap and R. J. Brulle (New York: Oxford University Press, 2015), 300–32.

41. Groups such as Americans for Prosperity, part of what is often termed the "Koch Network," have been influential as well as major contributors to congressional campaigns, and they have emphasized an antienvironmental and pro-fossil fuels agenda at odds with public preferences for environmental protection and action on climate change. See Eric Lipton, "G.O.P. Hurries to Slash Oil and Gas Rules, Ending Industries' 8-Year Wait," *New York Times*, February 4, 2017; Theda Skocpol and Alexander Hertel-Fernandez, "The Koch Network and Republican Party Extremism," *Perspectives on Politics*, 14, no. 3 (September 2016): 681–99.

42. Allan Freedman, "GOP's Secret Weapon against Regulations: Finesse," *CQ Weekly*, September 5, 1998, 2314–20; Charles Pope, "Environmental Bills Hitch a Ride through the Legislative Gantlet," *CQ Weekly*, April 4, 1998, 872–75.

43. See Fiorino, *The New Environmental Regulation*; Eisner, *Governing the Environment*.

44. See Sara R. Rinfret and Scott R. Furlong, "Defining Environmental Rulemaking," in *The Oxford Handbook of Environmental Policy*, ed. Sheldon Kamieniecki and Michael E. Kraft (New York: Oxford University Press, 2013), 372–94.

45. See Michael E. Kraft and Sheldon Kamieniecki, eds., *Business and Environmental Policy: Corporate Interests in the American Political System* (Cambridge, MA: MIT Press, 2007).

46. See Jackie Calmes, "Both Sides of the Aisle Say More Regulation, and Not Just of Banks," *New York Times*, October 14, 2008.

47. On the broad regulatory reform actions in Congress in 2017, see William W. Buzbee, "Regulatory 'Reform' That Is Anything But," *New York Times*, June 15, 2017.

48. David Malakoff, "Republicans Ready a Regulatory Rollback," *Science* 354 (November 25, 2016): 951; Philip A. Wallach and Nicholas W. Zeppos, "Is the Congressional Review Act About to Supercharge Deregulation?" *FIXGOV* [blog of Brookings Institution], April 4, 2017. The quotation is from Juliette Eilperin, "Trump Undertakes Most Ambitious Regulatory Rollback Since Reagan," *Washington Post*, February 12, 2017. See also Michael Grunwald, "Trump's Secret Weapon Against Obama's Legacy," *Politico*, April 10, 2017.

49. Michael D. Shear, "Trump Discards Obama Legacy, One Rule at a Time," *New York Times*, May 1, 2017.

50. John H. Cushman Jr., "G.O.P.'s Plan for Environment Is Facing a Big Test in Congress," *New York Times*, July 17, 1995.

51. Robert B. Semple Jr., "Concealed Weapons against the Environment," *New York Times*, July 31, 2011.

52. Pope, "Environmental Bills Hitch a Ride."

53. Andrew Revkin, "Law Revises Standards for Scientific Study," *New York Times*, March 21, 2002. See also Rick Weiss, "'Data Quality' Law Is Nemesis of Regulation," *Washington Post*, August 16, 2004; Paul Raeburn, "A Regulation on Regulations," *Scientific American*, July 2006, 18–19. Raeburn reported that by 2006, perhaps one hundred Data Quality Act petitions had been filed with dozens of different government agencies, most of them by industry groups.

54. Susan Zakin, "Riders from Hell," *Amicus Journal* (Spring 2001): 20–22. For recent use of riders, see New York Times Editorial Board, "The Dirty Little Deals That Would Foul the Environment," *New York Times*, February 19, 2018.

55. Carroll J. Doherty and the staff of *CQ Weekly*, "Congress Compiles a Modest Record in a Session Sidetracked by Scandal: Appropriations," *CQ Weekly*, November 14, 1998, 3086–87 and 3090–91.

56. Jeffrey Mervis and the *Science* News Staff, "Congress Trumps President in Backing Science," *Science* 356 (May 5, 2017): 470–71. See also Lewis M. Milford and Mark Muro, "A Senate Panel Speaks for Sound Clean Energy Policy—and Rebukes Trump," Brookings Institution, August 8, 2017, available at www.brookings.edu/blog/the-avenue/2017/08/08/a-senate-rebukes-trump/.

57. David Hosansky, "Rewrite of Laws on Pesticides on Way to President's Desk," *Congressional Quarterly Weekly Report*, July 27, 1996, 2101–03; Hosansky, "Provisions: Pesticide, Food Safety Law," *Congressional Quarterly Weekly Report*, September 7, 1996, 2546–50.

58. David Hosansky, "Drinking Water Bill Clears; Clinton Expected to Sign," *Congressional Quarterly Weekly Report*, August 3, 1996, 2179–80; Allan Freedman, "Provisions: Safe Drinking Water Act Amendments," *Congressional Quarterly Weekly Report*, September 14, 1996, 2622–27.

59. Rebecca Adams, "Pressure from White House and Hastert Pries Brownfields Bill from Committee," *CQ Weekly*, September 8, 2001, 2065–66.

60. Mary Clare Jalonick, "Healthy Forests Initiative Provisions," *CQ Weekly*, January 24, 2004, 246–47.

61. Juliet Eilperin, "Keeping the Wilderness Untamed: Bills in Congress Could Add as Much as Two Million Acres of Unspoiled Land to Federal Control," *Washington Post* (National edition), June 23–July 6, 2008, 35; Avery Palmer, "Long-Stalled Lands Bill Gets Nod from Senate," *CQ Weekly*, January 19, 2009, 128.

62. See Juliet Eilperin and Dino Grandoni, "The Senate Just Passed the Decade's Biggest Public Lands Package. Here's What's in It," *Washington Post*, February 12, 2019; Matthew Daly, "Congress Passes Sprawling Plan to Boost Conservation, Parks, *Washington Post*, July 22, 2020.

63. Chuck McCutcheon, "House Passage of Bush Energy Plan Sets Up Clash with Senate," *CQ Weekly*, August 4, 2001, 1915–17.

64. Rebecca Adams, "Politics Stokes Energy Debate," *CQ Weekly*, January 12, 2002, 108.

65. Carl Hulse, "House Votes to Approve Broad Energy Legislation," *New York Times*, April 22, 2005.

66. See Ben Evans and Joseph J. Schatz, "Details of Energy Policy Law," *CQ Weekly*, September 5, 2005, 2337–45.

67. Steven Overly, "Trump to Pull Back EPA's Fuel Efficiency Determination, Opening the Door for Reduced Standards," *Washington Post*, March 15, 2020.

68. Editorial, "An $80 Billion Start," *New York Times*, February 18, 2009.

69. See Editorial Board, "An Energy Bill in Need of Fixes," *New York Times*, April 20, 2016.

70. Steven Mufson and Dino Gandoni, "Senate Prepares to Vote on Big Energy Bill This Week," *Washington Post*, March 2, 2020; Steven Mufson and Dino Grandoni, "House Democrats Unveil Ambitious Climate Package, Steering Toward a Net-Zero Economy by 2050," *Washington Post*, June 30, 2020. The House proposals did not go as far as the Green New Deal, but nonetheless were remarkably comprehensive and pathbreaking. Some termed the package the Democrats' green infrastructure bill.

71. See Coral Davenport and Emmarie Huetteman, "Lawmakers Reach Deal to Expand Regulation of Toxic Chemicals," *New York Times*, May 19, 2016.

72. The education act was part of the Higher Education Opportunity Act of 2008. It authorized competitive grants to institutions and associations in higher education to promote the development of sustainability curricula, programs, and practices. It was the first new federal environmental education program in 18 years.

73. Cited in *Science and Environmental Policy Update*, the Ecological Society of America online newsletter, April 20, 2001.

74. The Gallup poll reports regularly on the public's approval of Congress and its trust and confidence in governmental and other institutions. In mid-April of 2020, 30 percent approved of "the way Congress is handling its job," while 66 percent disapproved. See "Congress and the Public," at https://news.gallup.com/poll/1600/congress-public.aspx. On approval of other institutions, see Gallup's report at https://news.gallup.com/poll/1597/confidence-institutions.aspx.

75. See one report on the assistance in Emily Cochrane and Nicholas Fandos, "Senate Approves $2 Trillion Stimulus After Bipartisan Deal," *New York Times*, March 25.

6 ENVIRONMENTAL POLICY IN THE COURTS

Kimberly Smith[1]

On January 2, 1900, the Chicago Sanitation District completed an ambitious civil engineering project: A drainage canal that reversed the course of the Chicago River. Instead of fouling Lake Michigan—the source of the city's drinking water—the city's sewage would now flow into the Des Plaines River and, eventually, the Mississippi River. This engineering spectacle was rivaled by the legal spectacle that followed: Missouri's Attorney General Edward Crow sued the state of Illinois in the US Supreme Court, arguing that the sewage created a nuisance by spreading filth and disease in St. Louis and other towns along the Mississippi River. The lawsuit took 6 years to resolve, mobilizing hundreds of expert witnesses and producing more than 100 exhibits. A 13,160-page report was presented to the Supreme Court in May 1905, which issued its final decision in February 1906.[2] Missouri lost. Despite the massive record, the Court was not convinced that the sewage was harming human health.

This lawsuit opened a new era of environmental litigation in federal courts. By the 1920s, the Supreme Court had become the principal forum for resolving, or trying to resolve, a host of interstate conflicts over natural resources based on common law causes of action like nuisance.[3] State attorneys general were the driving force behind these lawsuits; it took the resources of a state government to pursue them. They typically involved hundreds of hours of testimony, compiled records thousands of pages long, and dragged on for years or even decades. And they were hard to win. Given the complexity and uncertainty of environmental science, it was challenging to prove whose pollution was causing what sort of harm. But in the absence of federal pollution legislation, states had little choice but to ask the federal courts to decide these matters. As a result, for much of the twentieth century, federal courts decided how much and what kind of waste the states could emit into the air and water.

With the passage of the Clean Air Act (CAA), Clean Water Act (CWA), and other federal environmental statutes in the 1970s, the Supreme Court happily relinquished to Congress and the executive branch the responsibility for setting transboundary pollution policy.[4] But those days of judicial policymaking may not be safely in the past. With Congress hampered by gridlock, environmental activists—along with a growing contingent of state attorneys general—are reviving common law doctrines and asking federal courts once again to take the lead in addressing new environmental problems like climate change. It's not a role that judges are used to playing.

To be sure, courts are important players in the environmental policy system, but their primary role today is to supervise the administrative agencies that implement environmental statutes. In that role, courts may force reluctant policymakers to act by enforcing statutory mandates, but more commonly they slow down or moderate policy change. Judges affect the pace and scope of policy change by enforcing statutory requirements on agencies, interpreting legal rules, and reviewing the constitutionality of environmental policies.

Both state and federal courts can play this supervisory role, but the focus of this chapter is environmental policymaking in the federal courts. First, I will profile the US court system and explain the federal courts' role in implementing environmental policy and overseeing administrative agencies. Next, I will review how courts can force, delay, or shape environmental policymaking, as illustrated by several case studies. I end the chapter by considering how the president's appointment power and congressional gridlock may affect the courts' role in environmental policy in the coming years.

THE JUDICIAL SYSTEM

Although the US Supreme Court receives the most attention, most litigation in the federal courts begins and ends in the trial court, called the District Court. When questions of law arise during the trial, the question may be appealed to the intermediate appellate court (the Circuit Court). Questions of federal law, whether arising in a state or federal court, may ultimately be reviewed and decided by the US Supreme Court.

The authority to determine the meaning and application of regulations, statutes, and constitutional provisions gives the courts influence in the policy system. Moreover, they can exercise that influence at least somewhat independently of the other branches. The 860 federal judges staffing the regular federal courts enjoy life tenure. Although appointed by the president and confirmed by the Senate, they cannot be removed from office (barring serious malfeasance), nor can Congress lower their salaries. These features insulate judges from partisan pressure and make them attractive audiences for policy arguments that failed to prevail in the legislature. Thus, environmental advocates and their opponents spend at least as much time in court, litigating over the interpretation and implementation of environmental policy, as they do in lobbying Congress and the executive branch. Nevertheless, several features of courts make it difficult for this branch to exert consistent influence on policy (see Figure 6.1).

First, courts are largely passive institutions; they must wait for plaintiffs to bring cases to them rather than actively seeking out policy conflicts to resolve. Although a judge may (in an opinion, article, or public speech) indicate an interest in addressing certain legal questions, it is the plaintiffs who decide whether and, often, in which court to bring suit. Environmental organizations and their opponents typically have sophisticated litigation strategies, seeking out conflicts that are suitable for resolution in court, finding sympathetic plaintiffs and fact situations that favor their side. This is also the role often played by state attorneys general, since states may have standing to sue—that is, a legally cognizable interest—even when no private parties do.

Second, the sheer multiplicity of courts means that questions about a law may arise in many different forums and be decided by different panels of judges, leading to conflicting outcomes. Eventually, of course, some of those questions may be resolved by the US Supreme Court. But the Supreme Court, which has broad discretion over its docket, hears only a very small fraction of cases each year. Most decisions on federal law by state and federal lower courts will not be reviewed, so conflicting decisions among circuits can persist for years.

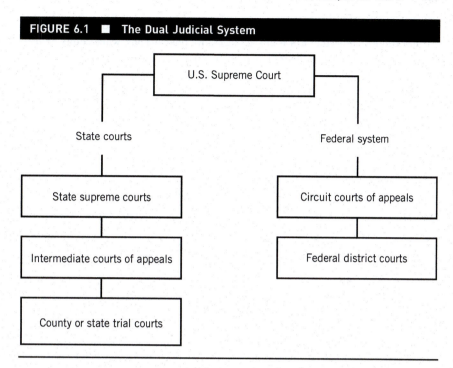

FIGURE 6.1 ■ The Dual Judicial System

Finally, courts have very limited power to enforce their decisions. They rely on the cooperation of the other branches, and their only means of persuading those actors is through well-reasoned legal argument and the general public respect for the courts. Without the support of the political branches, judicial decisions may be rendered meaningless simply by being ignored.

Despite these limitations, courts do influence environmental policy. First, the courts supervise how federal agencies implement environmental laws. Some statutes authorize judicial review directly, but most administrative decisions are also reviewable under the Administrative Procedure Act[5] (APA), which sets out the basic procedures by which federal agencies carry out their rulemaking and enforcement duties. Because it provides for judicial review of agency actions, the APA is the basis for a great deal of environmental litigation. It requires agencies to follow certain procedures (such as giving notice and collecting comments from the general public) before making a new rule to implement a statute. It also allows the court to strike down a rule as "arbitrary and capricious" (lacking adequate justification) or exceeding the agency's statutory mandate. In all these decisions, judges have some discretion in interpreting the APA's requirements and evaluating the agency's actions. However, the usual result of an adverse decision by the court is that the agency simply has to make another attempt at rulemaking. A determined agency can usually win judicial approval, if it is careful and persistent.

Courts also exercise discretion in interpreting statutes. But here, also, they do not have the last word. The legislature may respond to a judicial interpretation of a statute by amending the statute to clarify its meaning. Only when a court finds a rule or statute unconstitutional altogether can it block a legislative policy. Thus, while courts do have some influence on the direction of environmental policy change, they have much more influence on the *pace* of change. The following sections discuss four ways in which courts exercise that influence: they can force reluctant policymakers to act; delay the implementation of a new policy; shape policy directly through statutory interpretation; and put roadblocks in the way of policy change through constitutional review.

FORCING POLICY ACTION

Congress often instructs administrative agencies to address environmental problems but gives them no specific deadlines. A long list of mandates combined with no deadlines is a recipe for agency inaction. A good deal of environmental litigation is therefore aimed at getting agencies to carry out their statutory mandates. Climate change provides a good example of how courts can be useful, if not entirely successful, at prodding a reluctant agency into action.

In 1999, a group of environmental advocacy organizations petitioned the Environmental Protection Agency (EPA) to address climate change by regulating four greenhouse gases (carbon dioxide, methane, nitrous oxide, and hydrofluorocarbons) emitted by new motor vehicles. The petitioners argued that the EPA had authority, and indeed was mandated, to regulate greenhouse gases under the CAA.[6] The EPA—when it finally addressed the petition four years later—disagreed. The petitioners challenged that denial in court, and the result was *Massachusetts v. Environmental Protection Agency,*[7] the 2007 Supreme Court decision directing the EPA to take action on greenhouse gases under the CAA.

Since methane, nitrous oxide, and hydrofluorocarbons were already regulated under the CAA, the legal dispute centered on carbon dioxide. Section 202(a) (1) of the CAA provides that "the Administrator [of EPA] shall by regulation prescribe... standards applicable to the emission of any air pollutant from any class or classes of new motor vehicle... which in his judgment cause, or contribute to, air pollution which may reasonably be anticipated to endanger public health or welfare." The EPA argued that carbon dioxide wasn't "air pollution" within the meaning of the CAA, which was aimed at addressing ground-level pollutants rather than those that occur high in the atmosphere.[8] According to the EPA, Congress did not intend for the CAA to address climate change. On the contrary, it had considered and rejected a proposal to address climate change when that Act was amended in 1990. In short, the EPA and petitioners disagreed about how to interpret the CAA.

The petitioners appealed the agency's decision to the US District of Columbia Court of Appeals, which upheld the EPA's refusal. The petitioners, now joined by several state and local governments, successfully petitioned the Supreme Court to review the case. On April 2, 2007 (8 years after the original petition), the Supreme Court reversed the lower court.

The Supreme Court focused on two questions. First, EPA argued that the petitioners did not have standing to challenge its decision. The standing requirement derives from Article III of the Constitution, which authorizes federal courts to hear "cases and controversies." The Supreme Court has interpreted that provision to mean that the party bringing an action to court must have a real stake in the conflict. This stake, a particular injury that can be addressed by a legal remedy, gives the party "standing" to sue. Only one petitioner needs to satisfy the standing requirement, however, and the Court (in the majority opinion by Justice John Paul Stevens) concluded that the state of Massachusetts did so. Following the line of decisions that started with *Missouri v. Illinois*, discussed above, the Court held that Massachusetts had standing as a "quasi-sovereign" with an interest in defending "all the earth and air within its domain."[9]

On the main question, the Court concluded that Sec. 202(a) (1) clearly does cover carbon dioxide. The CAA has a "sweeping definition of 'air pollutant'" which includes "*any* physical, chemical... substance or matter which is emitted into or otherwise enters the ambient air."[10] It rejected EPA's argument that Congress didn't intend for EPA to address climate change, noting that the fact that Congress didn't address climate change in the 1990 amendments "tells us nothing about what Congress meant when it amended § 202(a) (1) in 1970 and 1977."[11] The Court concluded that "EPA has offered no reasoned explanation for its refusal to decide whether greenhouse gases cause or contribute to climate change. Its action was therefore 'arbitrary, capricious, ... or otherwise not in accordance with law'"—that is, with the APA.[12]

The Court therefore ordered EPA *not* to start regulating carbon dioxide immediately but to make a decision as to whether carbon dioxide "may reasonably be anticipated to endanger public health or welfare." Only if the agency decides that carbon dioxide does endanger the public is it required to regulate these emissions under the CAA. It took two more years (and a new administration) for the EPA to issue its finding that carbon dioxide and other greenhouse gases do endanger public health and welfare.[13] President Obama used that finding both to raise the Corporate Average Fuel Economy (CAFE) mileage standards for cars and light trucks, and to curtail stationary source carbon emissions from power plants in his Clean Power Plan (Chapter 4). However, the latter plan was enjoined by the Supreme Court in 2016 and was later repealed by the Trump administration.

Indeed, *Massachusetts v. Environmental Protection Agency* itself may be vulnerable to being overturned. Recently appointed Justices Neil Gorsuch and Brett Kavanaugh have expressed reluctance to allow agencies to address "major questions" of public policy under statutes that weren't clearly aimed at addressing those questions. For example, Justice Kavanaugh used a statement on the Supreme Court's decision *not* to review a case to state that he supports this "major question" doctrine. The statement has no authority as precedent; rather, it is a good example of how judges invite plaintiffs to bring cases that would allow them to develop doctrine in this way.[14] If the Trump administration's replacement of the Clean Power Plan—the "Affordable Clean Energy" rule—ends up in the Supreme Court after being challenged by attorneys general from Democratic states, it could serve as an opportunity to overturn *Massachusetts v. Environmental Protection Agency* on these grounds (see Chapters 2 and 7).

Even if *Massachusetts v. Environmental Protection Agency* stands, however, the pace of EPA action under the CAA has been very slow, and climate policy advocates are trying other legal strategies to force action on climate change. One high-profile case is *Juliana v. United States*, in which a group of young people argue that the federal government's failure to address climate change violates their "substantive due process rights to life, liberty, and property" (a constitutional claim) and "their obligation to hold certain natural resources in trust for the people and for future generations" (a common law claim invoking "public trust" doctrine).[15] The plaintiffs asked the federal court to order the US government to "develop a plan to reduce CO_2 emissions."[16]

Given federal courts' experience with common law environmental litigation in the early twentieth century, they are likely to be reluctant to encourage these lawsuits. In this case, the District Court agreed that plaintiffs have standing to bring the suit, but the Ninth Circuit reversed that decision, dismissing the case before even considering whether the constitutional or public trust arguments have merit.[17] Nevertheless, such litigation at least provides an incentive for Congress or state legislatures to take some action, in order to preempt the courts from wading into this complex and contentious policy area.

DELAYING POLICY ACTION

Where courts are most effective is in delaying changes in environmental policy. Presidents seeking to roll back environmental regulations often face substantial obstacles in the form of adverse judicial decisions. The Trump administration's deregulatory efforts offer a case in point. Most administrative regulations cannot simply be repealed by executive order; they must be replaced through the same rulemaking process by which they were created in the first place. Any effort to change an existing regulation may be challenged on the grounds that the agency failed to meet the APA's procedural requirements, that the agency's action is "arbitrary and capricious," or that the action exceeds statutory authority (see Chapter 7).

So far, most of the Trump administration's environmental regulations have either been struck down by federal courts or remain held up by litigation. The Institute for Policy Integrity is tracking the administration's regulatory efforts, and as of October 1, 2020, only 22 out of 138 agency actions challenged in court have survived judicial review.[18] Fifteen of the successful lawsuits concerned environmental regulation, but most of those have very limited effects, such as a postponement of compliance dates for restrictions on toxic metal wastewater discharges from power plants.[19] Other unsuccessful lawsuits have challenged individual permits or leases, and one actually upheld the designation of a Superfund site against an industry challenge.[20] The Trump administration's record in court was unusually poor during its first 3 years, however—an administration typically wins about 70 percent of its lawsuits.[21] And a loss in court doesn't always doom the deregulatory effort, since the agency can make another attempt to satisfy the rulemaking requirements. But sometimes a loss in court is enough to discourage or at least moderate an administration's efforts.

These dynamics are illustrated by *California v. United States Bureau of Land Management*.[22] The case concerns a regulation issued by the Bureau of Land Management (BLM) during the Obama administration, restricting the venting, flaring, and leaking of natural gas during oil and natural gas production on federal lands. The rule took effect in January of 2017 and was immediately challenged in the Wyoming District Court by some of the regulated companies.[23] At the same time, President

Trump issued Executive Order 13783, directing agencies to "suspend, revise, or rescind those that unduly burden the development of domestic energy resources beyond the degree necessary to protect the public interest or otherwise comply with the law."[24] Following that directive, in June 2017 the BLM gave notice that it was postponing compliance dates for some sections of the rule. The states of California and New Mexico immediately filed suit, asking the District Court for the Northern District of California to vacate the postponement.

The California Northern District Court agreed with California and New Mexico. The BLM had not complied with the usual notice and comment requirements in issuing the postponement, nor had it offered any new information that would justify the delay. Instead, the agency had relied on APA section 705, which allows it to postpone the effective date of a rule "when an agency finds that justice so requires." The BLM argued that this provision allowed it to postpone compliance dates as well, and that the postponement was necessary to "preserve the status quo" until the Wyoming case challenging the rule was decided.[25] The court rejected this reasoning. The APA does not authorize agencies to change compliance dates once a rule has gone into effect, and the postponement would not in fact preserve the status quo but bring to a halt efforts already underway to comply with the rule. The court noted that while "new presidential administrations are entitled to change policy positions," the APA requires them to "give reasoned explanations for those changes."[26] The BLM appealed the decision but then dropped the appeal, abandoning the effort.

This did not end the saga, however. The BLM continued to work on suspending the rule and replacing it with a different, more lenient one. In the meantime, the Wyoming case was proceeding. In April 2018, the Wyoming District Court issued an order to stay (prevent) the implementation of the original rule, in order to allow the BLM to complete its revision. While the court expressed frustration with its options, it decided that postponing the compliance dates in the original rule would "provide certainty and stability for the regulated community and the general public while BLM completes its rulemaking process, will allow the BLM to focus its limited resources on completing the revision rulemaking, and would prevent the unrecoverable expenditure of millions of dollars in compliance costs."[27] Delay, it turns out, can work both ways.

The BLM's plans for oil and gas extraction on public lands continue to run into judicial roadblocks, however. For example, in February 2020, an Idaho District Court suspended oil and gas leases in sage grouse habitat on the grounds that the BLM did not allow adequate public comment on its leasing plan, as required by the APA. That case is discussed in more detail in Chapter 9.[28]

SHAPING POLICY

Courts may have a more direct influence on environmental policy when asked to consider how a statute drafted several decades ago should be applied under new conditions. This sort of "updating" can look very much like the implementing agency crafting a new policy altogether—a task that courts may be reluctant to allow. That was the concern behind the 2018 Supreme Court decision *Weyerhaeuser Company v. United States Fish and Wildlife Service.*[29]

The United States Fish and Wildlife Service (USFWS) administers the Endangered Species Act[30] (ESA), and in 2001 it listed as endangered the dusky gopher frog, which lives in upland pine forests in coastal Alabama, Louisiana, and Mississippi. The frog's habitat needs are fairly specific: it breeds in ephemeral ponds in open-canopy forests. Open canopies ensure that there is sufficient ground-level vegetation to support the insect populations that the frog feeds on. Unfortunately, such open-canopy, longleaf pine forests have become rare in the Southeast, and the frog's numbers have declined precipitously. Shrinking habitat makes the frog more vulnerable to climate disruption and other threats. Accordingly, the USFWS plan for its conservation seeks to protect not only its current habitat but additional areas that the frog might be able to inhabit in the future. The agency faces similar challenges for many species, as climate change alters habitats and forces species to migrate into new territory. The *Weyerhaeuser* case raises questions about the agency's power to deal with those climate-induced changes under the ESA.

A provision of the ESA requires the USFWS to designate as "critical habitat" any geographical area within the species' habitat that is essential to its conservation, including "specific areas outside the geographical area occupied by the species ... upon a determination by [the USFWS] that such areas are essential for the conservation of the species."[31] Following this somewhat confusing statutory mandate, the USFWS designated as critical habitat not only the areas in which the dusky gopher frog is currently found but also a 1,544-acre site in St. Tammany Parish, Louisiana. That area, a closed-canopy timber plantation owned by Weyerhaeuser, contains some high-quality ephemeral ponds, and could be turned into the open-canopy forest needed by the frog with (in the USFWS judgment) "reasonable effort." Such a designation, however, might prevent Weyerhaeuser from logging the tract.[32] Weyerhaeuser therefore brought suit, asking the court to order the USFWS to remove the unit from its critical habitat designation.

Weyerhaeuser argued that the tract cannot be "critical habitat" because, in its current condition, the frog couldn't survive there. In other words, it argued that the USFWS misinterpreted the statute, adopting too broad a definition of "critical habitat" to include territory outside the frog's habitat altogether. The District Court and the Fifth Circuit Court of Appeals ruled in favor of the USFWS, deferring to its expertise, but the Supreme Court reversed the lower courts and rejected the agency's interpretation of the statute.

The central question, according to Chief Justice Roberts' majority opinion, was "whether 'critical habitat' under the ESA must also be habitat"—that is, whether the agency is limited to considering territory that is *already* a suitable habitat for the frog.[33] Put that way, he had no difficulty concluding that the term "critical habitat" is so limited. In short, he rejected the agency's attempt to give the key statutory term a broad reading that would reach *potential* habitat. However, he didn't give Weyerhaeuser a clear victory either. The USFWS claimed that the frog could in fact inhabit the area in question, so the Court remanded the case back to the lower court for further proceedings on that question.[34] (It also decided that the lower court improperly failed to review the USFWS's consideration of the economic impact of designating the area as critical habitat.)[35]

The USFWS's approach to interpreting the phrase "critical habitat" illustrates the challenge of applying the ESA in an era of climate change. Protecting endangered species may increasingly require agencies to become more creative in interpreting statutes—to extend key terms, for example, such as reading "critical habitat" to include new territories into which a species can migrate as their existing habitat is altered by climate change. But such extensions or reinterpretations may subtly reshape the policy itself. The issue raised by *Weyerhaeuser Company v. United States Fish and Wildlife Service* is whether the courts will defer to agencies as they shape policy through such creative statutory interpretations.

There is reason to believe they may not. As mentioned above, some conservative jurists like Justices Kavanaugh and Gorsuch are concerned to limit the scope of agency discretion, to ensure that major policy questions are resolved by elected officials and not relatively obscure agency staff. A key point of conflict in this debate focuses on the 1984 Supreme Court decision *Chevron v. Natural Resources Defense Council (NRDC)*.[36] The case concerned (once again) EPA's interpretation of the CAA. The EPA allowed states (which give permits to emitters) to treat all of the pollution-emitting devices within the same industrial groupings as though they were encased within a single bubble, as one "source" of pollution rather than several different sources. That approach would give the facility some flexibility to allow more pollution from one device by cutting down pollution from another. The NRDC, in contrast, argued that this was not a reasonable interpretation of the statutory term "source," and that such trade-offs shouldn't be allowed. The Supreme Court upheld the agency's policy, holding that where the language of the statute isn't clear, courts should defer to the agency's interpretation.

That principle of deference has come under attack recently by conservative critics of the administrative state. Judge (now Justice) Neil Gorsuch has complained that the *Chevron* rule "permit[s] executive bureaucracies to swallow huge amounts of core judicial and legislative power and concentrate federal power in a way that seems more than a little difficult to square with the Constitution of the framers' design."[37] One worry is that it seems to give (politically accountable) agency staff rather than (neutral, objective) courts the final word on the interpretation of statutes—even though the APA explicitly directs reviewing courts to interpret statutes.[38] The second worry is that the rule seems to delegate law-making power to administrative agency staff—now described as remote from public view and politically unaccountable—rather than in the (elected and accountable) legislature.[39] Obviously the two critiques are somewhat inconsistent: Are agencies too political or not political enough to exercise regulatory power? Either way, of course, one might wish to see them more firmly checked by the judicial branch. Environmental groups, in contrast, tend to worry that the courts are the institution to worry about because they lack the agencies' expertise and sense of mission. Under this view, refusing to defer to agency decisions may prevent those agencies from implementing statutory protections effectively.

In practice, however, the debate over the *Chevron* rule may have little effect on the court/agency relationship. As many scholars have noted, the rule has never prevented courts from overruling agencies. On the contrary, it is "a principle often honored in the breach."[40] Under the *Chevron* rule, deference is called for only when the statute is unclear or silent on the issue in question, which leaves the court a good deal of discretion in deciding whether a statute is clear. In *Weyerhaeuser*, both lower courts cited *Chevron* in upholding EPA's interpretation of the statute, but the Supreme Court did not; Justice Roberts found the statute crystal clear. Thus it may not matter much whether the *Chevron* doctrine is rejected. The larger issue is whether courts will give agencies the flexibility to extend statutory protections to address new environmental risks posed by climate change. If such judicial deference is not forthcoming, those policies will have to come from Congress—or from state government, which has its own constitutional authority to protect endangered species.

CONSTITUTIONAL REVIEW

The final area in which courts influence environmental policy is in determining the constitutional limits of government power to protect the environment. Traditionally, the American federal system has left natural resource management to state government as a part of the state's "police power," its general power to protect the health, safety, and welfare of the community. The federal Constitution contains no provisions explicitly empowering the federal government to protect or manage natural resources. However, beginning in the early twentieth century, two constitutional provisions have been interpreted to give Congress fairly broad power over many natural resources. The Property Power, found in Article IV of the US Constitution, gives Congress power "to dispose of and make all needful Rules and Regulations respecting the Territory or other Property belonging to the United States." This provision gives Congress authority to manage the extensive federal land holdings (including more than 400 national parks, 560 national wildlife refuges, and nearly 250 million acres of other public lands) and to enact regulations aimed at protecting that property. The Supreme Court has consistently given that power a very broad reading, permitting Congress to address threats to those resources even if they originate beyond the bounds of federal property. For example, in 1976 the Supreme Court upheld the federal government's power to protect wild burros and horses "as an integral part of the natural system of the public lands," even if they stray onto private property.[41]

The other power, however, has been more controversial. Article I section eight gives Congress the power to regulate interstate commerce, which has been interpreted to give Congress authority to manage and protect "navigable waters" (including lakes, rivers, and their tributaries).[42] That may have once seemed like a fairly limited power, but evolving ecological science challenges those limits. Given the complex and interconnected nature of hydrological systems—and their connection to the global climate system—Congress' power to protect navigable waters threatens to extend its authority over virtually the whole of industrial society.

Not surprisingly, both the courts and the opponents of federal environmental policy have sought for some principled limit to this power. The result is a number of Supreme Court cases considering how far the federal authority over water resources extends. The debate focuses on the CWA, which applies to "waters of the United States." But which waters are those? Strictly speaking, the cases addressing this question are about statutory interpretation rather than constitutional power; that is, they focus on whether the agencies implementing the CWA have interpreted the term "waters of the United States" properly. But the Supreme Court has signaled quite clearly that it is looking for a definition of that term that will limit Congress's constitutional power over water resources.

The most recent decision in this line of cases is *Rapanos v. United States.*[43] John Rapanos was fined for filling three wetlands in order to develop them. The wetlands lay several miles from the nearest navigable river, but each one was connected by either a man-made or natural drain to a river that eventually emptied into Lake Huron. The EPA and the Army Corps of Engineers share authority for implementing the CWA, and they concluded that the wetlands were covered by the Act because "there were hydrological connections between all three sites and corresponding adjacent tributaries of navigable waters."[44] To Justice Antonin Scalia, who wrote the opinion for the Court, the agencies' action were "a small part of the immense expansion of federal regulation of land use that has occurred under the Clean Water Act...during the past five Presidential administrations." The Corps and the EPA have interpreted their jurisdiction to cover "270-to-300 million acres of swampy lands" as well as "virtually any parcel of land containing a channel or conduit—whether man-made or natural, broad or narrow, permanent or ephemeral—through which rainwater or drainage may occasionally or intermittently flow." As he correctly noted, "the entire land area of the United States lies in some drainage basin, and an endless network of visible channels furrows the entire surface, containing water ephemerally wherever the rain falls." So "any plot of land containing such a channel may potentially be regulated as a 'water of the United States.'"[45] However, Scalia's opinion avoided considering whether Congress' powers over water resources could extend so far. Instead, he focused on interpreting the statutory term "waters of the United States," which in his view should be read to cover only relatively permanent standing or flowing surface waters and wetlands with a surface connection to those water bodies. He would thus exclude intermittent streams, man-made drains, and wetlands without a surface connection to a standing or flowing water body. The Court thus vacated the lower court's judgment for the government and remanded the case for further proceedings.

Scalia's opinion, however, was joined by only three other justices (Roberts, Alito, and Thomas). Justice Anthony Kennedy concurred with the result but offered his own rationale. This opinion has since been treated as the more important one because Kennedy served as the "swing vote" on this issue. (The liberal Justices would have upheld the agencies' actions.) Kennedy echoed an earlier ruling by Justice Stevens, arguing that "waters of the United States" could cover any water or wetland having a "significant nexus" to waters that are or were navigable in fact or that could reasonably be so made.[46] The key point, in his view, was whether the regulated waters actually have an impact on the quality of navigable waters. He would thus allow the agencies to regulate any wetlands adjacent to navigable waters (whether or not they have a surface connection), as well as intermittent streams. He believed that this approach would prevent federal regulation of those categories of waters "that appeared likely, as a category, to raise constitutional difficulties and federalism concerns."[47]

Rapanos' fractured decision did not end the controversy, however. Justice Roberts' concurrence noted that "[a]gencies delegated rulemaking authority under a statute such as the Clean Water Act are afforded generous leeway by the courts in interpreting the statute they are entrusted to administer" (citing *Chevron U.S.A., Inc. v. Natural Resources Defense Council, Inc.*).[48] He thus urged those agencies to define the term "waters of the United States" through the rulemaking process. The agencies accordingly set out on a 9-year effort that finally, in 2015, resulted in a formal rule defining "waters of the United States" (known as the WOTUS rule).[49] That rule largely followed Kennedy's guidance, including water bodies with a significant hydrological connection to navigable waters. However, the rule was repealed by the Trump administration, which issued a final rule reinstating the former (pre-2015) regulatory definition in October 2019. That was not the end of the story, however. The Trump administration also began a new rulemaking process that would more narrowly define "waters of the United States"—excluding many wetlands, ephemeral streams, and man-made water bodies. It remains to be seen whether that definition will survive judicial review or the next Presidential administration.

This debate has clearly generated intense partisan conflict, with liberals and conservatives both assuming that a decision limiting federal authority will result in fewer or weaker protections for water resources. But supporting state authority over environmental management doesn't necessarily undermine environmental protection. Consider, for example, *United Haulers Assn., Inc. v. Oneida-Herkimer Solid Waste Mgmt. Authority.*[50] This case arose over a waste management plan enacted by Oneida and Herkimer counties. The plan required that all solid waste generated in the counties had to be delivered to the Solid Waste Management Authority, a public agency that provided recycling services but also charged higher fees than other (private) waste processing facilities. United Haulers Association argued that the plan interfered with interstate commerce, invoking the "dormant Commerce Clause" doctrine. Under this long-standing doctrine, state (and local) governments are prohibited from unreasonably interfering with interstate commerce even if they are not violating any federal statute.

The counties' plan clearly did have some impact on interstate commerce, since it prevented people from contracting with private waste disposal companies that would take their waste out of state. One could argue that the counties were essentially "hoarding" their waste for the benefit of the Solid Waste Management Authority. But the Supreme Court, in an opinion by Justice Roberts, upheld the plan. Disposing of trash, he noted, has long been a traditional function of local government. The counties, he concluded, treated both in-state and out-of-state waste management companies the same, since they prevented residents from contracting with both in-state and out-of-state companies. The critical point was that, although the ordinances did have some impact on interstate commerce, that impact was outweighed by the benefits they confer on the county residents.[51]

The decision was a victory and possible bellwether for state and local environmental regulations. Such regulations often interfere with interstate commerce to some extent. For example, some states may attempt to combat climate change by restricting electric companies from contracting with out-of-state coal-fired power companies.[52] The *United Haulers Ass'n* decision suggests that the Court may look favorably on such policies, as long as they are aimed at environmental management and not simply protecting local industry from competition.

It is worth noting, however, that the case did not result in the standard conservative/liberal split. The dissenters were Justices Alito, Kennedy, and Stevens (who was ordinarily considered a liberal jurist). Justices Scalia and Thomas concurred in the judgment because they reject the "dormant Commerce Clause" doctrine altogether. That is, they believe state and local laws should not be struck down unless they conflict with a federal statute. The dissenters, in contrast, would strike down the ordinances on the ground that they unjustifiably discriminated against interstate commerce (even though no federal law explicitly prohibits such discrimination). This decision demonstrates the difficulty of predicting how the Supreme Court will rule on a given environmental policy. Federal judges may be less concerned with the goal of the policy than in which level of government is enacting the policy, how it was enacted, or the means it uses to achieve its goal.

On one constitutional issue, however, the conservative and liberal Justices usually do find themselves at odds: How far may any government, state or federal, restrict property owners' rights to develop their land? The federal Constitution prohibits the government from taking anyone's property "without just compensation."[53] This provision may come into play when a government enacts a regulation on private property that ends up diminishing its economic value. That was the claim in *Lucas v S. Carolina Coastal Council*,[54] a 1992 Supreme Court decision concerning a state law prohibiting David Lucas from erecting a permanent dwelling on two vacant lots on a barrier island near Charleston. The statute, aimed at protecting fragile coastal areas, was enacted after Lucas bought the property; it effectively ended a residential development project he was planning. Lucas sued, arguing that the regulation effectively "took" his property by prohibiting all reasonable economic use of the land. The state trial court agreed with Lucas, but the state Supreme Court reversed, holding that no compensation was required. The US Supreme Court, in turn, reversed the state supreme court.

The South Carolina Supreme Court had followed an earlier US Supreme Court decision, *Keystone Bituminous Coal Ass'n v. DeBenedictis*,[55] which held that when a regulation diminishes the value of someone's property, the court should consider a variety of factors in deciding whether the regulation constitutes a compensable "regulatory taking." Those factors include (1) the economic impact of the regulation; (2) the regulation's interference with investment-backed expectations; (3) the character of the government action (that is, whether there is a physical invasion of property); and (4) nature of the State's interest in the regulation. In the *Lucas* case, the South Carolina court weighed heavily the fact that the state was trying to protect a valuable natural asset—its beaches—in concluding that the regulation was a legitimate exercise of the state's traditional power to protect natural resources, and not a "taking."[56] Justice Scalia's opinion, however, took a different approach.

Scalia acknowledged that the state did have the power to protect the coastal area by prohibiting its development. However, he found it significant that the regulation in this case denied Lucas *all* economically beneficial or productive use of land. The situation would be different, he argued, if the regulation merely prohibited the owner from creating a public nuisance (by building a factory in a residential area, for example). Landowners do not have the right to create a nuisance. But here, the intended use of the land was authorized by law before the statute suddenly changed its status.[57] In this situation, where a regulation destroys all economic value of the land, the court should not balance the various factors listed in the *Keystone* decision. Instead, it should treat the regulation as though the government had actually taken the property away from the owner altogether, and order compensation.

Justice Blackmun, dissenting, complained that this new rule "launches a missile to kill a mouse."[58] He and many others worried that the decision favored private property owners, and would invite them to challenge any environmental or other health and safety regulations that impacted their development projects as a "regulatory takings." The cost of compensating all such properties owners could doom environmental protection. (South Carolina, for example, eventually had to settle with Lucas, buying his property and then selling it to another developer to recoup the cost.[59])

The impact of the *Lucas* decision on government regulation, however, has not been quite so dramatic. A 2016 empirical study examining over 2000 cases decided between 1979 and 2012 concluded that regulatory takings claims fail more often than they succeed because courts seldom find that a regulation has destroyed all economic use of property; on the contrary, they "almost always defer to the regulatory decisions made by government officials."[60] In fact, claims that regulation completely destroyed the property's economic value succeeded more often before the *Lucas* decision (18 out of 34 cases) than after (8 out of 36 cases).[61] Moreover, a recent Supreme Court decision[62] has made it even more difficult for landowners to claim that a regulation has destroyed all economic value of their property.[63] Nevertheless, litigating these claims may impose a burden, especially on local governments.

The same concern, and qualification, applies to the extension of a "regulatory takings" regime to international trade. The North American Free Trade Agreement was one of several trade agreements that provided for compensation when a nation imposes environmental regulations that "indirectly expropriate" the property of a foreign investor by reducing its investment value.[64] Legal commentators have noted that international arbitrators tend to draw on American regulatory takings doctrine in resolving disputes under such provisions. For example, Transcanada Corps sued the US government in 2016 under the NAFTA provision, seeking $15 billion in compensation for the Obama administration's rejection of a permit for the Keystone XL oil pipeline. The lawsuit was rendered moot when the Trump administration granted the permit in 2017, but the threat of such lawsuits hovers over attempts to address climate change through aggressive regulation.[65] Few governments would welcome the prospect of compensating investors for their investments in the fossil fuel infrastructure. Thus far, however, that prospect seems fairly remote. International tribunals usually expect investors to take the risk of regulation into account in making their investments, and the threat of greenhouse gas regulation is easily foreseeable at this point.[66]

CONCLUSION

In sum, the federal courts' chief role in environmental policy has been to slow down or moderate dramatic policy shifts, in either a more liberal or more conservative direction. That role is unlikely to change in the coming years. The president does, of course, have some influence on the judiciary by making appointments to the federal bench (which must be confirmed by the Senate). However, even two-term presidents are usually unable to make a substantial change to the partisan balance of the judiciary. There are 860 federal judgeships, and since President Clinton, each president has appointed between 300 and 400 judges (Clinton: 387; Bush: 340; Obama: 334).[67]

Presidents cannot fill seats until they become available, and federal judges often stay on the bench until a president of their own party is in power. As a result, most appointments merely replace one judge with another of the same party. For example, as of September 2020, Trump had 218 judges appointed and confirmed (and one of those has left the bench), with 59 additional vacancies.[68] He primarily replaced Republican-appointed judges.[69] His appointments have changed the partisan balance on some appellate circuits, but overall the federal appeals courts are still fairly balanced between Democratic- and Republican-appointed judges.

So far, President Trump has had the opportunity to appoint three Supreme Court Justices. One might expect these appointments to shift constitutional and other important areas of doctrine in a more conservative direction—but it is far from clear what that shift would mean for environmental regulation. As mentioned above in the discussion of *United Haulers Assn., Inc. v. Oneida-Herkimer Solid Waste Mgmt. Authority*, it's hard to predict how a liberal or conservative judicial philosophy will affect a given environmental dispute. Some conservatives, for example, support state and local protections against "dormant commerce clause" claims. Others may subject regulations repealing environmental protection to less deferential scrutiny, thus impeding deregulatory efforts. Moreover, judicial philosophies are not static over time, and neither are partisan principles. A judge who seems like a reliable conservative vote when appointed might end up joining the liberal wing of the court—as did Justices Earl Warren (appointed by President Eisenhower), Harry Blackmun (appointed by President Nixon), John Paul Stevens (appointed by President Ford), and David Souter (appointed by the first President Bush).

This is not to suggest that it's unimportant who is appointed to the federal bench. As discussed above, if congressional gridlock continues, federal courts may be asked to play an expanding role in updating policies to deal with new challenges like climate change. The courts may be reluctant to take up that invitation, of course. But if neither Congress, federal agencies, nor federal courts are willing to address pressing environmental policies, state governments will become even more important sources of environmental policy—which will give *state* judiciaries an expanded role in environmental policy. State courts can entertain suits based on state legislation, common law, and state and federal constitutional law. Similarly, international tribunals may become more important to policymaking as international trade and environmental law grows more extensive.

Finally, it is worth noting that courts also have the potential to shape political discourse. Arguments made in court often find their way into the news media, giving environmental advocates a chance to make their voices heard. Moreover, as a forum dedicated to reason and law, courts can require other branches of government to support their actions with scientific evidence and reasoned argument. And perhaps most importantly, as one of the most trusted government institutions, the courts can lend legitimacy to environmental policy by articulating the rationale for government protection of natural resources. In an era of partisan polarization, overheated rhetoric, and fake news, American courts can serve as an outpost of sensible, fact-based policy discussion.

SUGGESTED WEBSITES

Environmental Law Institute (www.eli.org) This organization's website provides expert, nonpartisan research and education on current issues of environmental law.

Legal Planet (www.legal-planet.org) This blog is a collaboration between faculty at UC Berkeley School of Law and UCLA School of Law, providing insight and analysis on energy and environmental law and policy.

Natural Resources Defense Council (www.nrdc.org) The NRDC site provides expert analysis of legal issues and reports relevant to ongoing policy debates.

NexisUni and Westlaw (www.nexisuni.com; www.westlaw.com) These commercial websites provide access to federal, state, and international laws, judicial opinions, legal scholarship, and other legal materials.

Regulations.gov (www.regulations.gov) This website is a portal to all federal regulations open for comments from the public.

US Environmental Protection Agency (www.epa.gov/laws-regulations) This section of the agency's website provides links to environmental laws and information about current and pending regulations.

NOTES

1. I would like to thank John C. Dernbach, a distinguished professor of law at Widener University and an environmental lawyer, for his assistance on the chapter.
2. *Missouri v. Illinois*, 200 U.S. 496 (1906); Robert Percival, "The Clean Water Act and the Demise of the Federal Common Law of Interstate Nuisance," *Alabama Law Review* 55 (2004): 717–74.
3. "Nuisance" is a common law cause of action in which the complainant charges that the defendant used his/her/their property in an unreasonable way, causing inconvenience or damage to the complainant (for example, by emitting noxious odors or fouling a water source).
4. *Ohio v. Wyandotte Chemicals Corp.*, 401 U.S. 493 (1971).
5. 5 U.S.C. Secs. 500–596.
6. 42 U.S.C. §7401 et seq. (1970).
7. 549 U.S. 497 (2007).
8. Notice of denial of petition, Control of Emissions From New Highway Vehicles and Engines, 68 FR 52922, September 8, 2003, at 52928.
9. 549 U.S. 581–519, quoting *George v. Tennessee Copper Co.*, 206 U.S. 230, 237 (1907).
10. 549 U.S. 528–529, quoting Sec. 7602(g) of the CAA (Stevens' emphasis.)
11. Ibid., 530.
12. Ibid., 534, citing APA, 42 U.S.C. Sec. 7607(d) (9) (A).
13. Endangerment and Cause or Contribute Findings for Greenhouse Gases Under Section 202(a) of the Clean Air Act; Final Rule, 74 Fed Reg. 66496 (December 15, 2009).
14. *Paul v. United States*, 589 U.S. ___ (2019) (statement by Kavanaugh).
15. *Juliana v. US*, Case No. 6:15-cv-01517-TC, Opinion and Order (ED Oregon November 10, 2016).
16. Ibid., 2.
17. *Juliana v. United States*, Docket No. 6:15-cv-01517-AA (January 17, 2020). As of this writing, the plaintiffs are petitioning for review of the dismissal by the full Ninth Circuit panel.
18. *Round-Up: Trump Era Agency Policy in the Courts* (Institute for Policy Integrity), https://policyintegrity.org/deregulation-roundup (accessed March 31, 2020).
19. *Clean Water Action et al. v. EPA et al.*, No. 18-60079, 2019 U.S. App. LEXIS 26001 (5th Cir. August 28, 2019).

20. *Meritor, Inc. v. EPA* 966 F.3rd 864 (D.C. 2020).
21. Connor Raso, *Trump's Deregulatory Efforts Keep Losing in the Court—And the Losses Could Make it Harder for Future Administrations to Deregulate* (Washington, DC: Brooking Institution, 2018), https://www.brookings.edu/research/trumps-deregulatory-efforts-keep-losing-in-court-and-the-losses-could-make-it-harder-for-future-administrations-to-deregulate/ (accessed December 10, 2019).
22. 277 F. Supp. 3d 1106 (N.D. Cal. 2017), appeal dismissed, 2018 WL 2735410 (9th Cir. March 15, 2018).
23. *Wyoming v. DOI*, 366 F.Supp. 3rd 1284 (D.Wyo. 2018).
24. Ibid., 1112.
25. Ibid.
26. Ibid., 1123.
27. *Wyoming et al v. DOI*, 366 F.Supp.3rd, 1292.
28. *Western Watersheds Project v. Zinke*, 441 F.Supp.3rd 1042 (D. Idaho 2020).
29. 139 S.Ct. 361 (2018).
30. 16 U.S.C. secs. 1531 et seq.
31. 16 U.S.C. sec. 1532(5) (A).
32. 139 S.Ct., 365–366.
33. Ibid., 368.
34. Ibid., 369. The USFWS subsequently settled the case by withdrawing the critical habitat designation for the unit in question. Laura Blies, "Frog Case Settled with No Decision on Habitat" (July 18, 2019), Wildlife.Org (https://wildlife.org/frog-case-settled-with-no-decision-on-habitat-definition/ (accessed April 1, 2020).
35. Ibid., 372.
36. 467 U.S. 837 (1984).
37. *Gutierrez-Brizuela v. Lynch*, 834 F.3d 1142, 1149 (10th Cir 2016) (Gorsuch, J., concurring).
38. 5 U.S.C. sec. 706; Gutierrez-Brizuela, 1151.
39. Similar concerns animate Justice Gorsuch's warnings concerning Congress' delegation of rulemaking authority to the executive branch. See *Gundy v. U.S.*, 588 U.S. ___ (2019) (J. Gorsuch, dissenting).
40. Michael Herz, "Chevron is Dead; Long Live Chevron," *Columbia Law Review* 115 (November 2015): 1867–909.
41. *Kleppe v. New Mexico*, 426 U.S. 529 (1976).
42. *Gibbons v. Ogden*, 22 U.S. 1 (1824).
43. 547 U.S. 715 (2006).
44. Ibid., 730.
45. Ibid., 722.
46. Ibid., 759.
47. Ibid., 776.
48. Ibid., 758.
49. Definition of "Waters of the United States," 80 Federal Register 37054 (June 29, 2015).
50. 550 U.S. 330 (2007).
51. Ibid., 334.
52. *North Dakota v. Heydinger*, 825 F.3d 912 (8th Cir. 2016).
53. U.S. Constitution, Amendments 5 and 14.
54. 505 U.S. 1003 (1992).
55. 480 U.S. 470 (1987).
56. *Lucas v. S.C. Coastal Council*, 304 S.C. 376 (S.C. 1991), 379.
57. 505 U.S. 1027.
58. Ibid., 1036.
59. William Fischel, *Regulatory Takings* (Cambridge, MA: Harvard University Press, 1995), 61.

60. James Krier and Stewart Sterk, "An Empirical Study of Implicit Takings," *Wm. & Mary Law Review* 58 (October 2016): 35–95.

61. Ibid., 60.

62. *Murr v. Wisconsin*, 137 S. Ct. 1933 (2017).

63. Stewart Sterk, "Dueling Denominators and the Demise of Lucas," *Arizona Law Review* 60 (2018): 67–90.

64. North American Free Trade Agreement, December 17, 1992, 32 I.L.M. 289 (chs. 1–9), 32 I.L.M. 605 (chs. 10–22), at Article 1110.

65. Hao Zhu, "Dynamically Interpreting Property in International Regulatory Takings Regimes," *Columbia Journal of Law and Social Problems* 51 (Fall 2017): 129–75.

66. Ibid., 173–4. On the environmental implications of the trade agreement replacing NAFTA, see Scott Vaughan, "USMCA versus NAFTA on the Environment," International Institute for Sustainable Development, https://www.iisd.org/blog/usmca-nafta-environment (accessed April 1, 2020).

67. U.S. Courts Administration website, https://www.uscourts.gov/judges-judgeships/author ized-judgeships (accessed December 10, 2019).

68. Ballotpedia, https://ballotpedia.org/Federal_judicial_appointments_by_president (accessed December 10, 2019).

69. Russell Wheeler, Judicial Appointments in Trump's First Three Years: Myths and Realities (Washington, DC: Brookings Institution, 2020), https://www.brookings.edu/blog/fixgov/2020/01/28/judicial-appointments-in-trumps-first-three-years-myths-and-realities/ (accessed April 1, 2020).

THE ENVIRONMENTAL PROTECTION AGENCY

7

Richard N. L. Andrews

The Environmental Protection Agency (EPA) is the lead US agency responsible for protecting the environment from air and water pollution and for protecting people from the health hazards of pollution and toxic chemicals in the environment. Created in 1970, just a few months after the first Earth Day and before most of today's major pollution control laws were enacted, it regulates air pollution from cars and smokestacks, water pollution from urban sewers and industrial outfalls, hazardous wastes and municipal landfills, drinking water contaminants, and pesticides and toxic chemicals; and under President Obama, it also began to regulate carbon emissions from motor vehicles as well as power plants and other industries that contribute to global climate change. Environmental and public health advocates see it as the government's champion for those widely shared values. Some critics, however, accuse it of imposing excessive regulatory burdens and unjustified costs on businesses, property owners, and state and local governments; and others accuse it at least of not using the most economically efficient and effective policy tools to achieve its goals.

The EPA itself pioneered the development of innovative policy tools to try to reduce environmental risks in the most cost-effective ways. Examples include "market-oriented" incentives, such as tradable emission allowances; information disclosure requirements, such as the Toxics Release Inventory which requires annual reporting by most industries of quantities of toxic chemicals released to the environment; and elaborate procedures for risk assessment. The EPA also provides subsidized loans for drinking water and wastewater treatment facilities, conducts research to reduce pollution, and provides technical assistance and enforcement cooperation to support state environmental protection programs (see also Chapter 2 on state environmental policymaking).

The EPA can use only the policy tools that Congress has authorized, however, and the primary tools Congress authorized for it are regulations, which inherently place restrictions and costs on influential businesses and state and local governments. This chapter discusses how the EPA makes decisions in the face of constant pressures—from advocates of environmental protection and the news media, from businesses, and from the president, members of Congress, state and local officials, and the courts—and in particular, how its decision-making processes are being altered by the Trump administration.

BACKGROUND[1]

The EPA was created in 1970 by President Richard Nixon, in the midst of a widespread public outcry for the federal government to "do something" about pollution.[2] It was created not by an act of Congress but through a presidential reorganization plan that pulled together a number of separate programs into a single new agency. Air pollution and waste management programs were transferred from the Department of Health, Education, and Welfare, water pollution programs from the Department of the Interior, pesticide programs from the Department of Agriculture, and some radiation protection programs from the Atomic Energy Commission.

Beginning in 1970, Congress passed a series of far-reaching new statutes to address pollution and other environmental health hazards, and assigned them to the EPA to carry out. These included the Clean Air Act, the Federal Water Pollution Control Act, the Safe Drinking Water Act, and laws regulating pesticides, toxic substances, and solid and hazardous wastes (see Chapter 1 and Appendix 1). Since the EPA itself was created only by a presidential reorganization plan rather than by an act of Congress, however, it functions largely as an umbrella organization managing separate programs to implement each of these laws, plus crosscutting units for enforcement, legal counsel, research and development, and more recently information and financial management. Its administrator has only limited authority to integrate, coordinate, or set priorities among its separate program units except through its annual budget requests. And Congress often does not grant its requests: in recent years, a gridlocked Congress often has passed only continuing resolutions to maintain specified levels of existing funding, not new appropriations bills—let alone new environmental laws.

Federal administrative agencies such as the EPA can act only under the authority of laws passed by Congress ("statutes"). In approving a new statute, Congress authorizes the EPA to implement and enforce the statute's requirements, but also limits the ways in which the agency may do so, either specifically or simply by not authorizing other options. Appropriations legislation also determines how much money the agency can spend each year to implement and enforce each of its statutes. A regulation, in contrast, is a standard or rule written by the agency to interpret a statute, apply it in particular situations, and enforce it.

The authority to issue regulations was one of the main tools Congress gave the EPA to carry out these laws. In short, Congress authorizes the EPA to issue regulations to solve an environmental problem, and the EPA must then write the details of the regulations to do so: a process that is often challenging, contentious, and politicized by those who are affected. If someone challenges a regulation, decisions by the courts can either confirm or overrule the agency's interpretation. Statutes, regulations, and judicial decisions all have the force of law.

In addition, Congress authorized the EPA to provide technical and compliance assistance to state governments; to provide grants to state and local governments to implement federal air and water quality standards, monitor public water supplies, and clean up hazardous wastes; and to provide low-interest loans to local governments to build new drinking water and wastewater treatment facilities. Many of these tools thus rely heavily on environmental federalism, discussed in Chapter 2.[3] In Fiscal Year 2019, the EPA had a budget of $8.8 billion, just two-tenths of 1 percent of the overall federal budget; but only a small fraction of this was used to pay its fourteen thousand staff members, while nearly half was spent on grants and assistance agreements to state and tribal governments.[4]

To reduce air pollution, for instance, Congress directed the EPA to set National Ambient Air Quality Standards (NAAQS) specifying how clean the air around us must be in order to protect public health; and it required states to produce state implementation plans (SIPs) for achieving these standards. The law also required all new stationary sources of air pollution (such as factories and power plants) to obtain EPA permits and to use the "best available control technology" to minimize their emissions. Important amendments in 1990 set drastically reduced limits on total emissions of sulfur from all large power plants, and allowed polluters to buy and sell their shares of that total—their "emission allowances"—so that companies that do better than the requirements could sell their allowances to companies that found it cheaper to buy more allowances and keep polluting while still achieving the specified overall level. This proved to be one of the most dramatically successful environmental policy innovations. The law also ordered the EPA to set tailpipe emission standards for cars and trucks that all manufacturers must meet on average across the "fleet" of new vehicles they sell each year.

The Federal Water Pollution Control Act authorized the EPA to regulate all "point sources" of water pollution (such as factory outfalls and municipal wastewater treatment plants), not just new sources as in the case of air pollution. All point sources of water pollution must get permits from the EPA (or in practice, from the state agency administering the EPA permit program), and must use the "best available technology" to purify their discharges. The EPA also provides low-interest loans to fund construction of new publicly owned wastewater treatment plants. However, the EPA was not authorized to regulate pollution from "nonpoint" sources such as farm runoff—due to the influence of the farm lobby—and as municipal and industrial sources have been reduced, these nonpoint sources have thus become some of the main remaining sources of water pollution today (see Chapter 1). All treatment, storage, and disposal facilities for hazardous wastes also must have EPA permits, as must municipal landfills and incinerators; and all shipments of hazardous wastes must be documented from the factory where they were generated as waste to their ultimate disposal in a permitted facility.

Each of these statutes addressed a particular environmental problem, but many of them affected the same industries, often with conflicting consequences. Many electric utilities complied with air pollution regulations in the 1970s, for instance, by building taller smokestacks to disperse their pollutants so that they would reduce health hazards immediately downwind; but this simply caused the pollutants to rain out further away as acid rain, damaging forests and fisheries.[5] More recently, the EPA tightened regulations on sulfur dioxide (SO_2) and mercury emissions to protect public health, and to comply, electric utilities put expensive "wet scrubbers" on their stacks to capture these pollutants before they were released into the air; but these materials then were piped into coal ash ponds, some of which later leaked and caused serious water pollution.[6] Similarly, sewage treatment improves water quality by removing contaminants from wastewater, but these materials then must themselves be managed, often by landfilling, incineration, or spraying them on farmlands where they may cause new hazards.

In an ideal world, the EPA would design an integrated set of policy incentives to promote pollution *prevention*: minimizing the use of polluting materials and energy all the way from the initial extraction of resources through production, consumer use, reuse and recycling, and eventual disposal. In practice, however, the EPA must use limited and sometimes expensive regulatory tools aimed at separate problems—such as technology-based standards for air and water pollution—to try to solve complex environmental problems whose outcomes are often environmentally interconnected. In a few cases Congress has allowed the EPA to adopt more flexible tools to reduce pollutants, such as cap-and-trade requirements, but these options are available only for the specific uses for which Congress has approved them, such as limiting sulfur emissions from power plants.

In writing its regulations, the EPA must satisfy a burden of proof to show that its regulations are within its statutory authority and are justified by a written record of substantial evidence: that is, that they are not arbitrary or "capricious." Throughout its history the EPA has met this burden of proof through two primary sources of evidence: scientific evidence that the proposed regulations accurately address environmental problems and their risks in appropriate ways, and economic evidence that their benefits to society are greater than their costs.

For example, in addition to air and water quality, the EPA regulates individual substances that have environmental health risks, such as pesticides, drinking water contaminants, and toxic chemicals used in manufacturing. Before doing so, it conducts an elaborate process of "risk assessment" to determine how serious a hazard a substance is and how many people might be exposed to it, and it then proposes a regulation that balances that risk against the economic benefits and costs of restricting it.

Finally, the EPA and the laws it administers have relied on a philosophy of "environmental federalism," under which the federal government sets minimum standards and permit requirements—especially for air quality and for water and waste management facilities—but allows the states to implement and enforce them, and to set tougher standards if they wish to, so long as they do not violate federal minimum standards or interfere with the ability of downwind or downstream states to meet the federal standards as well. With a few exceptions, the EPA itself remains the primary regulator for most products sold in interstate commerce, such as cars and trucks, pesticides, and other toxic chemicals.

The EPA and the laws it administers were created with broad bipartisan support, and the agency itself and most of its laws were even approved by Republican presidents. Supporters of the agency argue that the EPA has made valuable and cost-effective contributions, given the tools, funds, and limitations Congress has given it, toward cleaning up air and water pollution and hazardous wastes and preventing toxic chemicals from endangering public health and the environment.[7] Critics argue, however, that its regulations are burdensome to the industries, small businesses, and state and local governments that must comply with them, even though most have been documented as having greater overall benefits to society than their costs. Critics also charge that these regulations hurt jobs and profits; that the EPA sometimes fails to do its job effectively (for example, when it failed to protect drinking water in Flint, Michigan, from lead contamination); and that the agency should therefore be reined in or even abolished.[8]

THE TRUMP ADMINISTRATION: A NEW ERA?

Donald Trump was elected president in November 2016, after a campaign in which he promised to get rid of "burdensome" environmental regulations—especially those addressing climate change—and to drastically shrink or even abolish the EPA. He immediately rescinded all of President Obama's presidential directives on climate change, and ordered the EPA to review and rescind the Clean Power Plan (CPP), greenhouse gas (GHG) emission standards for cars and trucks, and other major EPA regulations. He also issued an executive order requiring that the agency must rescind at least two existing regulations before approving any new one, and that it should focus only on reducing compliance costs and disregard the benefits of regulations. And he appointed as EPA's senior administrators a series of former lobbyists for fossil fuel and other industries and former staff members to long-time congressional critics of the agency.

In his initial budget blueprint, President Trump proposed narrowing the EPA's scope to focus on its "core legal requirements," leaving many of its responsibilities to the states and "easing the burden of unnecessary regulations that impose significant costs on workers and consumers without justifiable environmental benefits." He proposed to cut the EPA's budget and staff by more than 30 percent (though Congress repeatedly declined to do so): under his proposal, support would continue for locally popular drinking water and wastewater treatment grants programs, but the agency's research funds would be cut by nearly half, its enforcement budget by nearly a quarter, and its hazardous waste cleanup funds by 30 percent. Trump also proposed eliminating funds for the EPA's climate change programs, its cleanup programs for major multistate resources such as the Great Lakes and Chesapeake Bay, and more than 50 other programs. Despite his verbal support for the states and for "environmental federalism," moreover, his budget proposed leaving many more responsibilities to the states while cutting the EPA's grants that support them by 44 percent.[9]

In short, the election of President Trump brought into office an administration whose appointees sought to radically redirect the agency's activities: to weaken or eliminate many of its regulations, to reduce its personnel and budget, to diminish its enforcement efforts, and even to alter the scientific and economic criteria used to justify its regulations. According to the *New York Times*, by December 2019 the Trump administration had either completed or initiated rollbacks of 95 environmental rules, 35 of them involving the EPA; seven of these were subsequently reinstated following legal challenges.[10] The agency's future effectiveness will be profoundly influenced by whether or not these initiatives endure and are upheld by the courts.

The following examples illustrate both the processes by which the EPA has historically reached its decisions, and some of the most significant changes initiated by the Trump administration.

Politics and "Grandfathering": Air Pollution From Electric Power Plants

Coal-fired power plants are one of the most significant sources of air pollution, including particularly SO_2, nitrogen oxides (NO_x), particulates, and mercury (as well as

GHGs—discussed later). As of 2008, the United States had 1,466 coal-fired generating units, most of them built before 1990; 58 percent of them were built even before the landmark Clean Air Act requirements of 1970. By 2019 more than a third of these facilities had been retired, mainly due to declining demand growth and the falling cost of natural gas as an alternative and more efficient fuel, but others continued to operate even after more than 40 years.[11]

When Congress passed the Clean Air Act in 1970, it directed the EPA to set air pollution emission standards for all *new* facilities: each must meet "new source performance standards," based on "the best emission reduction technology that had been adequately demonstrated, taking into account its cost." However, it exempted *existing* power plants so long as they were not modified: retrofitting existing plants would have been far more expensive, and Congress therefore preferred to assume that these old sources would gradually be phased out anyway. If an existing facility underwent "any" physical change or change in method of operation that would increase emissions, however, it would have to undergo review as a new source ("New Source Review," or NSR) and become subject to the new source standards.[12]

The EPA faced the question, therefore, of how to interpret this mandate. Did Congress really intend it to require costly new pollution controls for literally "any" physical or operational change in an existing facility? Or did Congress really mean to leave existing facilities alone so long as changes to them did not cause significant increases in air pollution? The EPA exempted "routine maintenance, repair, and replacements," as well as modifications that added only minimal amounts of pollutants. But once new power plants were more strictly controlled than existing ones, the utilities had a greater incentive to keep the old plants operating longer, and also to upgrade them as much as they could without triggering the NSR process.[13] The EPA tolerated these practices during the 1980s, under Presidents Ronald Reagan and George H. W. Bush, but President Bill Clinton's EPA administrator, Carol Browner, began an aggressive investigation into evasion of the NSR requirements by old coal-fired power plants and filed suit against thirteen electric utilities for violations at fifty-one plants in thirteen states.[14] The utilities fought back, arguing that the EPA was now trying to enforce a more restrictive definition of "routine maintenance" than in the past; and they also spent heavily to support George W. Bush's successful presidential campaign over Clinton's vice president, Al Gore.

Once Bush was elected, the utilities lobbied vigorously to loosen the NSR rules, and in 2003, Bush's EPA officials proposed a new rule that redefined "routine maintenance" as any upgrades that did not cost more than 20 percent of the plant's value—a huge loophole—and announced a weaker enforcement policy, dropping some seventy-five NSR enforcement investigations.[15] Environmental and public health groups objected strenuously, and fourteen states sued to block the rule changes. In 2003, a court ruled that one of the utilities had indeed violated the NSR rules eleven times at one of its plants; and in 2005 and 2006, the US Court of Appeals for the District of Columbia rejected the Bush EPA's rule changes, holding that when Congress had originally applied NSR to "any" physical or operational changes that would increase pollution, it did indeed mean "any," not just those costing more than 20 percent of the facility's value.[16]

In the closing months of the Bush administration, his EPA officials tried once more to weaken NSR by substituting more discretionary (and thus less enforceable) criteria for review. Once President Obama took office, however, an environmental advocacy group immediately petitioned the new EPA administrator to reverse this change, and she did. After a decade and a half of litigation and a reaffirmation of support for NSR by a new president, the EPA ultimately took enforcement actions against some 45 percent of the country's electricity generating units, leading to 22 major settlements covering 263 units.[17]

The issue did not end there, however. The Trump administration issued new regulations promising that the EPA would no longer review the accuracy of industries' documentation of NSR decisions, narrowing what counts as a "source," and proposing to require NSR only if a proposed modification would cause an increase both in total annual emissions and in the hourly rate of emissions: that is, it would not require NSR if an old plant were modified to run at the same emissions rate but for many more hours per day, causing an increase in total annual emissions. It also made a series of piecemeal changes in policies and definitions through "guidance documents" that had the effects of weakening the NSR requirements without formally issuing new regulations, thus making it more difficult for environmental advocacy groups to document their impact and bring lawsuits to challenge them.[18]

What lessons does this case offer about how the EPA makes decisions? First, the issues involved are complicated, and subject to political influence on the technical details even when the plain language of the law appears to be clear and the science persuasive. Old coal-fired power plants clearly contribute to air pollution–related illness and mortality, but upgrading them to the emissions standards of new facilities would be a major expense: a political as well as economic issue in regions such as the Midwest and the South, where manufacturing is in decline yet electoral influence is high.

Second, the EPA itself is rarely the final decision-maker. Almost any significant EPA decision will be challenged in lawsuits, either by the regulated businesses or by environmental advocacy groups or both. The courts thus play an essential role in EPA decision-making, often supporting protective interpretations of the environmental laws, but not always (see also Chapter 6 on environmental policy in the courts).

Finally, the EPA's decisions rarely remain settled. Its decisions change economic outcomes for businesses that are regulated, thus creating ongoing incentives for companies to challenge regulations rather than comply. These challenges include not only petitions and lawsuits but also attempts to reverse EPA policies by congressional legislation or budget provisions and by electing presidents with different philosophies.

Science and Politics: Particulate Matter as an Air Pollutant

EPA's use of science to justify its decisions also is vulnerable to attempts to manipulate and politicize it by those affected by the outcome. Ever since its creation the EPA has been directed to set National Ambient Air Quality Standards for airborne particles and to update these standards every 5 years. Airborne particulates include particles of acids, organic chemicals, metals, and soot as well as dust, pollen, mold, and soil. From 1970 to 1987 the agency set standards only for "total suspended particulates," but as new scientific studies showed that small particles had the greatest health risks, in 1987 it set standards specifically for particles less than 10 microns in size ("PM_{10}"), including both a maximum annual average (50 micrograms per cubic meter of air, or $\mu g/m^3$) and a 24-hour limit of 150 $\mu g/m^3$. By 1997 further research showed that the most serious health risks were associated with even smaller particles that could be inhaled deeper into the lungs, and so the EPA set new annual and 24-hour standards for particles less than 2.5 microns in size ("$PM_{2.5}$"). The 1997 standards were updated in 2006 (from 65 $\mu g/m^3$ over 24 hours and 15 $\mu g/m^3$ averaged over a year, to 35 and 15), and again in 2012 to 35 and 12.[19] The next mandatory review was due in 2017.

$PM_{2.5}$ particles are extraordinarily small—less than 1/30 the width of a hair—and originate from many sources, from fires and automobile exhaust to industrial chemical reactions in the atmosphere. If not controlled, they can cause premature deaths in people with heart or lung disease, asthma attacks, decreased lung function and other respiratory symptoms, as well as atmospheric haze and acid rain. The smaller the particles and the tighter the standards, however, the more costly and widespread the standards' impact, and so the more politically and economically controversial they become. As the scientific evidence justifying tighter and tighter standards became more compelling, therefore, the affected industries increasingly attacked not just the proposed regulations but also the scientific evidence and review processes that led to them.

The EPA's review process normally includes a comprehensive review of the scientific literature across all relevant fields of research, by EPA professionals in cooperation with respected independent scientific researchers, with particular attention to new findings since the previous review. The staff then drafts proposed findings and policy recommendations based on their judgment of the overall "weight of the evidence" of the pollutant's health effects, and of the effectiveness of the standard in protecting against them. These staff recommendations then are reviewed in detail by the agency's Clean Air Science Advisory Committee (CASAC)—a chartered committee of respected environmental science experts from outside the agency—and particularly by its Particulate Matter Review Panel (PMRP), a panel of 20 experts most knowledgeable about the relevant science. CASAC and the PM panel include both academic and industrial scientists, with explicit requirements for disclosure of any conflicts of interest and recusal if any exist. Due to this rigorous, independent scientific review process, EPA's proposed standards usually have been upheld when challenged in the courts.

The Trump administration, however, made several major changes to this process. First, Trump's EPA administrator announced that anyone who held EPA-funded research grants had an automatic conflict of interest and was therefore banned from serving on EPA review panels. Accordingly, he removed or did not reappoint many of the academic members of the CASAC, the PMRP, and other scientific review panels, and replaced them primarily with state air quality officials from Republican-governed states and with industry scientists and consultants, with no similar concern about their potential conflicts of interest. As chair of CASAC he appointed an industrial consultant whose publicly stated view was that standards could only be justified by statistical evidence of "manipulative causality": that is, proof of a direct relationship between the standard and an improvement in health outcomes, and not merely an "association" (as is often the case in epidemiological studies) rather than considering the full "weight of the evidence" across all relevant fields of research.[20] Several courts subsequently upheld these appointments on the grounds that the EPA administrator has wide discretion in appointing members of advisory panels.

Second, the administrator issued a new rule prohibiting the use of "secret science" as a basis for EPA's regulatory decisions: that is, excluding any studies that did not fully disclose their raw data and analytical methods to the public. This proposal appeared to exclude many of the long-term epidemiological studies that have documented the human health effects of pollutants, since they relied on confidential medical records that could not be disclosed (and would be too expensive and probably impossible to anonymize).[21] This proposal was widely condemned by the scientific community, but was nonetheless pushed forward by the EPA administrator.

Third, he abolished the PMRP, leaving the external review of the PM standards solely to the seven-member CASAC which did not include expertise even in such critical fields as air pollution epidemiology and health effects. Several members of the CASAC, acknowledging the limits of their own expertise, requested that the panel be reestablished, but the EPA administrator responded only by naming a pool of a dozen handpicked consultants—itself lacking expertise in air pollution epidemiology—with whom the CASAC could communicate only through the CASAC chair and in writing.

With these changes as background, in October 2018 the EPA staff issued a draft update of its *Integrated Science Assessment for Particulate Matter*, a comprehensive evaluation and synthesis of the policy-relevant science with input from many scientific experts both within and outside the agency.[22] The chair and a majority of the CASAC responded with a harshly critical review, asserting that the staff draft was based on "unverifiable opinion"; that it used inadequate evidence, and did not adequately review the available science; that it should have paid more attention to "discordant" evidence (for instance, articles by the chair of the panel himself); that it showed only "associations" rather than causal connections between exposures to fine particulates and health effects; and therefore that it had not shown any justification for tightening the standard beyond the 2012 level.[23]

In September 2019 the staff released its draft *Policy Assessment for the Review of the National Ambient Air Quality Standards for Particulate Matter*, which concluded that "the current $PM_{2.5}$ standards do not adequately protect public health" and that "at the levels of the current fine particulate standards, the risk of premature mortality is unacceptably high."[24] However, the majority of a divided CASAC responded that since the assessment depended on the *Integrated Science Assessment* which they had previously criticized, "most CASAC members conclude that the Draft PM PA does not establish that new scientific evidence and data reasonably call into question the public health protection afforded by the current 2012 $PM_{2.5}$ annual standard," and that the current 24-hour standard for $PM_{2.5}$ as well as the PM_{10} standards should also be retained.[25]

In response to these CASAC assertions, the twenty members of the disbanded PMRP—all recognized as distinguished experts on the health effects of airborne particulates—reconvened as an "Independent Particulate Matter Review Panel," and issued a consensus statement to the administrator that the 2012 standards did not adequately protect public health, and that the annual standard should be tightened from 12 to 8–10 $\mu g/m^3$, and the 24-hour standard from 35 to 25–30 $\mu g/m^3$.[26]

The ultimate outcome of this process remains to be seen, but what is clear is that the Trump administration is trying to radically change the EPA's use of science and of independent scientific review processes to justify its standards and regulations. It dramatically reduced the presence of subject matter research experts on its independent scientific review boards and substituted the judgments of a far smaller CASAC composed primarily of industrial consultants and state officials from conservative states. And the new chair of the CASAC proposed rejecting the EPA's well-established "weight of the evidence" basis for standard setting with a burden of proof so stringent that it would be far more difficult to use science to justify regulation. This new approach deeply divided the CASAC and the larger community of subject matter experts, and at least for the immediate future, seriously damaged trust in the CASAC to reach consensus conclusions about the relevant science. If it is upheld and continued, it may seriously diminish the EPA's ability to prevent exposure to environmental health hazards.

Authority to Address New Problems: Greenhouse Gas Emissions

Another case raises the question, how can the EPA deal with a newly identified environmental problem that was not fully anticipated when its regulatory statutes were enacted?

Climate change has been an important public policy issue since the late 1970s, when scientists proposed that global warming was increasing beyond its historic range due to carbon dioxide emissions from human activities: in particular, fossil fuel combustion by power plants, other industries, and motor vehicles. In 1987 Congress directed the EPA to develop a coordinated national policy on climate change, but most early policymaking focused on crafting an international agreement—the 1992 Framework Convention on Climate Change, which the United States adopted, and the 1997 Kyoto Protocol, which it did not—rather than on policies to control domestic emissions.

The Kyoto Protocol included binding targets for GHG emission reductions, which President Clinton agreed to but the US Senate in 1997 voted overwhelmingly to reject unless industrializing countries such as China were also held to them. With Clinton's support, therefore, the EPA began to assert a more active policy role. In 1998, its legal office issued an opinion that EPA had the authority to regulate GHG emissions under the Clean Air Act even though it had not previously done so. In 1999 a group of environmental organizations and renewable energy businesses, citing this opinion, petitioned the EPA to regulate GHG emissions from new motor vehicles under the Clean Air Act; and in January 2001, just as the Clinton administration left office, the EPA invited public comments on this petition, which produced nearly fifty thousand responses during the first five months of the incoming Bush administration.

While Bush was campaigning for the presidency against Al Gore in September 2000, he pledged that "[w]e will require all power plants to meet clean-air standards in order to reduce emissions of carbon dioxide within a reasonable period of time."[27] Once elected, however, he renounced this pledge and ended US participation in the Kyoto Protocol negotiations; and his new appointee as the EPA's legal counsel issued a reinterpretation which asserted that contrary to the agency's previously issued opinion, the EPA did *not* have authority to regulate GHGs as air pollutants under the Clean Air Act. Bush's EPA officials also rejected the petition that had called on the EPA to regulate GHG emissions from motor vehicles. Unlike other regulated air pollutants, they argued, GHGs were only significant at a global scale and therefore were not amenable to the national- and state-level regulations provided by the Clean Air Act. Moreover, they argued that such a far-reaching new regulatory initiative should only be undertaken with explicit direction by Congress and after more extensive research, and that in the 1990 Clean Air Act amendments, Congress itself had only directed the EPA to pursue research and nonregulatory solutions for GHG emissions.[28]

However, a group of states, cities, and environmental groups led by the state of Massachusetts challenged this reinterpretation, and asked the courts to require the EPA to regulate GHG emissions from motor vehicles. In 2005 an initial court upheld the Bush EPA's interpretation, but in 2007 the Supreme Court ruled that GHGs *did* fall within the Clean Air Act's definition of air pollutants, and that the EPA therefore did have authority to regulate them. Given that authority, the court said, the EPA also had a legal *responsibility* to determine whether they "cause, or contribute to, air pollution which may reasonably be anticipated to endanger public health or welfare" (an "endangerment finding").[29] An affirmative endangerment finding, in turn, would automatically trigger an EPA obligation to set GHG emission standards.

In 2009, following the election of President Barack Obama, a bipartisan group of congressional leaders tried but ultimately failed to enact new legislation that would have created a cap-and-trade regime to reduce GHG emissions. Given the Supreme Court's *Massachusetts* decision as a mandate and President Obama's commitment to address the problem, the EPA thus became the lead agency to try to reduce GHG emissions. The new administrator issued an endangerment finding in 2009, concluding that GHGs did indeed contribute to risks to public health and welfare and that emissions from new cars and trucks contributed to these effects. President Obama then announced a joint initiative by the EPA and the Department of Transportation to coordinate the regulation of GHG emissions and of motor vehicle fuel efficiency, and several of the major car manufacturers and other major corporations announced their support. The motor vehicle standards (the "Tailpipe Rule") were finalized in 2010 and tightened further in 2012 (the continuing story of the motor vehicle standards, and the federalism issues related to the battle over them between the Trump administration and the state of California, would require a whole additional case study, but see Chapter 2 for more detail).[30]

An additional implication of the endangerment finding, however, was that once an emission is regulated as an air pollutant under *any* part of the Clean Air Act (motor vehicle emissions, in this case), all major *stationary* sources automatically became subject to regulation as well.[31] This principle empowered EPA to regulate GHGs from stationary as well as mobile sources, but it also had unintended consequences: "major" stationary sources had been defined as those emitting more than one hundred tons per year of traditional air pollutants, which meant only large industrial facilities and power plants; but far more facilities might emit that amount of carbon dioxide, including many hospitals, schools, restaurants, office buildings, farm buildings, and other establishments. The number of sources required to have permits thus could potentially increase from fewer than fifteen thousand to over six million, even though most of them were relatively small contributors to total GHG emissions. Annual administrative costs would increase from $62 million to $21 billion, and the newly regulated sources would face permitting costs estimated at $147 billion.[32] In 2010, the EPA therefore issued a rule requiring permits for new or modified GHG sources but "tailoring" these regulations to focus only on the largest sources—the power plants and other industrial facilities that emitted more than one hundred *thousand* tons of GHGs per year—while excluding the many smaller and nonindustrial facilities.[33] In 2012, the EPA also proposed performance standards requiring that all new or modified power plants use the "best system of emission reduction" for GHGs, and in 2013, it issued a revised version of this standard.[34]

These standards relied on a crucial assumption: that carbon capture and storage (CCS) technologies had been "adequately demonstrated" as a "best system of emission reduction," even though only a handful of CCS facilities were actually operating so far, or that other options such as energy efficiency improvements could be used to achieve the reductions. The EPA's rationale was that few new coal-fired power plants were planned before 2020 and that new plants after that were already being designed to include CCS technology. Businesses and states opposed to the rule argued, however, that these technologies had not yet been adequately demonstrated at a commercial scale.

Two further complications also followed. First, issuing a performance standard for *new* power plants also triggered a requirement for the EPA to develop guidelines that *states* must use to reduce emissions from *existing* facilities, as part of their required "state implementation plans" for complying with the Clean Air Act. In June 2014, the EPA issued the Obama administration's Clean Power Plan (CPP), a proposed rule that set state-specific goals based on each state's power plant GHG emission rates.[35] The goal of the rule was to reduce overall power plant GHG emissions by 30 percent from 2005 levels, while allowing each state to adopt flexible strategies for doing so that best suited its circumstances. If a state did not submit a plan that satisfied the EPA's guidelines, EPA could write a plan for the state itself. The final CPP rule was issued in October 2015; initial state plans were to be completed by September 2016, and final ones by September 2018.

Second, because only some of the necessary emission reductions would likely be achieved by the existing power plants, the states would have to use additional measures to achieve the goals, such as increasing the substitution of natural gas and renewable energy for coal, avoiding the retirement of existing nuclear plants, increasing energy efficiency, adopting market-based incentives, and perhaps joining multistate cap-and-trade programs. Critics immediately argued that the CPP stretched the EPA's use of planning guidelines under environmental federalism beyond previous precedents and without legal authority.

In 2013 a group of utilities, other carbon-intensive industries, and some states and public officials (the "Utility Air Regulatory Group," or UARG) petitioned the Supreme Court to overrule the EPA's regulation of GHG emissions from stationary sources, arguing that CCS technology had not yet been adequately demonstrated and that trying to regulate GHG emissions from stationary sources would thus expand EPA regulation far beyond what Congress had intended in the Clean Air Act (the *Massachusetts* decision had only addressed motor vehicle emissions, not the effects of this decision in triggering the regulation of other sources as well). In its decision in this case in June 2014, the Supreme Court confirmed that the EPA had the authority to regulate GHG emissions from stationary sources that also emitted other pollutants specifically covered by the

Clean Air Act. It also ruled, however, that the EPA could not regulate other sources solely for GHGs without congressional approval, nor did it have administrative discretion to change the specific tonnage triggers in the law (from one hundred to one hundred thousand tons per year) to solve the awkwardness of trying to "tailor" these provisions to "major" GHG sources. In effect, under this ruling the Supreme Court confirmed the EPA's authority to regulate sources of 83 percent of GHG emissions but protected many other smaller sources from becoming newly subject to EPA regulation without congressional action.[36]

In the same month as the court's *UARG* decision, the EPA announced its proposed CPP rule, and the coal industry, along with twenty-six state governments and hundreds of other businesses and business organizations, immediately sued the EPA again to try to block it. The EPA, they argued, was trying to force the states to reduce GHGs by taking actions far beyond the actual performance of the regulated power plants themselves, using the Clean Air Act not just to reduce the power plants' own emissions but to try to create a whole new low-carbon energy economy. The EPA responded—supported by eighteen other states, many local governments, businesses, and environmental groups—by arguing that the flexibility it was offering to the states was in fact a classic case of cooperative environmental federalism: consistent with the original Clean Air Act and similar to previous environmental regulations, it set a national goal to reduce emissions endangering public health and welfare and allowed states broad flexibility to develop plans to achieve it.

The rule was finalized in August 2015, but in February 2016, a 5–4 majority of the Supreme Court put the rule on hold until the court case was decided, and in September 2016, the US Court of Appeals for the District of Columbia Circuit heard oral arguments on the case.[37] In November 2016, meanwhile, Donald Trump was elected president, and on taking office he immediately issued an executive order directing the EPA to review and rescind the CPP rule. In October 2017 his new EPA administrator issued a proposed rule repealing it, and in June 2019 the EPA issued its proposed substitute—the "Affordable Clean Energy" (ACE) rule—which would be far weaker, leaving it to each of the states to decide not only the methods by which they would reduce GHG emissions but even how much they would choose to do so, and requiring only improvements in efficiency at existing plants (a change that might even increase GHG emissions by making older plants more efficient so that they could run more frequently). In August 2019 ten environmental groups filed suits to overturn the new rule and reinstate the CPP; a month later their suit to reinstate the CPP was dismissed as moot on the grounds that the CPP had already been repealed, but the suit challenging the ACE remained active and was scheduled for oral arguments in October 2020.

This case thus offers several further lessons. First, as in both the previous cases, the EPA's decisions are driven by both internal and external forces, including its own staff, environmental and business advocacy groups, presidential policy preferences, and court decisions. Key decision points in this case included a memo from the EPA's legal staff under a supportive Clinton administration; a petition by environmental groups, based on this memo, asking the EPA to regulate carbon emissions from motor vehicles; a contrary memo by President Bush's EPA officials, disavowing that interpretation and rejecting the petition; a Supreme Court decision reaffirming the previous interpretation; a series of subsequent EPA rules under the Obama administration implementing that interpretation; a further Supreme Court decision affirming the EPA's authority to regulate but specifying more clearly the rationale and limits of that authority; an EPA rule implementing that authority but delayed by the Supreme Court pending final judicial review; and, following the 2016 election, a presidential executive order directing the EPA to review and rescind the rule.

Second, as in the previous two cases, the EPA's policies clearly are influenced by presidential politics. Its first administrator, William Ruckelshaus, sought vigorously to establish its independence as a regulatory agency responsible first and foremost to faithfully execute the laws, based on the best science and economics available, even when that mission conflicted with President Nixon's business supporters. Beginning with the Reagan administration in the 1980s, however, the agency's policies became much more subject to change based on presidential politics, though those changes remained constrained by judicial oversight. The EPA's position on regulating GHG emissions changed significantly from the Clinton to the Bush administration and again under Obama, and yet again under President Trump; a critical question still unanswered is the extent to which President Trump's appointees will be upheld by the courts in their initiatives to change regulations put in place by their predecessors.

Third, what may appear to be relatively straightforward choices, such as whether the EPA should protect the environment by regulating GHG emissions from cars and trucks, can trigger far more complex consequences. In this case, it triggered regulatory consequences for thousands of stationary sources as well as for state governments. Many observers believe that a broad-based carbon tax or cap-and-trade system would be more effective and more workable, whether or not they support that goal; but either of those tools would require new legislation, which Congress so far has been unable to pass. State cap-and-trade programs may be another option, but a more effective national solution would require congressional action (see also Chapter 2 on state environmental policymaking).

Economic Benefits and Costs of Regulations: Manipulating the Analysis

A final illustration of how the EPA normally makes decisions, and how the Trump administration is attempting to change it, is a sequel to the previous case: the Trump administration's attempt to use changes in economic criteria to justify the "Affordable Clean Energy" rule, an attempt that has potential implications far beyond the climate change rule itself and would affect the agency's economic-analysis protocols for all future regulations. In addition to the CPP, Trump's EPA administrators also used these criteria changes to try to justify rescinding rules regulating methane emissions, defining EPA's jurisdiction over the "Waters of the United States," and the long-established standards for mercury and air toxics emissions, among others.[38]

Since at least the 1980s, the White House's Office of Management and Budget has directed the EPA to document the economic costs and benefits of proposed regulations that would significantly impact the US economy, and to show that their benefits to society are greater than their costs. In practice this requirement poses difficult and sometimes contentious technical questions. The costs of an environmental regulation, for instance, include the costs of compliance to the businesses that are regulated, but historical examples suggest that the compliance costs that are predicted in advance—as they must be, for a proposed regulation—sometimes turn out to be far lower in reality, since once a regulation goes into effect, regulated businesses tend to find cheaper ways to comply.[39] The benefits to society of regulating an environmental health hazard, in turn, often are far greater but also are harder to measure than the private costs of compliance that are claimed. These benefits include the economic benefits of lower mortality and morbidity, and of people living longer and healthier and thus more productive lives: that is, reducing the "social cost of pollution." They also include "cobenefits," such as the benefits of cleaning up one pollutant that also contributes to cleaning up others. They include improvements in the overall efficiency of the economy, as they create incentives to substitute cleaner and more efficient production methods for the dirtier ones that become more expensive. In the case of a hazard such as climate change, they also include benefits to the entire population of the world, not just to people in the United States.

Finally, the economic analysis must calculate an appropriate relationship between present costs and benefits and those in the future, by using an appropriate discount rate: a high discount rate places greater value on costs and benefits in the immediate future, whereas a low discount rate places a higher value on long-term considerations.

Up until the Trump administration, the EPA has used well-established and widely accepted economic procedures and assumptions for calculating benefits and costs, including cobenefits, benefits to all who in fact are expected to benefit, and discount rates appropriate to the full period over which the standard's benefits and costs are expected to occur. In rescinding the CPP and other regulations, however, the Trump administration introduced several significant changes. First, Trump signed an executive order directing all agencies to eliminate two regulations for every one new one they proposed, and to base their analysis solely on reducing compliance costs: that is, to ignore the benefits of the regulations involved, such as reduction of the social costs and health damage caused by the pollution that was regulated.[40]

Second, Trump's EPA administrator proposed that calculation of the benefits of the CPP include only benefits to residents of the United States, ignoring the fact that its purpose was to reduce a global environmental hazard to which the United States was one of the largest contributors. This change of definition by itself significantly lowered the rule's calculated economic benefits. In addition, however, he proposed that only direct economic benefits to climate mitigation should be included, excluding "cobenefits" such as the fact that cleaning up GHG emissions from power plants would also greatly reduce emissions of other pollutants and thus improve health and other outcomes. This proposal violated not only common sense but also a provision of the Office of Management and Budget's guidance on benefit–cost analysis, which specifically called for inclusion of cobenefits. Finally, he called for using a high discount rate—focusing only on short-term costs and benefits—despite the fact that climate change and its consequences were inherently a long-term process.[41]

These proposals were contested and their outcome remained uncertain, but what was clear was that they represented an attempt to manipulate the EPA's procedures for economic analysis in ways that would lower the calculated benefits of EPA regulations, by substituting assumptions that were inconsistent both with established EPA practice and with the principles accepted by most economists.

ACHIEVEMENTS AND LIMITATIONS

Over the 50 years of the EPA's existence to date, the EPA has accomplished a great deal in making the environment cleaner using the regulatory tools it has been given (see Chapter 1). Air pollution has been dramatically reduced, and regulations for cars and trucks have produced major improvements in motor vehicle design to reduce air pollution and increase fuel efficiency. The EPA's permit requirements also have greatly reduced water pollution from wastewater treatment plants and industrial discharges, although runoff from farms, construction sites, and other nonpoint sources continues to cause serious problems. Solid and hazardous wastes are now managed far more safely: the EPA's regulations closed more than five thousand open-burning dumps in the 1970s, and municipal and commercial landfills are now far better managed by professionals under permit standards set by the EPA.[42] The EPA itself has pioneered some of the most important innovations in environmental policy, such as emissions trading and information disclosure requirements.

The EPA's risk-based, substance-by-substance regulations of hazardous chemicals have had far more limited success. They have banned or restricted a few highly visible and controversial toxic chemicals, but due to scientific uncertainties, limited staff and resources, political and legal resistance, and the heavy burden of proof it must sustain, the EPA has actually studied and regulated very few. One hope for the future was a rare bipartisan vote by Congress in 2016 to pass the Lautenberg Chemical Safety Act, which reformed the EPA's least effective mandate—the Toxic Substances Control Act—by making all commercial chemicals subject to EPA review, prioritizing those that posed the greatest risks, and giving the EPA broader options to manage these risks and to require more safety testing. However, the Trump administration assigned implementation of this law to a chemical industry scientist who rewrote its implementation rules to apply only to the risks of the manufacturer's intended uses (rather than all potential uses), and made other changes sought by industry to limit its effect.[43]

Both business advocates and policy scholars have sometimes criticized the EPA's regulations. Businesses often criticize them because of the additional costs of compliance, sometimes disregarding the "external" or "social" costs of the health effects and other economic damage that their own pollution imposes on others. Policy scholars often criticize the regulations as economically inefficient and discouraging of innovation: technology-based permit requirements, they argue, sometimes impose extra costs on firms that could reduce pollution more cheaply in other ways, and "best available technology" requirements tend to "lock in" the best existing technology rather than encouraging the discovery of more innovative solutions. Policy scholars have repeatedly recommended the use of markets, environmental taxes, information disclosure, and other innovative policy tools to achieve environmental protection more efficiently and effectively (see Chapter 10).[44]

In reality, however, the EPA can use only the tools that Congress has authorized, which often do not yet include many innovative solutions that have been proposed.[45] It therefore has tried to use the regulatory powers it does have to address new problems, such as allowing trading of emission allowances, redefining animal feedlots as point sources, regulating genetically modified organisms as pesticides, and regulating GHGs as dangers to public health and welfare. These are sometimes awkward substitutes, however, for designing more effective policies by statute; yet the Congress has become too polarized and gridlocked to pass most new environmental statutes that have been proposed.

THE EPA'S FUTURE

What then is the likely future of the EPA? One outcome could indeed be the radical diminution of the agency that Trump proposed. Its staff was demoralized and diminished by retirements and buyouts,[46] and its scientific advisory boards no longer included many of the most respected academic research scientists. Some of its most visible regulations were weakened or rescinded, and recommendations to strengthen them to better protect public health were ignored: these included particularly its initiatives to diminish the risks of global climate change, such as the CPP; the automotive standards for fuel efficiency and GHG emissions; and proposals to eliminate particularly hazardous pesticides. And its commitment to enforcement declined precipitously: between 2017 and 2018, EPA investigations under the Clean Air and Clean Water Acts dropped by one-third to half, and under its hazardous waste programs by 17–23 percent. Its civil enforcement actions declined by 20–50 percent, to the lowest levels in 10 years; referrals for criminal prosecution dropped to a 30-year low, just 166 cases in 2018; criminal convictions fell from more than one hundred cases in each year since 1997 to just sixty-two cases in 2018;[47] and the amounts of financial penalties collected were the lowest since 2006.[48]

President Reagan tried once before to "deregulate, defund, and delegate" the EPA's responsibilities, but it did not end well for him: the public and many members of Congress mobilized to "save EPA," he was forced to retreat and replace most of the agency's leadership, and in the next elections, the public voted in congressional majorities supportive of the agency. The second President Bush also tried to weaken environmental regulations by reinterpreting them, but he too was then confronted with a more environmentally supportive Congress during his second term and with rejection of his proposals by the courts.

Trump could perhaps be more successful. First, his appointees to the EPA were more politically experienced, and the Republican majority of senators was both more ideologically conservative and more fearful of crossing him than those in the 1980s. Second, he and the Senate's Republican leadership focused single-mindedly on replacing retiring judges with lifetime appointments of young conservatives vetted by the Federalist Society for their ideological commitment to limited government; their philosophies might reshape the courts' receptiveness to the agency's initiatives for a generation.

However, the extent to which Trump would ultimately succeed in weakening the EPA was by no means certain. He was easily able to cancel rules that had not yet been finalized, but once rules have been adopted, they cannot be rescinded simply by the stroke of the president's or administrator's pen. They can only be replaced through the full process that was required to approve them in the first place, and proposed changes can be overturned by the courts, which happened to a large number of his proposals. Rewriting the EPA rules requires convincing new evidence to justify changing rules that were previously justified. It also is a time-consuming process, not only for the EPA and its staff but also for the courts.

In addition, a serious attempt to gut the agency could well rekindle a backlash of public and even business opposition. Environmental protection often is not a major issue in elections, but frequently becomes one in the face of any serious threat to it. The EPA's mission and many of its programs are still popular with the public, with some influential businesses, and with many members of Congress, including even some Republicans; note that even the Senate Republican majority refused to seriously consider Trump's proposed budget cuts to the agency. Even with unanimous Republican support, moreover, the Republican majority in the Senate was less than would be needed to override a filibuster, and by 2019 the House was controlled by Democrats. Some business interests supported radically weakening the EPA's regulations, but others benefited from them: they provided a single set of national standards rather than different ones in each state, they protected companies' reputations and compliance investments against less principled competitors, and they provided predictable requirements consistent with global as well as domestic expectations. In 2019, for instance, even the US Chamber of Commerce began to moderate its hostility to climate change policy and to recognize it as an issue that needed to be addressed.[49]

The most likely outcome, therefore, was that at least during the remainder of Trump's first term of office, his EPA appointees would continue to issue few if any new regulations but would work energetically to weaken or rescind current ones, hoping to push them through court review before his term ended.

If Trump were to be reelected, however, the future of the EPA could be very different than its past 50 years. One of the most serious long-term dangers was the Trump administration's attempt to radically defund the scientific research activities that are needed to understand the problems the agency is charged with addressing. A particular danger, for instance, is the attempt to eliminate the federal capacity to address climate change, including not simply the EPA's regulations but also climate science research, mitigation, and adaptation capacities throughout the federal agencies. In October 2020, for instance, he appointed two outspoken climate change critics to senior science and policy positions in the National Oceanic and Atmospheric Administration, which is the lead agency for the nation's periodic official assessments of climate change; these changes could do serious damage to both America's and the world's understanding of important environmental forces and trends and significantly hinder our ability to respond to them.

RAYS OF HOPE

Despite attacks by some businesses and other critics, however, as of 2020 the basic statutory frameworks and scientific foundations of the EPA's decisions remained largely intact. Even during unsupportive presidencies, the courts have frequently upheld the EPA's statutory responsibilities to protect the environment and have overruled attempts to reinterpret them in less protective ways. Both environmental advocacy groups and some state governments—and even some supportive businesses—have played key roles in bringing such lawsuits.

There is additional hope in the proliferation of state-level policy innovations (see Chapter 2). California and New Jersey led in developing hazardous chemical "right-to-know" laws in the 1980s, which led to the EPA's nationwide Toxics Release Inventory. More than half the states passed renewable energy mandates, and twenty-four passed tax credits for renewable energy. North Carolina passed a Clean Smokestacks Act in 2001, a state-level cap-and-trade requirement that forced its electric utilities to clean up or close down old coal-fired power plants that had been "grandfathered" under federal law: this act reduced the state's sulfur emissions by more than 80 percent.[50] California and a coalition of northeastern states (the Regional Greenhouse Gas Initiative, or RGGI) became leading voices in climate change policy, both within the United States and in cooperation with other countries, with California in particular passing strong policies to reduce GHG emissions and to promote energy efficiency and renewable energy (see Chapters 2 and 12). Other states also moved to promote renewable energy, including even "deep red" states such as Texas and Iowa. At the same time, however, polluting industries and wealthy individuals opposed to regulation began pouring money into state election campaigns as well as national ones, and proposing "model" state legislation to try to block strong environmental policy initiatives at the state level as well.[51]

There was also hope in coalitions between environmental advocacy groups and businesses that would prosper in a greener economy. Many leading businesses identified ways in which good environmental management could be good business, and in cooperation with some environmental organizations, positioned themselves to prosper in a more environmentally sustainable economy (see Chapter 14). Market forces favoring greener outcomes, such as the rapidly decreasing prices of solar and wind energy and rising global expectations demanding mitigation of climate change, also became important drivers of business decisions, and these influences would continue whether or not Trump's deregulation agenda succeeded. But the EPA's regulations remained an important motivator for other businesses, many of which were still polluting either because they lacked the will to modernize old facilities or because it was more expensive for them to control pollution than to pay for lawyers, lobbyists, and politicians to resist regulation.

The EPA thus remains an essential institution and—so far, at least—an important force within the limits of its authority, resources, and politically unstable leadership. The limits and imperfections of its policy tools are real, but more often than not, they result not from any unwillingness by the EPA to use more effective ones but from the inability of Congress—hamstrung by partisan and ideological gridlock and by the power of entrenched interests—to authorize better alternatives. Any hope of truly fundamental policy improvement must lie in broader political reforms and in broader coalitions of environmental and public health advocates with those businesses and state governments that are supportive of environmental protection.

SUGGESTED WEBSITES

US Environmental Protection Agency (www.epa.gov) There are numerous links on this site that provide access to all major activities and issues of concern at the EPA, including environmental laws and regulations.

US Environmental Protection Agency—EPA History (www2.epa.gov/aboutepa/epa-history) This section of the EPA site explores the agency's history and the significant changes over time in environmental laws and regulations.

US Environmental Protection Agency Clean Power Plan and Affordable Clean Energy Rule, see https://19january2017snapshot.epa.gov/cleanpowerplan_.html for developments under the Clean Power Plan and other EPA actions to address carbon pollution under the Obama administration. For the Trump administration's proposed substitute for it (the Affordable Clean Energy Rule), see https://www.epa.gov/stationary-sources-air-pollution/affordable-clean-energy-rule. For recent developments, see https://eelp.law.harvard.edu/2017/09/clean-power-plan-carbon-pollution-emission-guidelines/

US Environmental Protection Agency New Source Review (www.epa.gov/nsr) This part of the EPA site is dedicated to explaining the New Source Review provisions under the Clean Air Act. For recent developments, see https://eelp.law.harvard.edu/2018/12/new-source-review/

US Environmental Protection Agency Particulate Matter (PM) Pollution (https://www.epa.gov/pm-pollution) This part of the EPA website provides links to all aspects of its regulation of particulate matter pollution. For recent developments, see https://eelp.law.harvard.edu/2020/07/national-ambient-air-quality-standards-for-pm-and-ozone/

NOTES

1. For a more detailed history of the EPA, see Richard Andrews, *Managing the Environment, Managing Ourselves: A History of American Environmental Policy*, 3rd ed. (New Haven, CT: Yale University Press, 2020), https://yosemite.epa.gov/sab/sabproduct.nsf/0/FE50D8FD06EA9B17852583B6006B7499/$File/03-07-19+Draft+CASAC+PM+ISA+Report.pdf, Chapters 12 and 17.
2. Russell Train, "The Environmental Record of the Nixon Administration," *Presidential Studies Quarterly* 26 (1996): 185–96.
3. For more detail on EPA budgets and staffing, see Chapter 1 and Appendixes 2 and 3.
4. U.S. Environmental Protection Agency, *Fiscal Year 2018 Agency Financial Report* (Washington, DC: EPA, 2018), https://www.epa.gov/sites/production/files/2018-11/documents/epa-fy-2018-afr.pdf.; Sarah Gibbens, "This Is How the EPA Uses Its Budget," *National Geographic*, March 2017, https://www.nationalgeographic.com/news/2017/03/environmental-protection-agency-budget-cuts/.
5. Philip Shabecoff, "E.P.A. Aims to Curb Tall Smokestacks," *New York Times*, June 28, 1985.
6. See, for instance, Trip Gabriel, "Utility Cited for Violating Pollution Law in North Carolina," *New York Times*, March 3, 2014.
7. Brian Clark Howard and Robert Kunzig, "5 Reasons to Like the U.S. Environmental Protection Agency," *National Geographic*, December 9, 2016, http://news.nationalgeographic.com/2016/12/environmental-protection-agency-epa-history-pruitt/.
8. Coral Davenport, "E.P.A. Faces Bigger Tasks, Smaller Budgets and Louder Critics," *New York Times*, March 18, 2016.
9. U.S. Office of Management and Budget, *America First: A Budget Blueprint to Make American Great Again*, March 2017.

10. Nadja Popovich, Livia Albeck-Ripka, and Kendra Pierre-Louis, "95 Environmental Rules Being Rolled Back Under Trump," *New York Times*, December 21, 2019.

11. "Existing U.S. Coal Plants, Age Comparison," *SourceWatch*, http://www.sourcewatch. org/index.php/Existing_U.S._Coal_Plants#cite_note-14.; U.S. Energy Information Administration, "Most Coal Plants in the United States Were Built Before 1990," *Today in Energy*, April 17, 2017, https://www.eia.gov/todayinenergy/detail.php?id=30812.; U.S. Energy Information Administration, "More U.S. Coal-fired Power Plants are Decommissioning as Retirements Continue," *Today in Energy*, July 26, 2019, https:// www.eia.gov/todayinenergy/detail.php?id=40212.

12. Clean Air Act of 1970, Public Law 91-604, Section 111; U.S. Environmental Protection Agency, *New Source Review*, www.epa.gov/nsr.

13. Larry Parker, *Clean Air: New Source Review Policies and Proposals*, CRS Report No. RL31757 (Washington, DC: Congressional Research Service, 2003); *Clean Air and New Source Review: Defining Routine Maintenance*, CRS Report No. RS21608 (Washington, DC: Congressional Research Service, 2005).

14. James E. McCarthy, "Clean Air Act: A Summary of the Act and Its Major Requirements," in *Clean Air Act: Interpretation and Analysis*, ed. James P. Lipton (New York, NY: Nova Science, 2006), Chapter 1. See pages 41–43 on New Source Review.

15. Christopher Drew and Richard Oppel, Jr., "Air War—Remaking Energy Policy: How Power Lobby Won Battle of Pollution Control at E.P.A.," *New York Times Magazine*, March 6, 2004.

16. John Shiffman and John Sullivan, "EPA's Court Follies Sow Doubt, Delay," *Philadelphia Inquirer*, December 8, 2008; *New York v. EPA*, 413 F. 3d 3 (D.C. Cir. 2005); *New York v. EPA*, 443 F. 3d 880 (D.C. Cir. 2006).

17. Thomas McGarity, "When Strong Enforcement Works Better Than Weak Regulation," *Maryland Law Review* 72 (2013): 1204–94.

18. Joseph Goffman, Janet McCabe, and William Niebling, "EPA's Attack on New Source Review and Other Air Quality Protection Tools," *Environmental & Energy Law Program, Harvard Law School*, November 1, 2019, http://eelp.law.harvard.edu/wp-content/ uploads/NSR-paper-EELP.pdf.

19. U.S. Environmental Protection Agency, "Particulate Matter Pollution," https://www.e-pa.gov/pm-pollution/particulate-matter-pm-basics.; "Table of Historical PM NAAQS," https://www3.epa.gov/ttn/naaqs/standards/pm/s_pm_history.html, both accessed January 29, 2020.

20. Louis Anthony Cox, Jr., "Should Health Risks of Air Pollution be Studied Scientifically?" *Global Epidemiology* 1 (2019): 100015; Gretchen Goldman and Francesca Dominici, "Don't Abandon Evidence and Process in Air Pollution Policy," *Science* 363 (2019): 1398–1400.

21. Juliet Eilperin and Brady Dennis, "Pruitt Unveils Controversial 'Transparency' Rule Limiting What Research EPA Can Use," *Washington Post*, April 24, 2018.

22. U.S. Environmental Protection Agency, *Integrated Science Assessment for Particulate Matter (External Review Draft)*, Report No. EPA/600/R-18/179, October 2018.

23. Clean Air Scientific Advisory Committee (CASAC) Draft Report, Letter to EPA Administrator Andrew Wheeler, March 7, 2019, https://yosemite.epa.gov/sab/sabproduct.nsf/0/FE50D8FD06EA9B17852583B6006B7499/$File/03-07-19+Draft+CASAC+ PM+ISA+Report.pdf, accessed February 7, 2020.

24. U.S. Environmental Protection Agency, "Policy Assessment for Review of the National Ambient Air Quality Standards for Particulate Matter, External Review Draft," *Federal Register* 84 (September 11, 2019): 47944.

25. Clean Air Scientific Advisory Committee (CASAC) Draft Report, Letter to EPA Administrator Andrew Wheeler, November 13, 2019, https://yosemite.epa.gov/sab/sabproduct.nsf/WebCASAC/0A46BDBE59C86531852584B10077B0F6/$File/11-13-19+Draft+CASAC+PM+PA+Report.pdf, accessed February 4, 2020.

26. Independent Particulate Matter Review Panel, Letter to EPA Administrator Andrew Wheeler, October 22, 2019, https://ucs-documents.s3.amazonaws.com/science-and-democracy/IPMRP-FINAL-LETTER-ON-DRAFT-PA-191022.pdf, accessed February 4, 2020.

27. Seth Borenstein, "Bush Changes Pledge on Emissions," *Philadelphia Inquirer*, March 14, 2001.

28. Robert Fabricant (EPA General Counsel), "EPA's Authority to Impose Mandatory Controls to Address Global Climate Change under the Clean Air Act" (August 29, 2003), cited in *Federal Register* 68 (173): 52922–32 at page 52925.

29. *Massachusetts v. EPA*, 549 U.S. 497 (2007).

30. James McCarthy, *Cars, Trucks, and Climate: EPA Regulation of Greenhouse Gases from Mobile Sources*, CRS Report No. R40506 (Washington, DC: Congressional Research Service, 2014).

31. The "timing rule" (also known as the "PSD Triggering Rule") was issued in 1980; it made any source emitting more than one hundred tons of any regulated air pollutant subject to requirements for an EPA permit for "prevention of significant deterioration" of air quality (PSD). The Supreme Court unanimously reconfirmed the EPA's authority to regulate GHGs from power plants in *American Electric Power Co. v. Connecticut*, 564 U.S. 410 (2011).

32. *Utility Air Regulatory Group v. EPA*, 134 S. Ct. 2427 (June 23, 2014).

33. *Federal Register* 75: 31514, June 3, 2010; Robin Bravender, "EPA Issues Final 'Tailoring' Rule for Greenhouse Gas Emissions," *New York Times*, May 13, 2010.

34. *Federal Register* 77 (2012): 22392; James McCarthy, *EPA Standards for Greenhouse Gas Emissions from Power Plants: Many Questions, Some Answers*, CRS Report No. R43127 (Washington, DC: Congressional Research Service, 2013).

35. Environmental Protection Agency, *Carbon Pollution Emission Guidelines for Existing Stationary Sources: Electric Utility Generating Units*, 40 CFR Part 60, Fed. Reg. 79: 34830 (2014).

36. *Utility Air Regulatory Group v. EPA*, 134 S. Ct. 2427 (June 23, 2014).

37. Linda Tsang and Alexandra Wyatt, *Clean Power Plan: Legal Background and Pending Litigation in West Virginia v. EPA*. CRS Report No. R44480 (Washington, DC: Congressional Research Service, March 8, 2017).

38. Niina Heikkinen, "Crunching New Numbers for a Potent Greenhouse Gas," *Climatewire*, November 20, 2017.

39. Elizabeth Kopits, Al McGartland, Cynthia Morgan, Carl Pasurka, Ron Shadbegian, Nathalie B. Simon, David Simpson, and Ann Wolverton, "Retrospective Cost Analyses of EPA Regulations: A Case Study Approach," *Journal of Benefit-Cost Analysis* 5 (2014): 173–93.

40. Alan Krupnick, "Trump's Regulatory Reform Process: Analytical Hurdles and Missing Benefits." Resources for the Future, *Resources* 194 (Spring 2017).

41. Chris Mooney, "New EPA Document Reveals Sharply Lower Estimate of the Cost of Climate Change," *Washington Post*, October 11, 2017.

42. Richard Andrews, *Managing the Environment, Managing Ourselves: A History of American Environmental Policy* (New Haven, CT: Yale University Press, 2006), 245–49.

43. Eric Lipton, "Why Has the E.P.A. Shifted on Toxic Chemicals?" *New York Times*, October 21, 2017.

44. At least one scholar, however, has argued that this conventional wisdom is overstated and often wrong and that "there are solid reasons to suspect that an emissions trading program does a poorer job of stimulating innovation than a comparably designed traditional regulation." See David Driesen, "Does Emissions Trading Encourage Innovation?" *Environmental Law Reporter* 33 (2003): 10094–108.

45. Richard Andrews, "The EPA at 40: An Historical Perspective," *Duke Environmental Law and Policy Forum* 21 (2011): 223–58 at 229–34.

46. Brady Dennis, "EPA Plans to Buy Out More Than 1,200 Employees This Summer," *Washington Post*, June 20, 2017, https://www.washingtonpost.com/news/energy-environment/wp/2017/06/20/epa-plans-to-buy-out-more-than-1200-employees-by-the-end-of-summer/?utm_term=.a7a9c9ba72ef.

47. Ellen Knickmeyer, "EPA Criminal Action Against Polluters Hits 30-Year Low," Associated Press, January 15, 2019.

48. Leif Frederickson, Marianne Sullivan, Christopher Sellers, Jennifer Ohayon, Ellen Kohl, Sarah Lamdan, Alissa Cordner, Alice Hu, Katarzyna Kaczowka, Natalia Navas, Linda Wicks, *A Sheep in the Closet: The Erosion of Enforcement at the EPA*, Environmental Data and Governance Initiative, November 2018, https://www.eenews.net/assets/2018/11/16/document_gw_01.pdf, accessed November 19, 2018.

49. Jennifer Dlouhy, "U.S. Chamber to Re-Examine Climate Policy That Cost It Members," *Bloomberg News*, September 24, 2019.

50. Richard Andrews, "State Environmental Policy Innovations: North Carolina's Clean Smokestacks Act," *Environmental Law* 43 (2013): 881–940.

51. Suzanne Goldenberg, "Conservative Lobby Group Alec Plans Anti-Environmental Onslaught," *The Guardian*, December 2, 2014, https://www.theguardian.com/us-news/2014/dec/02/alec-environmental-protection-agency-climate-change.

PUBLIC POLICY DILEMMAS

ENERGY POLICY

8

Sanya Carley

Energy, as both a commodity and service, fundamentally affects the day-to-day lives of all individuals. Energy is also essential to the functionality of modern economies and a critical source of economic growth in both the United States and elsewhere around the world. Primary energy resources (e.g., coal, natural gas, oil, nuclear, wind, solar, hydroelectric, geothermal, biomass) are used to generate—or reduce in the case of energy efficiency—different secondary energy services (e.g., electricity, transportation, heating, and cooling) that support both human well-being and economic prosperity. Each of these energy resources, and the policies that encourage them, however, is also accompanied by a set of benefits and drawbacks, institutional challenges, vested interests, and difficult trade-offs.

Energy policy refers to policy, legal, and political decisions about the production and consumption of these energy resources. Energy policy thus also requires making decisions under complex conditions that necessitate difficult trade-offs. The direction of energy policy, at any given point in time, is shaped by those who make policy decisions—and the concerns, values, personal and professional interests, and partisanship that they uphold—as well as the conditions of the time. In the 1970s, when faced with oil shortages and a drastic increase in the price of oil, US policymakers pursued energy policy—in the form of the Energy Policy and Conservation Act of 1975—aimed at improving the efficiency of vehicles and building strategic petroleum reserves. During the George W. Bush presidency, the country's energy policies tended to focus on enhancing energy security and domestic production of energy through nuclear, ethanol, and so-called clean coal incentives. In 2009, a year into the Great Recession, the Obama administration urged Congress to pass the American Recovery and Reinvestment Act, which allocated approximately $90 billion toward energy innovation, research and development, and energy projects.[1]

Most recently, the Trump administration has sought to shift the energy policy paradigm once again, in favor of fossil fuels and away from renewable energy, energy conservation, and efficiency. Among many regulatory changes, the administration has prioritized removing barriers to fossil fuel extraction; lessening or abdicating any responsibilities to reduce greenhouse gas (GHG) emissions; changing the way in which analyses that support regulatory decisions are conducted; and propping up the coal industry (see Chapters 4 and 7). All of these efforts are accompanied by a growing political divide on energy issues within the United States. The administration's policy efforts mark a sharp contrast with policies pursued during the Obama administration and highlight just how deeply political energy policy decisions are.

A major condition facing energy markets in the modern era is the generation of GHG emissions. The production and consumption of energy, primarily through fossil fuel combustion to satisfy transportation and electricity needs, is the single largest contributor to US GHG emissions.[2] It is widely recognized, based on scientific evidence from a large contingent of prominent scientists, that the accumulation of GHG emissions in the atmosphere has the potential to affect the stability of ecosystems through climate change and compromise human health and well-being.[3] The majority of Americans also believe that climate change is happening, specifically perpetuated by anthropogenic sources,[4] and that policymakers should address climate change through public policy (see Chapters 3 and 12).[5]

Another primary condition facing energy markets is the rapidly changing price of different energy resources, due in part to past energy policies, but also to various market conditions and political factors. Between 2009 and 2018, for example, the cost per kilowatt-hour of onshore wind declined on average by 69 percent and the cost of solar photovoltaic (PV) panels declined on average by 88 percent.[6] The decrease in the price of renewable energy sources has, in many cases and in many locations across the United States and the world, led to cost parity with more conventional fossil fuel resources. Whereas the most common energy resource consumed just a decade ago was coal, which is highly carbon-intensive and also the source of significant air and water pollution,[7] low- to no-carbon, efficient, and advanced energy resources are penetrating energy markets across the world.

In some jurisdictions, these penetration levels are significant, such as in California, which produced more renewable energy in 2017 than any other energy source, with 16 percent of its electricity generation coming from solar PV or solar thermal.[8] Iowa also produced 29 percent of its electricity from renewable sources.[9] Consider also Germany, which operated on 85 percent renewable energy for a full day in May 2017.[10] Several US cities, such as Aspen, Colorado, and Burlington, Vermont, are also able to claim 100 percent renewable electricity consumption based on their energy mix, energy efficiency measures, and purchases of tradable credits.[11] These market changes, as well as others discussed below, reveal that an energy transition is underway.[12]

In light of these conditions, US energy policies—led by a combination of federal, state, and municipal governments as well as regional partnerships, but especially state and municipal governments during the Trump administration—are driven increasingly by three objectives: (1) reduce GHG emissions; (2) deploy more low-carbon and efficient technologies; and (3) either prepare for or fight against the energy transition. Some jurisdictions are pursuing all three at once, while others focus predominantly on one objective. Some seemingly seesaw between clean energy technology diffusion and protecting vested interests against the energy transition. Several red states such as Ohio and Indiana, for example, have repealed policies put in place to incentivize renewable energy industry expansion and instead promoted fossil fuel interests, despite a general interest in expanding industries with a growing number of jobs and general support from their constituents for clean energy.

In this chapter, I first provide more context about the energy transition. I then provide a detailed account of several energy policies that uphold the three aforementioned objectives and set each in context. While it is impossible to cover all US energy policies, the examples included represent a range of policy instruments, case studies, and levels of government. Throughout the narrative, in addition to presenting the intricacies of various energy policies, I highlight political tensions, trade-offs, and transitioning norms and institutions, and subsequently return to a discussion of these themes in the conclusion.

SETTING THE CONTEXT: A US ENERGY TRANSITION

The term "energy transition" is increasingly used to describe the state of energy markets across the world. An energy transition, by loose definition, is a switch from dependence on one energy resource or a set of energy resources, to another.[13] By one account, and as a much stricter definition, it is an increase in the prevalence of a specific energy resource from 5 to 80 percent.[14] There are several examples of previous energy transitions, at least by the loose definition of the term, such as the transition from whale oil to kerosene and wood to coal, both during the nineteenth century. The present energy transition is marked by a movement away from carbon-intensive fossil fuels toward lower-carbon, more efficient, and advanced energy resources, albeit not without a great deal of resistance by the fossil fuel industry and conservative supporters. Figure 8.1 displays these trends in the United States between 1990 and 2017 as the electricity mix has shifted from the left-most bar of 3 million megawatt-hours (MWh) to the right-most bar of 4 million MWh, with resource changes tabulated in between.

Although experiential evidence is limited due to the low number of historical examples, such transitions are rarely rapid. The current energy paradigm is solidified and continually reinforced by almost a century of technological path dependency (i.e., where history tends to determine the decisions that are made about the future) and

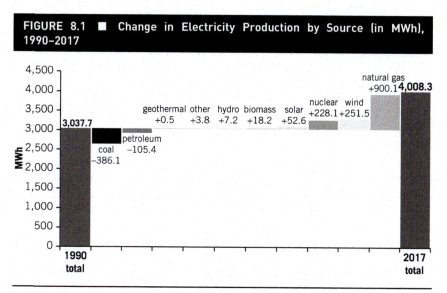

FIGURE 8.1 ■ Change in Electricity Production by Source (in MWh), 1990–2017

Source: "Table 7.2a Electricity Net Generation: Total (All Sectors). Monthly Energy Review September 2018," *U.S. Energy Information Administration* (2018), from https://www.eia.gov/totalenergy/data/monthly/pdf/sec7_5.pdf.

so-called "carbon lock-in," which refers to the continual reinforcement of a fossil fuel–based economy through technological and institutional norms.[15] For example, energy systems also tend to operate on long-term planning horizons, in which new power plants or other infrastructure tend to last decades before replacement. Thus, any changes in energy systems tend to be slow and gradual; and it is difficult for new, advanced, efficient, or low-carbon technologies to penetrate the market and gain consumer and producer acceptance.[16] Arguably, policy and other institutional norms tend to evolve even slower than the technologies.

Scholars continue to debate the pace of the current energy transition.[17] Those who claim it to be slow tend to abide by the strict definition. They observe the percentage of low-carbon technologies out of total demand over time and note that the percentage is small but growing gradually. Those who claim it to be rapid tend to consider the annual change in energy resources, and point out that low-carbon energy resources have provided the majority of new capacity in energy markets over the past decade, as well as most new capacity that is projected into the future.[18] Assumptions about uncertain future political circumstances also influence perceptions of the pace of the energy transition. Those who anticipate significant policy backlash against renewables as such technologies mature and penetrate previously well-established markets, for example, are more likely to assume a slower energy transition. As new energy technology penetration tends to increase, so too does political resistance;[19] and such political turmoil can slow the pace of the transition.

What is not debated is that the energy transition is occurring—this appears undeniable.[20] Projections from 13 different long-range energy modeling scenario studies published between 2017 and 2018—including those published by ExxonMobil, the Organization of Petroleum Exporting Countries (OPEC), and the United States Energy Information Administration—all predict a decline in coal of, on average, 28 percent of the global 2015 mix to between 12 and 17 percent in 2040, and an increase in renewables from 14 percent to between 16 and 31 percent over the same time period.[21] These studies do not agree about the future of nuclear: some predict a decline from the 5 percent globally in 2015, while others predict a modest increase. These estimates also predict a growth of energy consumption of between 20 and 30 percent between 2015 and 2040, which means that the percentage of any resource is out of a much larger total. Another source predicts that fossil fuel demand will peak by 2020 and 2027.[22] These various findings all suggest that the energy transition will continue to unfold, and at an accelerating pace.

Energy experts also do not debate the fact that the energy transition is accompanied by changes that extend beyond the resource mix, including new institutional actors, markets for exchanging energy commodities, and arrangements between consumers and producers. These changes are forcing traditional energy actors, such as utilities and car manufactures, to reevaluate their business models and, in the case of energy consumers, to modify the ways in which they interact with energy goods and services. For example, not only are centralized energy operations transitioning toward lower-emitting sources, but the prevalence of smaller-scale, low-carbon energy resources is also increasing. In many cases, the owners of such systems are not the traditional utility: for larger-scale wind turbines and solar panels, the owners are often independent power producers; for small-scale distributed generation, the owners are often consumers turned "prosumers" (when a consumer also serves as a producer of energy).

These new actors and market arrangements require a reevaluation of institutional rules and regulations and present a host of technical, economic, equity, and policy challenges. Such changes inevitably inflict growing pains on the entire energy system as well as on individual entities and subsectors and markets within the system. For example, energy producers that work in legacy industries such as coal—and possibly in the future natural gas and nuclear—face declining economic opportunities, a loss of employment, and, in some cases, stranded assets. As another example of such challenges, wholesale power markets are working to better integrate higher penetration levels of renewable energy, including energy storage and powerline construction. In addition, US energy policy decision-makers operate within a federal system, and new market operations require some degree of coordination across all levels within the system or decisions can become bottlenecked or caught in lengthy court processes.

While these various changes and challenges are welcomed by some, they are opposed—sometimes fiercely—by others. Incumbent and heavily invested interests, for example, may be reluctant to strand assets—such as in an already built and fully functional coal power plant—or diversify business operations in new directions. Resistance to change can be observed through lobbying efforts, where companies pay significant sums of money to influence legislation—such as to resist a carbon policy[23] or to support the continuation or removal of energy subsidies—that would require the company to drastically change its investment strategy or render the company's product outdated or noncompliant. Between 2000 and 2016, the fossil fuel industries spent approximately $370 million on climate lobbying (compared to $48 million spent by renewable energy industries).[24] These expenditures reached a climax in 2009 and 2010, right before onshore wind energy, the lowest cost and most commonly deployed form of renewable energy, reached cost parity with traditional forms of fossil fuels.[25] The timing of this climax also coincided with Congressional consideration of the Waxman-Markey Bill, which proposed an emissions trading plan, as well as the drafting of the Obama administration's American Recovery and Reinvestment Act of 2009.

The public is observing these various developments with mixed reactions. As of 2019, 67 percent of those polled believe that the government is not doing enough to reduce the effects of climate change,[26] up from 46 percent in 2010 who agreed with the statement that global warming is a problem requiring immediate government action.[27] While these numbers have shifted in recent years toward a greater share of those who hold this belief, strong political divides still remain. Of those who self-identified as Democrat in 2019, 90 percent believe that the government is not doing enough, compared to 39 percent who self-identified as Republican. In recent years, the public strongly supports renewable energy. In 2016, 89 percent of polled citizens said that they favor expanding solar panels and 83 percent favor wind turbine expansion, compared to 41 percent who favor coal mining expansion.[28] (See Chapter 3 for a more in-depth discussion of public opinion.)

This unfolding energy transition is at least partly attributable to energy policies aimed at technology deployment and GHG emission abatement that have been enacted and implemented over the past several decades, as well as due to market dynamics, mechanization, and other trends.[29] Most of the policies have been adopted and implemented at the state level, where economic and social interests tend to be more consistently aligned, but the federal and municipal levels have contributed substantively as well (see Chapter 2). The energy transition also instigates new energy policies, so as to either accommodate evolving energy markets or address difficult challenges that the energy transition poses, or to entirely resist the energy transition and its effects on vested interests. In the next section, I discuss some of the policies that have contributed to the energy transition, as well as recent modifications of these policies as the energy transition has evolved.

POLICY PRIORITIES: GHG EMISSIONS AND TECHNOLOGY DEPLOYMENT

State Renewable Portfolio Standards

One of the most popular US energy policies is the state-level renewable portfolio standard (RPS), first adopted by a small subset of states in the late 1990s as an outgrowth of electricity deregulation and in response to a lack of leadership at the national level on energy and climate policy. An RPS is a mandate that a specific percentage of total state electricity sales, or generation, is sourced from renewable energy (e.g., wind or solar), often including hydroelectricity. A more recent form of an RPS discussed in greater detail below, often called a clean energy standard, is a mandate that a percentage of total electricity comes from clean energy (e.g., a form of energy that produces limited or no GHG emissions). (See Chapter 2 for further discussion.)

As an example, Illinois has an RPS mandate that requires that 25 percent of all electricity sales be sourced by renewable energy by 2025. Eligible resources that count toward this requirement include solar PV, solar thermal, wind, biomass, hydroelectricity, landfill gas, other anerobic digestion gas, and biodiesel.[30] RPS policies, as a strict command-and-control type instrument, are almost always accompanied by renewable energy certificate (REC) markets. Those utilities required to comply with RPSs may buy (or sell) credits to satisfy their mandates, each of which counts as 1 MWh of renewable electricity. Typically, RECs only account for the environmental attributes of the renewable energy, not the actual electricity as well (one can pay separately for the electricity that comes from the renewable energy). The electricity is then sold through a separate energy market.

As of early 2020, 29 states and the District of Columbia have an RPS policy, as displayed in Figure 8.2. These states account for approximately 55 percent of total US retail electricity sales.[31] An additional 8 have voluntary policies. Democratically leaning states more commonly adopt these policies, but both red and blue states have done so.[32] In fact, one of the biggest RPS success stories with a significant penetration of wind energy—with an annual increase in wind consumption of 37 percent between 2000 and 2017—and fairly consistent support across key stakeholders is the state of Texas, which voted for a Republican presidential candidate every election between 1980 and 2016.[32] The majority of Southeastern states, however, where the cost of electricity is low and the endowment of renewable sources such as wind are low, do not have an RPS

FIGURE 8.2 ■ States With Binding Renewable Portfolio Standards in 2020

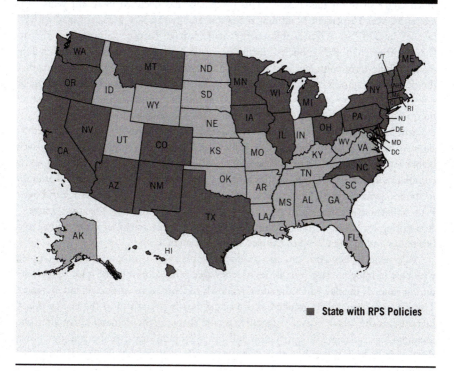

■ **State with RPS Policies**

Source: Cale Jaffe, "Melting the Polarization around Climate Change Politics," *Georgetown Environmental Law Review* 30 (3) (2018): 455–498.

policy. As of 2018, 32 other countries also have RPS policies, or what is more commonly referred to as quota policies.[33] The RPS policy is politically more feasible than other energy policies because the gradual nature of the requirement allows policymakers to set their goals into the distant future, with a minimal immediate commitment, but with a signal today that they are prioritizing clean energy. Voluntary policies send the signal that the state is interested in developing renewable energy, even if the policy has no "teeth."

In the early years of RPS implementation, states tinkered frequently with their RPS policy designs, with some states revising their policies up to five times. Common amendments included increasing policy stringency, adding new requirements for specific resources (e.g., solar PV or distributed generation), expanding the list of eligible resources (e.g., adding energy efficiency or fossil fuel technologies), adjusting geographic restrictions on RECs, and adding cost recovery and planning mechanisms.[34]

The vast majority of studies on RPS policies confirm that they have facilitated renewable energy growth successfully across the United States. Utility reporting documentation reveals that utilities that face RPS standards tend to achieve compliance with their state's interim benchmarks, with only some exceptions.[35] RPS policies were accountable for about 34 percent of all new renewable energy capacity in 2017, with a much larger share on the coasts.[36] This percentage has declined in recent years—for reference, over 75 percent of all new renewable energy capacity was attributed to RPS policies in 2007[37]—as renewable energy markets have become more robust due to factors other than RPS policies, including other policies, voluntary credit trading markets, long-term contracts, and other market dynamics. Research also confirms that the design of a state RPS policy is essential to its success; for example, more stringent standards lead to more renewable energy development and placing geographic restrictions on the exchange of RECs leads to more in-state renewable growth.[38]

Studies also confirm that state RPS policies could collectively reduce GHG emissions. A modeling study that accounted for all RPS standards in place by 2016—well before the recent explosion of 100 percent clean energy commitments—found that the cumulative 2015–2050 GHG savings from the collection of RPS policies could amount to about 6 percent. With a stronger set of standards that match the former Clean Power Plan obligations, states could reduce GHG by 23 percent over this time frame.[39] Accounting for all new 100 percent clean energy pledges across all states, cities, and businesses would increase this estimate quite considerably.

RPS policies, however, can also produce complications. Some states have grappled with how to integrate large quantities of renewable generation into the grid,[40] especially during times of the day when the sun goes down concurrent to when demand increases. This phenomenon, when graphed over a single day, is referred to in the energy policy literature as the "duck curve," suggestive of the shape of the demand curve after one accounts for available solar generation and the steep ramp-up after the sun goes down that needs to be satisfied by other energy resources that quickly penetrate the grid.[41] Another challenge is how to account for spillovers, such as when one state changes its energy mix as a result of a surrounding state's RPS policy.[42] A third challenge that electricity market actors are increasingly confronting is how to integrate individual states' RPS mandates into regional electricity markets, or what are titled Independent System Operations (ISO) or Regional Transmission Operators (RTO).[43] A fourth challenge is overcoming political and personal opposition to energy infrastructure siting.

In recent years, leading energy policy states began to modify their policies again, this time with more ambition. Hawaii adopted a 100 percent clean energy mandate in 2016. Following this precedent, several other states began to seriously pitch the idea of 100 percent clean energy. Many increased their targets in 2018 (e.g., California, Connecticut, Massachusetts, and New Jersey), followed by several more in 2019 (e.g., Maryland and Nevada), including those with 100 percent commitments (e.g., California, the District of Columbia, New Mexico, Washington, and the territory of Puerto Rico). These new, highly ambitious mandates not only double or triple the amount of low-carbon energy that states must achieve, but they also: (1) introduce greater complexity into the definition of clean energy and (2) are accompanied by other policy features aimed at efficiency and equity.

While all of these jurisdictions previously had aggressive RPS policies, or were at least early movers in the RPS policy diffusion process, these new, bolder mandates were, arguably, less of an effort to one-up their standards than an effort to increase their targets to a maximum in response to the Trump administration's efforts to derail the

US's commitments to climate mitigation. In this sense, these mandates are both symbolic—sending a signal to the federal government and the rest of the world that climate change is still a leading priority—and potentially significant for climate change mitigation, since these jurisdictions contain a significant proportion of the US population.

A case for consideration is the state of Washington and its Clean Energy Transformation Act, passed in 2019. This policy sets several benchmarks for the state:

1. by 2025, the state can no longer produce electricity from coal;

2. by 2030, the electricity mix must be 100 percent carbon neutral, at least 80 percent of which comes from state-generated carbon-free sources (e.g., renewables, nuclear, or natural gas with carbon capture and storage) and 20 percent can come from RECs, carbon taxes, or allowable energy project expenditures (e.g., electric vehicle infrastructure or weatherization); and

3. by 2045, 100 percent carbon-free and in-state generation.[44]

This policy also requires that the state switch from a traditional rate design to a performance-based design in order to encourage utilities in the state to be more innovative and supportive of energy efficiency and similar investments. Finally, the policy calls for equitable distribution of the benefits and burdens of policy outcomes, financial support for low-income energy assistance, grants for clean energy projects, and a requirement that utilities consider the social cost of carbon when making investment decisions.[45]

Federal Fuel Economy Standards

In contrast to the exclusively state-led RPS policy, fuel economy standards in the United States have been led by the national government, but with influence from the states. Fuel economy standards are arguably the most ambitious and long-lasting energy policies that the federal government has enacted over the past several decades. Corporate average fuel economy (CAFE) standards were first introduced in response to the 1975 Energy Policy and Conservation Act, during the term of President Gerald Ford, which authorized the National Highway Traffic Safety Administration (NHTSA) to set mileage standards for new cars and light-duty vehicles. The original standard was 27.5 miles per gallon (MPG) by 1985. After two decades of only slight modifications to the standards, but significant advancements in the state of California, the George W. Bush administration increased standards more substantially beginning in 2005, and then again in 2007, working with a Democratically controlled Congress, through the Energy Independence and Security Act (EISA). The Obama administration followed suit with executive authority granted through the EISA: in 2011, it increased standards from 2011 to 2016; in 2012, it issued a rulemaking that established a new goal of 54.5 MPG by 2025.

Just a few months after President Trump was sworn into office, his administration announced a plan to revisit the standards and reopen the midterm review of the program. The administration first released a draft plan, led by the NHTSA, that was riddled with calculation errors and in which many of the modeling scenarios did not even pass a basic benefit–cost test.[46] In March 2020, more than three years after Trump first announced his intent to change the standards, the administration released its final Safer Affordable Fuel-Efficient (SAFE) Vehicles Rule. Instead of an approximately 5 percent increase in efficiency each year, as the Obama administration rules were designed, the new regulation requires an annual increase of approximately 1.5 percent. Several scholars have raised similar questions about the underlying assumptions and conclusions within the regulatory impact assessment as they did about the proposed rule (for additional coverage of this topic, see Chapters 4 and 15).

There are three particularly noteworthy aspects of the history of fuel economy standards. First, US fuel economy standards are a case of dynamic federalism, in which the states, led by California, have influenced the federal government to continually increase the stringency of the standards. Under a provision offered to the state of California in the 1970 Clean Air Act, the state is able to seek a waiver of federal preemption over emissions regulations for motor vehicles. Since it is not practical to have two separate sets of vehicle standards—one for California and those states that adopt California's standards wholesale, and the rest of the country—the federal government has tended to match California's standards. That is, until recently. The Trump administration, as part of the rollout of the SAFE rule, revoked California's tailpipe emissions waiver, a decision that immediately resulted in judicial action when the state of California and more than 20 other states sued the administration. (For more extensive coverage of this topic, see Chapters 2 and 7.)

Second, fuel economy standards have received overwhelming bipartisan support through the years—after all, improving efficiency should presumably result in cost savings, unless the cost of efficiency measures outweigh the savings—and have proven durable across changes in administration and have been regularly amended by Congress since 1990. That is, again, until recently, when the Trump administration released its SAFE rule. They did so despite overwhelming support among the American public for advancements in vehicle efficiency, and even some pushback from the automobile industry in response to the Trump administration's declaration that it would revise standards. Automobile companies initially asked for greater flexibility in fuel economy and GHG emissions compliance, but not a substantial reduction to the standards themselves, variation in emissions standards across United States due to the California waiver, nor an increase in regulatory uncertainty. The oil and gas industry, however, one of the biggest sources of federal lobbying expenditures, stands to benefit significantly from the new rule due to consumer purchases of gasoline.

Third, fuel economy standards are part of what is referred to as the Joint National Program, which is the first major federal policy initiative that directly combines energy and climate goals.[47] This marriage grew out of the 2007 United States Supreme Court decision, Massachusetts versus EPA (see Chapter 6), that the Environmental Protection Agency (EPA) has the authority to regulate GHG emissions, and the subsequent 2009 endangerment finding issued by the EPA that GHG emissions from mobile and stationary sources threaten human health and the environment. In 2012, the EPA established GHG limits for cars and light trucks for model years 2017–2025 that correspond to the NHTSA MPG standards; and now both agencies, jointly, regulate vehicle efficiency.[48] This combined Joint National Program, with a set of dual objectives, may help garner greater or more diverse political support; of course, it could also divide political support for the program, given the extreme polarization on the topic of climate change.[49]

Modern fuel economy and GHG emissions standards have been marked by mixed success. The standards have achieved full manufacturer compliance,[50] although different companies comply in different ways. For example, some manufacturers bank or borrow credits toward compliance; some build new vehicle fleets, such as electric vehicles; and some change the vehicle materials, such as using aluminum instead of steel in the Ford F-150. Many manufacturers have even complied early.[51] As a result, the program has increased steadily the average MPG of new vehicles. The program also, however, has several drawbacks. As a result of focusing on improving the efficiency of the vehicle, manufacturers have neglected to improve other vehicle attributes that consumers value, such as horsepower and torque.[52] Fuel economy standards are also likely not the most efficient policy if one's objective is GHG abatement.[53] Scholars have also raised concerns about the rebound effect (i.e., when one drives more because the cost of owning a vehicle decreases) diminishing the benefits of the policy,[54] and the likelihood that the increased price of a new vehicle will cause people to hold onto their used cars longer.[55]

Zero-Emissions Vehicle Policy

Another joint GHG abatement and technology diffusion policy is the zero-emissions vehicle (ZEV) program. The objective of the ZEV program is to increase the percentage of vehicles on the road that produce low to no tailpipe emissions during operation. Over the long term, this has the potential to both reduce GHG emissions[56] and help advanced vehicles reach economies of scale and thus decline in price. The policy operates as a mandated percentage of ZEVs that must be purchased by a specified date, as tracked by the number of ZEV credits earned by each regulated manufacturer. Battery electric vehicles and fuel cell vehicles count for full credit. Vehicles that produce low levels of emissions, such as hybrid battery electric, hydrogen, natural gas, and hybrid vehicles, can receive partial credit. Similar to REC markets for renewable electricity, vehicle manufacturers can bank and borrow ZEV credits over time and exchange the credits with each other.

The ZEV program first originated in California in 1990 when the California Air Resources Board (CARB) introduced it as a way to combat smog and soot in Los Angeles. The original mandate was 2 percent ZEV vehicles on the road by 1998; it was subsequently raised to 10 percent by 2003. In 2012, CARB folded the ZEV program into its Advanced Clean Cars program and extended the ZEV program to focus on GHG abatement. It also increased the mandate of required ZEV credits earned by each manufacturer to amount to a projected 15.4 percent of ZEV sales by 2025.[57] This estimate has since declined to about a 7.5 percent penetration of ZEV vehicles due to consumers buying a higher than originally anticipated number of long-range battery electric vehicles as well as automakers' use of banked credits.[58]

Other states are allowed to adopt California's vehicle emissions standards wholesale without seeking EPA approval, as granted in section 177 of the Clean Air Act. Since 2004, ten states and the District of Columbia have adopted California's ZEV program, most recently Colorado in 2019: Colorado, Connecticut, Maine, Maryland, Massachusetts, New Jersey, New York, Oregon, Rhode Island, and Vermont. These 11 states and Washington, DC account for a sizeable portion of total US automobile demand. These states do not, however, produce the majority of vehicles; it is the manufacturing states, such as Michigan, Ohio, and Indiana, that will bear the majority of costs and benefits from these manufacturing adjustments.

Since 2010, when the first modern electric vehicles were released to US markets, the number of purchased electric vehicles has grown steadily, and crested over one million in 2018, or roughly 2 percent of the total automobile market.[59] These sales are not, however, exclusively in ZEV states and some ZEV states are doing better than others at working toward their ZEV mandates, as displayed in Figure 8.3, which plots cumulative electric vehicle sales between 2011 and 2019 relative to the final 2025 ZEV goal for each state. Over this same time period, the cost of electric vehicles has also declined significantly. All major manufacturers now offer at least one electric vehicle model, if not several, with variations in driving distance on a single battery charge and amenities. The ZEV is, arguably, the most important factor in these developments, since auto manufacturers develop cars for the entire US market, and not, for example, one set of cars for California and other ZEV states, and another set of cars for the non-ZEV states.

A brief history of automobile industry reactions to the ZEV program reveals mixed political reactions within the industry. In 1990, when California first introduced the ZEV, automobile manufacturers resisted it through focused lobbying efforts.[60] Their

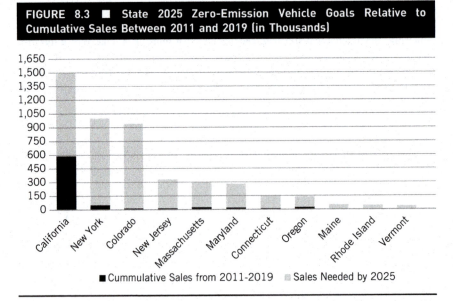

FIGURE 8.3 ■ State 2025 Zero-Emission Vehicle Goals Relative to Cumulative Sales Between 2011 and 2019 (in Thousands)

■ Cummulative Sales from 2011-2019   Sales Needed by 2025

Sources: Alliance of Automobile Manufacturers (2019). Advanced Technology Vehicle Sales Dashboard. Data compiled by the Alliance of Automobile Manufacturers using information provided by IHS Markit (2011–2018) and Hedges & Co. (2019). Data last updated 8/20/2019. Retrieved 02/18/2020 from https://autoalliance.org/energy-environment/advanced-technology-vehicle-sales-dashboard/

resistance was to protect their investments in the incumbent technology, the internal combustion engine. Over time and with each subsequent revision to the ZEV mandate, however, the industry became less resistant, tempered their lobbying efforts, and even engaged in more proactive lobbying. Examples of such lobbying efforts include those to reduce compliance costs through more flexible credit rules and other policy design modifications,[61] to increase regulatory certainty, and to assure compatibility between the ZEV and other vehicle regulations such as fuel economy standards.[62]

ZEV mandates have the potential to reduce significantly GHG emissions and US demand for oil.[63] Like RPS policies and fuel economy standards, however, ZEV mandates also have their challenges. There is no guarantee that consumers will demand enough electric or fuel cell vehicles by 2025 for manufacturers in ZEV states to comply with their mandates.[64] There is also extensive evidence that consumers are concerned about some of the functional and financial drawbacks to electric vehicles—such as concerns about the range of the vehicle and the upfront cost to purchase it—and are still hesitant to adopt this new technology.[65] These conditions suggest the need to couple the ZEV policy with other electric vehicle programs that build demand and reduce consumer concerns, such as providing tax credits, charging station infrastructure, and making use of creative ways to get people to experience electric vehicles (e.g., fleet purchases, test-drive opportunities, or leasing structures).[66] States are, to varying degrees, pursuing all of these activities.

Boston, Massachusetts, is an example city that has offered a variety of electric vehicle incentives and, as a result, experienced significant electric vehicle growth rates. Massachusetts offers a tax credit of up to $2,500 for battery electric or fuel cell vehicles and $1,500 for hybrid electric vehicles and helps coordinate an electric vehicle advocacy network. These incentives are on top of the federal tax credit for electric vehicles, which can provide up to $7,500 depending on the type of electric vehicle. Boston residents have benefited from these incentives, as well as the activities facilitated specifically by the city: personal charging station incentives, public charging stations throughout the city, requirements that new developments are electric-vehicle ready, electrician training for charger installers, and a streamlined charger inspection process, among others.

POLICY PRIORITIES: PREPARING FOR OR PROTECTING AGAINST THE ENERGY TRANSITION

State Pushback in the Form of New Policies and Retrenchment From Old Ones

State RPS and ZEV policies, and federal CAFE standards, are all examples of policies designed to advance new, low-carbon energy technologies. Through technology deployment, these policies also reduce GHG emissions in energy sectors. As discussed above, such policies are also contributing to the energy transition, as these clean technologies mature, their costs decline, and they capture a larger share of energy markets.

Yet, while some jurisdictions are planning for the energy transition and using public policy to promote it, others are pushing back and seeking ways to slow the transition. Technological and business changes facilitated by the energy transition have the potential to threaten incumbent industries and local or regional economies that rely heavily on fossil fuel–based resources or manufacturing. Shareholders, politicians, and other entities that rely on these industries and economies are more likely to respond with resistance. Such protection of economic or political interests has led to political battles in some places, clean energy policy retrenchment in others, and fossil fuel–propping policies in yet others.

Several states have recently repealed or frozen their RPS policies. In 2009, Kansas adopted a standard of 20 percent renewable energy by 2020 but, in 2015, changed the policy from mandatory to voluntary through Senate Bill 91.[67] In 2008, Ohio adopted one of the earliest and most aggressive portfolio standards of 25 percent by 2025.[68] Ohio then froze its standard for two years in 2014 and extended the terminal year from 2024 to 2026. Although House Bill (HB) 554 threatened to freeze the standard even longer, the governor vetoed it and the standard became binding again. Then, in 2019, HB 6 eliminated the RPS through a gradual phase out (i.e., 8.5 percent renewable energy by 2026 and then no standard thereafter) as well as the state's energy efficiency resource standard. The efficiency standard was originally set as a 22 percent reduction in consumer energy by 2027. By one estimate, this efficiency standard saved Ohio consumers an estimated $5.1 billion in energy savings between 2009 and 2017.[69] HB 6 also removed surcharges on customer electric bills to support renewable energy and energy efficiency, and introduced new surcharges to instead cover $1 billion of subsidies to bail out two coal and two nuclear plants.

These changes were accompanied by significant lobbying investments by electric utilities across the country,[70] as well as by political pressure within the specific states. Utilities in Ohio, for example, pushed the state legislators to repeal the RPS because the policy threatened the viability of long-term power purchase agreements for their coal generation. Over a year after HB 6 passage, the Federal Bureau of Investigation arrested Ohio House Speaker Larry Householder, as well as four others, for the exchange of $60 million in a bribery scheme involving FirstEnergy Solutions, an Ohio utility that owns power plants poised to benefit from the policy.[71] Across the country, companies representing coal, utility, and nuclear interests have also contributed over $3 million to political campaigns in each election year since 2010, and up to $4.3 million in the 2010 election.[72] These various efforts were accompanied by local resistance to wind turbine siting, and new wind turbine setback rules that challenged the financial viability of wind entirely.[73]

Another example of policy retrenchment is Nevada's changes to its net energy metering (NEM) policy in 2015. NEM policies allow a consumer to connect small-scale residential energy systems, such as solar PV panels, to the electric grid and to give and take electricity from the grid as needed. This policy allows a residential electricity customer to act as a prosumer. Under a traditional NEM design, consumers are compensated for the solar electricity that they give to the grid at the same rate that they pay for electricity, that is, the retail rate. State NEM policies typically set limits on the size of the residential energy system—to ensure that the system is sufficiently small—and the total capacity that a state can accept as NEM systems.

Nevada first adopted its NEM policy in 1997 and amended it several times between 2001 and 2014. A benefit–cost analysis provided by an independent consulting firm in 2014 found that the state NEM program provided millions of benefits in net present value.[74] Despite this, in December 2015, the state's largest utility, NV Energy, made the case through its own market study that consumers without residential solar were subsidizing those with solar. In response to this case, the state's utility commissioners sided with the utility and voted unanimously to abolish the state's NEM program and replace it with a program that compensates residential energy owners at the wholesale rate of electricity and adds fixed charges for these customers.[75] No residential consumers were grandfathered in from the old system. In other words, everyone who made the decision to invest in solar panels under the former policy were then forced to accept completely different rules, and rules that would make the payback period for their panels much longer, if a positive payback at all.

This case was particularly noteworthy due to the activists involved. Leonard DiCaprio and Mark Ruffalo appeared on the scene to protest the utility commission decision and advocate for the solar industry. Ruffalo declared publicly the utility commission as "anti-Robin Hood,"[76] while 2016 presidential candidate Hillary Clinton criticized the decision over social media[77] and candidate Bernie Sanders declared in a speech that the utility commission decision was "just about the dumbest thing I have ever heard."[78]

After significant public protests and solar job losses as solar companies fled the state, the NEM was reinstated in 2017 to allow systems up to 25 kW to recover 95 percent of the retail price of electricity for all excess generation, with declining recovery rates for all total state NEM generation beyond 80 MW.[79] Arguably, however, the damage was already done, as solar investors observed the political uncertainty of energy policies and those in the energy industry observed how different assumptions within a benefit–cost analysis can yield wildly different outcomes, outcomes by which policymakers may make key decisions such as whether to backtrack policies after initial adoption and implementation.

Grid Modernization Efforts and Energy Sector Planning

As the energy transition evolves, states are increasingly pursuing efforts to facilitate grid modernization, a term used to capture a number of investment, planning, analysis, and regulatory activities to prepare the electricity grid for technological advancements and resilience. Such efforts are a direct response to the evolving economic, technical, social, and policy challenges—and opportunities—precipitated by the energy transition, as discussed above. Grid modernization involves, among other activities:

1. Studying and planning for how to integrate renewable energy, including storage, into electricity grids;

2. Extending smart metering and advanced metering infrastructure to customers;

3. Setting rules and regulations for the use of data on customer electricity consumption, after gathering such data with smart infrastructure;

4. Including the social cost of carbon—a price on each unit of carbon that accounts for all unpriced impacts of carbon emissions—in integrated resource planning and other planning activities;

5. Using rate design and other incentives to encourage utilities to innovate and pursue energy efficiency, or to encourage customers to be more mindful of their energy use with prices that vary by time and actual production costs;

6. Performing value of solar studies to determine both the private and the social net benefits of small-scale solar PVs;

7. Identifying who will be harmed by the energy transition, understanding the source and implications of possible inequities, and devising approaches to compensate or otherwise help those disadvantaged;

8. Coordinating regional transmission operations and state electricity efforts; and

9. Planning and preparing for large shifts in the revenue base for local economies (for example, from coal power plant closures, where the coal plant formerly accounted for a large share of the local tax base).

In 2019, the majority of states, 46 in total, as well as Washington, DC, took some form of action related to grid modernization, with the most cases focusing on energy storage.[80] As such, grid modernization efforts are clearly appealing to state governments that exhibit a range of political, socioeconomic, and energy resource mix characteristics. A set of state energy policy leaders, however, including New York and California and a handful of other blue states, tend to pursue the more progressive grid modernization efforts, such as offering incentives for advanced infrastructure and addressing some of the deeper social justice issues associated with energy developments.

New York's Reforming the Energy Vision (REV), as an example of a grid modernization effort, was a direct response to Hurricane Sandy and the damage that it inflicted on the state's energy infrastructure.[81] The REV program was introduced by Governor Andrew Cuomo and other stakeholders in 2014 with an objective to make the electric grid more resilient, to guide the way that energy is produced and consumed in the state, and to facilitate the integration of renewable energy and energy efficiency into both electricity and transportation sectors.[82] The program includes over 40 initiatives that run the gamut from a Clean Energy Community competition for cities to a series of low-carbon building initiatives. Energy infrastructure modernization initiatives under the REV include the establishment of an Integrated Smart Operations Center, which uses analytical software to forecast equipment failures in power plants and transmission lines across the state[83] and Energy Manager, a product that is used to build energy management strategies for individual buildings.[84] The REV program has been lauded in the literature as having the potential to change the way that consumers interact with both their utilities and more generally with new services (e.g., smart energy technologies) and new prosumer opportunities.[85]

Another example is the state of Arizona and its policy efforts to distribute "smart meters" and roll out dynamic electricity pricing, which is when the price charged for electricity changes over time, depending on the cost of energy generation, as well as over space. One of the state's largest utilities, Salt River Project, for example, received approximately $57 million in federal American Recovery and Reinvestment Act of 2009 funds to defray the costs of smart meter—meters that have two-way information sharing from utility to customer and from customer to utility—installations to approximately one million customers by 2013.[86] Along with the smart meters, the utility allows customers to opt into different pricing plans, one of which offers different rates per season, another offers different prices at any given moment, depending on the cost of electricity at that moment (i.e., dynamic pricing).[87] Reports on this program find that it has saved significant labor and gasoline costs due to the reduced need for the utility to drive around and check consumer meters,[88] and saved consumers money on their electricity bills.[89]

CONCLUSION

As discussed in the introduction, modern US energy policy is driven by three primary objectives: reduce GHG emissions; advance the deployment of innovative, low-carbon, and efficient energy technologies; and help perpetuate or resist the evolving energy transition, and the effects of the transition on institutions, organizations, and people. While some jurisdictions are motivated by one of these objectives, others are motivated by all three at once, with little discernment among the three. For instance, with increasing prevalence of extreme weather patterns and forest fires that span multiple states if not entire countries, among other impacts, policymakers are confronting the modern challenge of global climate change. In the absence of policy tools that directly price or commodify GHG emissions—which may not work for certain levels of government or be a political nonstarter for others—policymakers' primary policy levers are to influence energy technology deployment and build resilient and smart energy infrastructure.

This chapter covers a range of energy policies that share these objectives. These policies represent variation in technological focus (e.g., electric vehicles, solar PVs, and smart meters), the level of government responsible for the adoption and implementation, degree of political influence or confrontation, and economic sectors. All policies discussed here share several common traits. First, each policy presents its own set of benefits and drawbacks. The new RPS variant of 100 percent clean energy standards, for example, sends a clear and ambitious signal to electricity market participants, but with the potential downside of overpromising before the technological, economic, and even social factors are in place to support 100 percent clean energy on the grid (perhaps similar to the ZEV program, as Figure 8.3 suggests may be the case). As another example, the ZEV program mandates the deployment of electric vehicles, which has the potential to reduce total GHG emissions, but with the potential downside of manufacturers focusing exclusively on building out electrification plans and not advancing efficiency in the rest of their vehicle fleets.

Second, and closely related, each policy presents its own set of trade-offs in design and implementation. Policymakers must make difficult trade-offs between specific technological advancement versus picking promising favorites, maximizing efficiency versus equity, keeping prices low versus aggressive innovation, and ensuring policy flexibility versus adaptability over time. Although, of course, decisions are not always as cut and dried as this list suggests. States also have very different strategies, priorities, and technical agencies to design, help implement, and oversee all of these policies.

Third, each policy is accompanied by vested interests and, thus, politics. Such politics tend to increase most significantly when new technologies or business models mature enough that they begin to threaten vested interests. Whereas, in early stages of technological advancement, incumbent industries tend to underestimate the potential for long-term market maturity and thus do not apply as much political opposition.[90] As an example, electric utilities did not oppose NEM policies when they were first introduced. It was only after the costs of residential solar PVs declined significantly, and a fraction of each utility's customers turned prosumer, that utilities began to oppose NEM policies. Vested interests also have perpetuated low-carbon energy technologies as well, such as through support of and pressure for policies like the production tax credit and the investment tax credit for wind. These policies, once in place, are hard to remove and, through political force, have been extended incrementally over time.[91] One may assert that a similar trend exists for RPSs, fuel economy standards, and ZEV standards, all of which policymakers have incrementally ratcheted up over time—with substantial support from political interests—for most jurisdictions, although repealed through acts of retrenchment for a few.

Finally, these policies, and the energy transition that is evolving concurrent to them, all introduce new institutional challenges, at least some of which need to be confronted with revised rules and regulations, market operations, human behavior, and institutional norms. To list just a few examples, as the deployment of smart meters increases, consumers may need to modify their energy consumption behavior, and utilities and others will need to consider how to analyze and protect consumer electricity data. As battery storage technologies mature and decrease in price, electricity markets will need to adapt to include options for market participation. As the energy transition continues to evolve, it will be necessary to also evaluate these new challenges continually, and devise new or amend existing policy solutions to address them.

I began this chapter by explaining that energy policies require decision-making under complex and evolving conditions and are shaped fundamentally by those who make these decisions and their values and interests. Policy priorities, the types of policy instruments deemed favorable, and even the approaches and methodologies used to evaluate a policy set often change with an administration and with shifting markets. As such, energy policies do not always progress in a linear fashion and can zigzag or proceed forward at times and retreat at others. The modern energy era is marked by an evolving energy transition away from dependence on carbon-intensive fossil fuels and toward low-carbon and efficient alternatives, as well as by intense partisanship over energy sources and policy. Policymakers must choose whether to continue to reinforce these trends through policy prescriptions or to stall the transition and prop up the declining industries. The Trump administration, through a series of executive orders and other regulatory processes, has focused on the latter. As we proceed through the 2020s, we will observe the implications of such policy setbacks, as well as policy responses made by other levels of government and future administrations.

SUGGESTED WEBSITES

Database of State Incentives for Renewables and Efficiency (www.dsireusa.org) The North Carolina Clean Energy Technology Center maintains an up-to-date database of all state renewable energy and energy efficiency policies, including information on how the policies are designed, and when and how they have been amended over time.

Resources for the Future (https://www.rff.org/) Resources for the Future is a nonprofit think-tank based in Washington, DC, that produces rigorous, objective research on environmental, energy, and economic policy.

U.S. Energy Information Administration (EIA) (https://www.eia.gov/ and https://www.eia.gov/tools/) The EIA collects a range of different energy statistics such as utility-level data on generating units, state-level production and consumption, and oil prices. The EIA also produces regular reports such as the Annual Energy Outlook, which contains energy modeling forecasts.

NOTES

1. Sanya Carley, "Energy Programs of the American Recovery and Reinvestment Act of 2009," *Review of Policy Research* 33, no. 2 (2016): 201–23.
2. "Sources of Greenhouse Gas Emissions," U.S. Environmental Protection Agency, 2019, https://www.epa.gov/ghgemissions/sources-greenhouse-gas-emissions.
3. V. Masson-Delmotte, P. Zhai, H. O. Pörtner, D. Roberts, J. Skea, P. R. Shukla, A. Pirani, W. Moufouma-Okia, C. Péan, R. Pidcock, S. Connors, J. B. R. Matthews, Y. Chen, X. Zhou, M. I. Gomis, E. Lonnoy, T. Maycock, M. Tignor, T. Waterfield, eds., "An IPCC Special Report on the Impacts of Global Warming of 1.5°C Above Pre-industrial Levels and Related Global Greenhouse Gas Emission Pathways, in the Context of Strengthening the Global Response to the Threat of Climate Change, Sustainable Development, and Efforts to Eradicate Poverty," *IPCC* 2018: Global Warming of 1.5°C, in press.
4. Anthony Leiserowitz, Edward Maibach, Connie Roser-Renouf, Seth Rosenthal, Matthew Cutler, and John Kotcher, "Climate Change in the American Mind: March 2018," April 2018, https://climatecommunication.yale.edu/publications/climate-change-american-mind-march-2018/.
5. J. De Pinto, F. Backus, and A. Salvanto, "Most Americans Say Climate Change Should Be Addressed Now – CBS News Poll," *CBS News*, September 15, 2019.
6. "Lazard's Levelized Cost of Energy Analysis-Version 12.0," *Lazard*, 2018, https://www.lazard.com/media/450784/lazards-levelized-cost-of-energy-version-120-vfinal.pdf.
7. Ralph L. Langenheim, Jr., "Coal (Mineral Resource)," *Salem Press Encyclopedia of Science*, 2019, https://search.ebscohost.com/login.aspx?direct=true&db=ers&AN=88806479&site=eds-live&scope=site.
8. "California: State Profile and Energy Estimates," U.S. Energy Information Administration, 2019, https://www.eia.gov/state/?sid=CA#tabs-3.
9. "State Energy Data System (SEDS): 1960–2017," United States Energy Information Administration, 2019, https://www.eia.gov/state/seds/seds-data-complete.php?sid=US#StatisticsIndicators.

10. Steve Hanley, "Germany Breaks a Solar Record – Gets 85% of Electricity from Renewables," *CleanTechnica*, May 8, 2017, https://cleantechnica.com/2017/05/08/germany-breaks-solar-record-gets-85-electricity-renewables/.

11. "Cities Are Ready for 100% Clean Energy: 10 Case Studies," *Sierra Club*, https://www.sierraclub.org/sites/www.sierraclub.org/files/blog/RF100-Case-Studies-Cities-Report.pdf.

12. Robert C. Allen, "Backward Into the Future: The Shift to Coal and Implications for the Next Energy Transition," *Energy Policy* 50 (November 2012): 17–23.

13. S. Carley and M. Graff, "A Just U.S. Energy Transition," in *Handbook of U.S. Environmental Policy*, ed. D. Konisky (Cheltenham: Edward Elgar Publishing, 2020) (forthcoming).

14. Roger Fouquet, "Historical Energy Transitions: Speed, Prices and System Transformation," *Energy Research & Social Science* 22 (December 2016): 7–12.

15. Gregory Unruh, "Understanding Carbon Lock-In," *Energy Policy* 28, no 12 (October 2000): 817–30.

16. Arnulf Grubler, "Energy Transitions Research: Insights and Cautionary Tales," *Energy Policy* 50 (2012): 8–16, https://doi.org/10.1016/j.enpol.2012.02.070.

17. See, e.g., Roger Fouquet, "Historical Energy Transitions: Speed, Prices and System Transformation," *Energy Research & Social Science* 22 (2016): 7–12; or Arnulf Grubler, "Energy Transitions Research: Insights and Cautionary Tales," *Energy Policy* 50 (2012): 8–16.

18. C. Lampe-Onnerud and J. Kortenhorst, "The Speed of the Energy Transition: Gradual or Rapid Change?" *World Economic Forum White Paper*, September 2019.

19. Hanna Breetz, Matto Mildenberger, and Leah Stokes, "The Political Logics of Clean Energy Transitions," *Business and Politics* 20 (2018): 492–522.

20. Robert C. Allen, "Backward Into the Future: The Shift to Coal and Implications for the Next Energy Transition," *Energy Policy* 50 (November 2012): 17–23.

21. Richard Newell, Daniel Raimi, and Gloria Aldana, "Global Energy Outlook 2019: The Next Generation of Energy," *Resources for the Future Report 19-06*, July 2019.

22. "2020 Vision: Why You Should See the Fossil Fuel Peak Coming," *Carbon Tracker*, 2018, https://www.carbontracker.org/reports/2020-vision-why-you-should-see-the-fossil-fuel-peak-coming/.

23. Matto Mildenberger, *Carbon Captured: How Business and Labor Control Climate* (MIT Press, 2020).

24. Robert J. Brulle, "The Climate Lobby: A Sectoral Analysis of Lobbying Spending on Climate Change in the USA, 2000 to 2016," *Climatic Change* 149 (2018): 289, https://doi.org/10.1007/s10584-018-2241-z.

25. "Lazard's Levelized Cost of Energy Analysis-Version 13.0," *Lazard*, 2019, https://www.lazard.com/media/451086/lazards-levelized-cost-of-energy-version-130-vf.pdf.

26. Cary Funk and Meg Hefferon, "U.S. Public Views on Climate and Energy," *Pew Research Center*, 2019. [PDF file], https://www.pewresearch.org/science/2019/11/25/u-s-public-views-on-climate-and-energy/.

27. "Little Change in Opinions about Global Warming," Pew Research Center U.S. Politics & Policy, *Pew Research Center*, 2010, https://www.people-press.org/2010/10/27/little-change-in-opinions-about-global-warming/.

28. "Americans' Opinion on Renewables and Other Energy Sources," Pew Research Center Science & Society, *Pew Research Center*, 2016, https://www.pewresearch.org/science/2016/10/04/public-opinion-on-renewables-and-other-energy-sources/.

29. Susan F. Tierney, "The U.S. Coal Industry: Challenging Transitions in the 21st Century," 2016.

30. "State Incentives for Renewables & Efficiency," North Carolina Clean Energy Technology Center, 2019, accessed October 18, 2019.

31. Galen Barbose, "U.S. Renewables Portfolio Standards: Preliminary 2018 Annual Status Report," *Lawrence Berkeley National Laboratory*, April 26, 2018.

32. "State Energy Data System (SEDS): 1960–2017," United States Energy Information Administration, 2019, https://www.eia.gov/state/seds/seds-data-complete.php?sid=US#StatisticsIndicators.

33. "Renewables 2019 Global Status Report," REN21, ISBN 978-3-9818911-7-1, 2019, https://www.ren21.net/wp-content/uploads/2019/05/gsr_2019_full_report_en.pdf.

34. Sanya Carley, Lincoln Davies, David Spence, and Nikolaos Zirogiannis, "Empirical Evaluation of the Stringency and Design of Renewable Portfolio Standards," *Nature Energy* 3 (July 2018): 754–63.

35. Galen Barbose, "U.S. Renewables Portfolio Standards: 2018 Annual Status Report," *Lawrence Berkeley National Laboratory*, November 2018.

36. Ibid.

37. Ibid.

38. Sanya Carley, Lincoln L. Davies, David B. Spence, and Nikolaos Zirogiannis, "Empirical Evaluation of the Stringency and Design of Renewable Portfolio Standards," *Nature Energy* 3 (July 2018): 754–63.

39. Ryan Wiser, Trieu Mai, Dev Millstein, Galen Barbose, Lori Bird, Jenny Heeter, David Keyser, Venkat Krishnan, and Jordan Macknick, "Assessing the Costs and Benefits of US Renewable Portfolio Standards," *IOP Publishing*, September 2017, https://iopscience.iop.org/article/10.1088/1748-9326/aa87bd/.

40. Eric Martinot, "Grid Integration of Renewable Energy: Flexibility, Innovation, and Experience," *Annual Review of Environment and Resources* 41 (November 2016): 223–51.

41. Paul Denholm, Matthew O'Connell, George Brinkman, and Jennie Jorgenson, "Overgeneration from Solar Energy in California: A Field Guide to the Duck Chart," *National Renewable Energy Laboratory*, NREL/TP-6A20-65023, November 2015, https://www.nrel.gov/docs/fy16osti/65023.pdf.

42. Alex Hollingsworth and Ivan Rudik, "External Impacts of Local Energy Policy: The Case of Renewable Portfolio Standards," *Journal of the Association of Environmental and Resource Economists* 6, no. 1 (January 2019), https://www.journals.uchicago.edu/doi/abs/10.1086/700419.

43. Sara K. Adair and F. T. Litz, "Understanding the Interaction Between Regional Electricity Markets and State Policies," *Nicholas Institute Primer*, November 2017.

44. S.B. 5116, 66th Legislature for the State of Washington, 2019, https://app.leg.wa.gov/billsummary?BillNumber=5116&Year=2019&initiative=#documentSection

45. Ibid.

46. Antonio M. Bento, Kenneth Gillingham, Mark R. Jacobsen, Christopher R. Knittel, Benjamin Leard, Joshua Linn, Virginia McConnell, David Rapson, James M. Sallee, Arthur A. van Benthem, and Kare S. Whitefoot. "Flawed Analyses of U.S. Auto Fuel Economy Standards," *Science* 352 (2018): 1119–21.

47. Sanya Carley, Nikolaos Zirogiannis, Denvil Duncan, Saba Siddiki, and John D. Graham, "Overcoming the Shortcomings of U.S. Plug-in Electric Vehicle Policies," *Renewable and Sustainable Energy Reviews* 113 (2019): 1–10, https://doi.org/10.1016/j.rser.2019.109291

48. Ibid.

49. Cale Jaffe, "Melting the Polarization Around Climate Change Politics," *Georgetown Environmental Law Review* 30, no. 3 (2018): 455–98.

50. Joshua Linn and Virginia McConnell, "The Role of State Policies under Federal Light-Duty Vehicle Greenhouse Gas Emissions Standards," *Resources for the Future Report*, June 2017.

51. Ibid.

52. Thomas Klier and Joshua Linn, "The Effect of Vehicle Fuel Economy Standards on Technology Adoption," *Journal of Public Economics* 133 (January 2016): 41–63.

53. See note 50.

54. Kenneth Gillingham, David Rapson, and Gernot Wagner, "The Rebound Effect and Energy Efficiency Policy," *Review of Environmental Economics and Policy* 10, no. 1 (2016): 68–88.

55. Mark R. Jacobsen and Arthur A. van Benthem, "Vehicle Scrappage and Gasoline Policy," *American Economic Review* 105, no. 3 (March 2015): 1312–38.

56. Jay Apt, Scott B. Peterson, and J. F. Whitacre, "Battery Vehicles Reduce CO_2 Emissions," *Science* 333, no. 6044 (August 2011): 823.

57. Sanya Carley, Nikolaos Zirogiannis, Denvil Duncan, Saba Siddiki, and John D. Graham, "Overcoming the Shortcomings of U.S. Plug-in Electric Vehicle Policies," *Renewable and Sustainable Energy Reviews* 113 (2019): 1–10, https://doi.org/10.1016/j.rser.2019.109291.

58. "Summary Report for the Technical Analysis of the Light Duty Standards," *California Air Resources Board [CARB] California's Advanced Clean Cars Midterm Review*, 2017.

59. "Advanced Technology Vehicle Sales Dashboard," *Auto Alliance*, Data compiled by the Alliance of Automobile Manufacturers using information provided by HIS Markit, August 20, 2019, https://autoalliance.org/energy-environment/advanced-technology-vehicle-sale=dashboard/, accessed December 13, 2019.

60. J. H. Wesseling, J. C. M. Farla, D. Sperling, M. P. Hekkert, "Car Manufacturers' Changing Political Strategies on the ZEV Mandate," *Transportation and Environment* 33 (2014): 196–209.

61. Ibid.

62. Saba Siddiki, Sanya Carley, Nikolaos Zirogiannis, Denvil Duncan, and John D. Graham, "Does Dynamic Federalism Yield Compatible Policies? A Study of Federal and State Vehicle Standards," *Policy Design and Practice* 1, no. 3 (2018): 215–32.

63. David L. Greene, Sangsoo Park, and Changzheng Liu, "Public Policy and the Transition to Electric Drive Vehicles in the U.S.: The Role of the Zero Emission Vehicles Mandates," *Energy Strategy Reviews* 5 (2014): 66–77.

64. Sanya Carley, Nikolaos Zirogiannis, Denvil Duncan, Saba Siddiki, and John D. Graham, "Overcoming the Shortcomings of U.S. Plug-in Electric Vehicle Policies," *Renewable and Sustainable Energy Reviews* 113 (2019): 1–10.

65. Ibid.

66. David L. Greene, Sangsoo Park, and Changzheng Liu, "Public Policy and the Transition to Electric Drive Vehicles in the U.S.: The Role of the Zero Emission Vehicles Mandates," *Energy Strategy Reviews* 5 (2014): 66–77; and Julio C. Zambrano-Gutierrez, Sean Nicholson-Crotty, Sanya Carley, and Saba Siddiki, "The Role of Public Policy in Technology Diffusion: The case of Plug-in Electric Vehicles," *Environmental Science & Technology* 52, no. 19 (September 2018): 10914–22.

67. "Database for State Incentives for Renewables and Efficiency," *North Carolina Clean Energy Technology Center, Renewable and Clean Energy Standards* (2020), accessed July 1, 2020.

68. Ibid.

69. "Energy Efficiency: A Good Investment for Ohio," *Midwest Energy Efficiency Alliance*, n.d., http://www.mwalliance.org/sites/default/files/media/Ohio-State-Fact-Sheet.pdf?current=/taxonomy/term/11.

70. "Industry Profile: Electric Utilities," *Center for Responsive Politics*, 2019, https://www.opensecrets.org/lobby/indusclient.php?id=E08&year=2018&filter=P.

71. National Public Radio, "Ohio House Speaker Arrested in Connection with $60 Million Bribery Scheme," 2020, https://www.npr.org/2020/07/21/893493224/ohio-house-speaker-arrested-in-connection-to-60-million-bribery-scheme.

72. Kathiann M. Kowalski, "Campaign Contributions Pay Off for Ohio Utilities and Coal Interests," *Energy News Network*, March 5, 2020.

73. Hana Breetz, Matto Mildenberger, and Leah Stokes, "The Political Logics of Clean Energy Transition," *Business and Politics* 20, no. 4 (2018): 492–522, https://doi.org/10.1017/bap.2018.14.

74. "Nevada Net Energy Metering Impacts Evaluation," *Energy and Environmental Economics*, July 2014, http://puc.nv.gov/uploadedFiles/pucnvgov/Content/About/Media_Outreach/Announcements/Announcements/E3%20PUCN%20NEM%20Report%202014.pdf?pdf=Net-Metering-Study.

75. Sanya Carley and Lincoln L. Davies, "Nevada's Net Energy Metering Experience: The Making of a Policy Eclipse?" *Brookings Mountain West Report*, 2016.

76. Noah Buhayar, "Who Owns the Sun?" *Bloomberg Business Week*, January 28, 2016.

77. Katie Fehrenbacher, "Why Nevada Brought Back Favorable Rates for Existing Solar Customers," September 2016, http://fortune.com/2016/09/16/nevada-solar-grand-fathering/, accessed October 5, 2016.

78. Scott Lucas, "In NLV Speech, Sanders Calls PUC Solar Decision 'Just about the Dumbest Thing I Ever Heard,'" *Las Vegas Sun*, December 28, 2015, http://lasvegassun.com/news/2015/dec/28/in-nlv-speech-bernie-sanders-calls-puc-solar-decis/, accessed October 5, 2016.

79. "Database for State Incentives for Renewables and Efficiency," *North Carolina Clean Energy Technology Center, Renewable and Clean Energy Standards*, 2020, accessed July 1, 2020.

80. "North Carolina Clean Energy Technology Center," *The 50 States of Grid Modernization: 2019 Review and Q4 2019 Quarterly Report*, February 2020.

81. "About REV," *Reforming the Energy Vision (REV)*, https://www.ny.gov/sites/ny.gov/files/atoms/files/WhitePaperREVMarch2016.pdf.

82. "REV Initiatives," *Reforming the Energy Vision (REV)*, https://www.ny.gov/sites/ny.gov/files/atoms/files/WhitePaperREVMarch2016.pdf.

83. "ISOC: Integrated Smart Operations Center," *NY Power Authority*, www.nypa.gov/services/digital-energy-services/isoc.

84. "NY Energy Manager," *NY Power Authority*, www.nypa.gov/services/digital-energy-services/ny-energy-manager.

85. Joseph Nyangon and John Byrne, "Diversifying Electric Customer Choice: REVing Up the New York Energy Vision for Polycentric Innovation," *Energy Systems and Environment* (2018): 4–19, http://dx.doi.org/10.5772/intechopen.76023.

86. Elizabeth Doris and Kim Peterson, "Government Program Briefing: Smart Metering," *National Renewable Energy Laboratory*, NREL/TP-7A30-52788, September 2011, https://www.nrel.gov/docs/fy11osti/52788.pdf.

87. Matthew S. Stern and Kevin B. Jones, "Salt River Project: Delivering Leadership on Smarter Technology & Rates," *Institute for Energy and the Environment*, June 2012, https://www.vermontlaw.edu/sites/default/files/Assets/iee/SRP-Report-Final-120618.pdf.

88. "The Smart Grid Done Right: Phoenix's Salt River Project," *Elster*, 2011, https://www.elstersolutions.com/assets/downloads/WP42-6003A_SaltRiverProject_DoneRight.pdf.

89. Judith Schwartz, "Case Study: Salt River Project—The Persistence of Customer Choice," Presented at the *National Forum of the National Action Plan on Demand Response*, June 2012, https://www.ferc.gov/industries/electric/indus-act/demand-response/dr-potential/napdr-srp.pdf.

90. Leah Stokes and Hanna L. Breetz, "Politics in the U.S. Energy Transition: Case Studies of Solar, Wind, Biofuels and Electric Vehicles Policy," *Energy Policy* 113 (February 2018): 76–86.

91. Ibid.

NATURAL RESOURCE POLICIES IN AN ERA OF POLARIZED POLITICS

William R. Lowry and John Freemuth

Sage-grouse, a bird a bit larger than a crow but with a larger, roundish body, smallish head, and long tail, have been affected since the beginning of settlement of the West and, by 1995, were in a serious decline. Conservation efforts began that year and the US Fish and Wildlife Service (USFWS), the Bureau of Land Management (BLM), and the US Forest Service (USFS) have worked together to develop a conservation strategy over the range of the species. However, resolution of the issue has been subject to increased polarization, politics, and policy reversals (Figure 9.1).

In 2002, the George W. Bush administration's USFWS received the first of three petitions to list the greater sage-grouse as threatened over its 11-state range. In 2004, some of its actions leaked to the press, including a key document containing inappropriate edits made by a higher-level political appointee in the Department of the Interior (DOI). The Inspector General for DOI found that the appointee had inappropriately changed the assessment document. Although she resigned, the Bush administration issued a "not warranted" decision on listing. Federal district Judge B. Lynn Winmill ruled that the determination not to list the sage-grouse was "arbitrary and capricious." The agency, in a settlement agreement, agreed to issue a new finding by May 2009. In 2010, USFWS ruled that the sage-grouse was at risk of extinction, but declined to list it, asserting that the listing was "warranted but precluded" because the agency had dedicated its resources to actions with a higher priority. However, the agency and others pledged to take steps to restore sagebrush habitat and in a court settlement, agreed to issue a listing decision by 2015.

In 2011, the Obama administration launched a concerted effort with federal and state agencies, nongovernmental organizations, and private landowners to develop plans to avoid an Endangered Species Act (ESA) listing. Policymakers in California, Colorado, Idaho, Montana, Nevada, and Wyoming all developed plans for conserving sage-grouse habitat, while the USFS and BLM revised 98 land use plans in 10 states. The US Department of Agriculture provided funding for voluntary conservation actions on private lands.

In 2015, Interior Secretary Sally Jewell, joined by a bipartisan group of Western governors, announced that these actions had reduced threats to sage-grouse habitat so effectively that a listing was no longer necessary. Despite the good feelings, some conflicts remained unresolved. The plan belatedly created zones called Sagebrush Focal Areas—zones deemed essential for the sage-grouse to survive—and proposed to bar mineral development on 10 million acres within those areas. Some Western governors viewed these as surprises and developers objected.

FIGURE 9.1 ■ Sage-Grouse Habitat

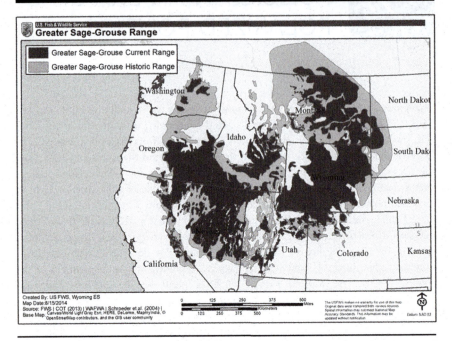

Sources: FWS | COT (2013) | WAFWA | Schroeder et al. (2004).

In March 2017, the United States District Court for Nevada ruled that the Obama administration violated the National Environmental Policy Act (NEPA) by not preparing a supplemental Environmental Impact Statement (EIS) about the designation of the Focal Areas, that EIS would have allowed for public comment on the focal area additions.

Subsequently, the Trump Interior Department decided not to create Focal Areas and allowed development in these zones to expand. Agency records show that as officials reevaluated the sage-grouse plan in 2017, they worked closely with oil, gas, and mining industry representatives, but not with environmental advocates. The administration then revised the plans developed by the previous administration. However, in October 2019, US District Judge Winmill granted a preliminary injunction that stopped the implementation of the Trump administration's sage-grouse plan revisions, asserting that environmental groups that sued to stop the plans would likely prevail in court. This court action puts the Obama-era plans back in place. Winmill's decision will be appealed, but it stands for now.

INTRODUCTION

Our thesis regarding natural resource policies is straightforward. As the sage-grouse case illustrates, politics matters. More specifically, natural resource policies are often based on short-term electoral or partisan decisions more than science, economics, the resource, or any long-term goals based on information or facts. This has always been true for natural resource policies to some extent, but it is even more apparent in recent years. The divergence between the Democratic and Republican parties in this policy area has grown to a chasm, and differences on policies, such as those on the sage-grouse, are often rigid and intense. We will cite some counterexamples, but in most cases, these days the two parties pursue very different policies regarding natural resources. And while the COVID-19 pandemic affected natural resources, partisan differences over governing policies on these issues neither changed nor diminished. Thus, policies are usually short term and inconsistent, and their impact problematic for natural resources that depend on long-term, sustainable management.

The chapter is organized as follows. We first provide some broader context with discussion of key statutes and then the factors that make political resolution of natural resource conflicts contentious. The result is often gridlock in Congress and unilateral actions by presidents. Thus, the second section examines the impact of politics on the public agencies that carry out agendas of presidents when implementing natural resource policies. The third section discusses some underlying trends in natural resource policies that continue across presidential administrations and shape policies currently and in the future.

EVOLUTION OF NATURAL RESOURCE POLICIES

North America was blessed with an abundance of natural resources. However, as those resources became scarcer over time, controversies and conflicts increased. This led to the development of key institutions and policies to govern them.

Historical Development

For centuries, humans used natural resources with little concern of exhausting them, but that changed in the nineteenth century when Americans intensified development of natural areas. People would move into an area, use up available resources, and move on. In the late nineteenth century, user groups as well as conservationists requested the creation of federal agencies to provide some oversight and balance between short-term usage and long-term needs.[1] Achieving this balance in natural resource policies has always been a challenge and policies almost inevitably favor one priority over another. Through the late 1960s, the two major parties did not automatically line up on different sides of an issue and important policies such as the Wilderness Act of 1964 passed by strong, bipartisan majorities.

By the early 1970s, most of the major formal institutions governing natural resource policies were in place. Four federal agencies administer most American federal lands. The exact amounts are always changing, but in 2017, the federal government owned nearly 650 million acres (about 28 percent of the total land mass). While these agencies have their own priorities, as discussed later, certain key statutes affect them all. Rulemaking on federal lands and species is supposed to take place according to the notice-and-comment rulemaking procedures established by the Administrative Procedure Act of 1946. If actions potentially affect the environment, they are subject to review under the NEPA of 1970. Wilderness areas established according to the process specified in the Wilderness Act of 1964 are to be set aside and maintained in pristine, roadless condition. The ESA of 1973 governs the protection of threatened and endangered species. Provisions in the Clean Air and Clean Water Acts affect all federal lands.

While collaboration between the two major political parties shaped the formation of many of these governing institutions, events in the middle and latter half of the 1970s reduced the chances of future consensus. As discussed in Chapter 8 on energy policy, oil shortages and higher gasoline prices shook American assumptions of continuing abundance of oil at cheap prices and were replaced by debates about conservation, production, and development of domestic supplies.[2] Much of the readily accessible crude oil and natural gas (this is before fracking opened other possible pools) in the country existed on public lands. The Republican Party began to demand more access and usage of these lands for fossil fuels, whereas the Democrats increasingly preferred the development of alternative sources of energy rather than accelerated tapping of public lands. This fissure in the relations between parties in natural resource grew.

Increasingly Contentious

Natural resource policies have become increasingly contentious since the 1970s. The divergence between the two parties in American politics in recent decades has been well documented and is discussed in Chapters 4 and 5. In short, the two parties have established different priorities, aligned with different interest groups, and mobilized different bases to get elected. Polarization is seen in nearly every issue area, including environmental, energy, and natural resource policies. Several factors make natural resource issues even more subject to partisan tensions.

One factor is that natural resource issues are not decided just at the federal level. Natural resources are also subject to policy impacts from state and local governments. Compared to the 650 million acres owned by the federal government, state governments own about 200 million. Every state has its own state park system, varying widely in terms of degree of development and protection.[3] The vast majority of states also have their own version of ESAs. Finally, much of the mineral and oil and gas development occurring today is potentially subject to state and local oversight as well as federal subsurface management by BLM. Politics at the state level add another level of potential partisan conflict.

A second factor is that many natural resource issues have a regional hue. Hostility towards the federal presence in the form of public lands is occasionally quite high in the western states, as evidenced by the recent takeover of the Malheur Wildlife Refuge in Oregon that we discuss in more detail later. At the federal level, this regional hostility takes on a partisan dimension. For instance, lawmakers introduced over 100 bills in Congress between 2011 and 2016 that called for the privatization or diminution of federal lands, nearly all involving Republican legislators from western states. The dominance of the Republican Party in many western states, in contrast to Democratic dominance on the coasts, only intensifies party differences on natural resource issues.

A third factor making natural resource policies contentious is that people have strong feelings about these issues. Most Americans value their public lands quite highly, expressing their support through public opinion polls regarding national parks and ballot measures for funding at state and local levels.[4] Broadly stated support does not always translate, however, to consensus on what to do with those lands and species. People have widely different views as to what purposes those lands should serve. Thus, a multitude of interest groups advocate for different policies. Organizations on both sides of natural resource issues have been intense and willing to use both legal and extralegal tactics to achieve their goals. Since the 1970s, the Republican Party aligned with development organizations while the Democratic Party aligned with traditional environmental groups.

Events during these decades also pushed the parties and their supportive groups even further apart. Controversies over endangered species motivated Republican efforts to alter the ESA that continue to this day. The Exxon Valdez oil spill off the coast of Alaska in 1989 strengthened the resolve of environmental groups and Democrats in Congress to stop some oil exploration in Alaska. Perhaps the ultimate issue driving the parties apart, at least since 1988, has been climate change. The impacts on natural resources, from glaciers to species, inspire debates that typically form along party lines. In all three of those cases, partisan divisions have prevented legislative resolution. Indeed, in general, Congress has rarely been able to pass significant natural resource policies in recent decades.

Two pieces of legislation ten years apart seem to provide exceptions to that rule. First, in 2009, Congress passed the Omnibus Public Land Management Act that provided protection for more than 2 million acres of wilderness and over 1,000 miles of wild and scenic rivers. The bill provided considerable benefits for many states and districts, including wilderness areas in nine different states and ten national heritage areas. Supporters of the bill used those distributional effects to overcome ideological gridlock and get overwhelming majorities for passage. Similarly, in 2019, Congress passed the John Dingell Conservation, Management, and Recreation Act. The legislation combined over 100 other pieces of legislation into one bill, the most important provision being to reinvigorate the Land and Water Conservation Fund (LWCF). Since 1964, the LWCF has used royalties from offshore oil and gas drilling to fund numerous projects such as matching grants for state and local parks. The LWCF has been inconsistently utilized but the 2019 legislation attempted a permanent reauthorization. The legislation also provided protection for millions of acres of land and hundreds of miles of rivers.

Closer examination of both these pieces of legislation, however, reveals numerous compromises such as allowing hunting in natural areas. These provisions facilitated support from constituencies, such as hunters, as part of a broader coalition. Broad coalitions do not, however, translate to support for controversial measures and neither act took meaningful action on issues such as regulation of greenhouse gas emissions on public lands or the protection of endangered species.[5] Instead, as one analysis summarized the 2019 act, "the hodgepodge bill offered something for nearly everyone, stretching across the country."[6] Congress finalized the legislation as the Great American Outdoors Act and President Trump signed it into law in August 2020.

Congress is much more capable of building support across party lines if it focuses on distributional policies such as spending money for individual projects even while it remains gridlocked when attempting to provide significant legislation on controversial issues.

NATURAL RESOURCE AGENCIES AND POLICIES

While Congress and the president set broad goals and priorities, putting policies into effect on the ground (literally in the case of natural resources) is the responsibility of public agencies. With Congress usually stuck in partisan gridlock, presidents have shown an increasing tendency to act unilaterally, typically through public agencies, rather than pursuing broad-based legislation. We describe the actions of key natural resource agencies below.

Before discussing specific agencies, several differences between the two most recent administrations affect all natural resource agencies. First, the administrations have pursued different priorities. The Obama administration emphasized science in appointments and actions, including politically controversial natural resource topics such as the permitting of coal mines and signing on to the Paris Climate Accord. President Trump and his administration, to the contrary, substantially deemphasized science by appointing lobbyists rather than scientists to head agencies, suppressing scientific studies that undermined political goals, politicizing the grant-making process and restricting scientific communication on numerous issues, particularly climate change.[7] Second, as Chapter 4 makes clear and Appendix 1 shows, budgets for recent administrations differ between Democratic and Republican presidents. The Trump administration cut budgets for all natural resource programs other than those involving recreation. Third, perhaps in no area are the administrations more different than in the use of regulations. The Democratic Party under President Obama supported the use of regulations on the protection of public lands, oil and gas drilling in federal areas, and rules for offshore oil drilling, among others that are discussed elsewhere in this volume. The offshore oil reforms followed the Deepwater Horizon oil spill of 2010, a disaster magnified by the failures of several federal agencies. The Trump administration reversed Obama-era regulations in these areas as well as more long-term rules such as protections for endangered species. The administration's deregulatory efforts have prompted numerous legal challenges as discussed in Chapter 6.

Table 9.1 lists the five agencies most responsible for natural resource policies. Other agencies that affect these policies are discussed fully elsewhere in this volume, notably the EPA in Chapter 7 and the Department of Energy in Chapter 8.

Department of the Interior

The DOI plays a major role in shaping natural resource policies. Created by a reorganization of several federal agencies in 1849, the DOI is a collection of different agencies that have some relation to American lands. Indeed, even people at the DOI refer to the agency as the "Department of Everything Else" because it has so many diverse responsibilities. Specific to natural resources, the DOI oversees about 75 percent

TABLE 9.1 ■ Natural Resource Agencies		
Agency	Mission	Controversies
Department of the Interior	Overall management of public lands and resources	Conflicting mandates and missions; inconsistent priorities
Bureau of Land Management	Manage uses of rangelands sustainably	Politics determines grazing and mining fees; extensive oil + gas development
National Park Service	Protect the nation's parklands for the future but allow for use	Politics affects balance between preservation and use; creation of monuments
Fish and Wildlife Service	Manage refuges and protect endangered species	Low funding; political pressures on listing species and allowing uses on refuges
Forest Service	Multiple-use doctrine to manage national forests for perpetuity	Politics determines priorities of usage; species protection and roadless areas

of federal public land as well as nearly 2 billion acres offshore. Three of the agencies discussed below are located in DOI. Over its history, the DOI has seen controversy as a result of the conflicting mandates of the agencies within as well as inconsistent priorities from different administrations.

In recent years, the influence of partisan politics has been notably evident in terms of White House appointments. The secretaries of the DOI under President Obama were a moderate Senator from Colorado, Ken Salazar, whose appointment received a mixed reception among conservation groups, and Sally Jewell, a former oil engineer. The first head of DOI under President Trump, Ryan Zinke, was a former representative from Montana who resigned under pressure in 2019, while under investigation for expenditures of federal money for private purposes. Zinke's replacement, David Bernhardt, had been a key lobbyist for the oil and gas industry before becoming Zinke's assistant.

Bureau of Land Management

The nation's largest landlord, the BLM, administers over 200 million acres of range and grassland, largely available for grazing and mining and energy development. The BLM was created through a reorganization of existing land agencies in 1946 at the behest of cattle ranchers who could not control their own tendencies to overgraze the low-vegetation lands of the American West. The agency did not receive a clear mission statement and, over time, it came to serve its one primary client, livestock interests.[8] As the agency's responsibilities grew to include governance of mining claims on many federal lands, critics often derisively referred to BLM as the "bureau of livestock and mining."

The title derives largely from the fact that both grazing and mining on federal lands are governed by systems that charge low user fees. Grazing fees for animals on federal land are many times lower than fees on state or private lands. Mining fees are still determined by the 1872 Mining Law wherein people or companies finding hard-rock minerals on federal land patent their claims at minimal prices and then pay no royalties. As a result, the lands under BLM control can be overgrazed or damaged by large mining operations.[9] To a significant extent, the BLM is not to blame for this situation but rather operates under mandates from Congress. Grazing fees are set each year by Congress, and members, particularly from western states, traditionally refuse to substantially increase those fees. Members from western states, including Democrats such as Harry Reid of Nevada, have also refused to significantly revise the 1872 Mining Law.

Being somewhat powerless then, the BLM is subject to the demands and preferences of whatever President is in power. Thus, under the Trump administration, the BLM has offered lease sales on millions of acres of land for oil and gas development. The acting head of BLM in 2020, William Perry Pendley, was a Deputy Secretary under James Watt in the 1980s; Watt was removed from office under allegations of offering coal leases at below-market prices. The case study at the outset of this chapter on the sage-grouse is illustrative of the conflicts facing the BLM.

National Park Service

The National Park Service (NPS) manages about 85 million acres in 419 units, including 61 national parks as well as national monuments, battlefields, and recreation areas. The agency was established in 1916 with a mandate to provide for short-term use as well as long-term protection, a perpetual source of tension.[10] Partisan fights over agency operations have intensified in recent years.

At the level of the entire system, partisan fights over budgets have severely affected needed resources. The operating budget for the NPS has actually declined in terms of constant dollars. Total funding in nominal dollars between 2005 and 2014 increased from $2.27 billion to $3.1 billion (15 percent) but, when adjusted for inflation, this was an actual decline of 3 percent.[11] Since 2011, the number of full-time NPS employees has declined by 16 percent.[12] These cuts occurred at the same time that the agency was dealing with increased visitation and a backlog of needed repairs to infrastructure. Visitation increased by over 20 percent, from 274 million in 2005 to a record 331 million in 2017.[13] The NPS currently has an estimated $12 billion maintenance backlog of needed repairs.[14]

Within individual units, including the most famous sites, polarized politics and the swings that come with different partisan priorities preclude significant progress on long-term plans. For example, partisan disagreements over priorities and budgets have resulted in only incremental progress on long-standing plans to reduce automobile usage in Yosemite.[15] Buffeted by polarized interests and parties, Yellowstone park managers have struggled to formulate and implement a plan for snowmobiles in the winter that makes ecological sense.[16] At the Everglades, an ambitious restoration plan for increasing freshwater flows signed into law by President Clinton has suffered from inconsistent funding and broken promises.[17] These and other cases affecting every park unit from Shenandoah to the Grand Canyon warrant much further discussion, but suffice it to say that even the nation's most precious lands are not immune to impacts from partisan politics.

Politicization of the NPS has increased during the Trump years. Personnel have been shuffled or replaced, based on their policy positions, the most visible being Yellowstone Superintendent Dan Wenk, a 43-year veteran, who was pressured to retire after taking positions protecting park bison that were contrary to the views of nearby ranchers.[18] In early 2018, nine of the twelve members of the NPS Advisory Board resigned in protest of administration policies. In 2019, the NPS was put in the center of one of the most visible controversies of recent years. Unable to get congressional funding for a wall on the Mexican border, President Trump ordered NPS rangers from as far away as the Great Smoky Mountains (NC) and Wrangell-St. Elias (AK) to relocate to Arizona to work with the US Border Patrol.[19] During the early part of the COVID-19 pandemic, the Trump administration displayed confusion over whether to close popular units of the park system, at first promoting visitation and then allowing them to be closed as criticism increased over the health of park employees and visitors.[20]

Perhaps the most visible partisan conflict involving the NPS during the Trump administration has centered on national monuments created under the Antiquities Act (Box 9.1).

BOX 9.1 CASE STUDY OF THE ANTIQUITIES ACT

Congress passed the Antiquities Act in 1906 to conserve the stunning archaeological treasures of the American Southwest. Sites include dwellings such as the Cliff Palace and Spruce Tree House in what is now Mesa Verde National Park in Colorado. Spurred on by vandalism and looting in such places, Congress passed the "Act for the Preservation of American Antiquities" on June 8, 1906. Almost every president except Richard Nixon, Ronald Reagan, and George H. W. Bush has used the Antiquities Act. Many well-known national parks were first designated as national monuments, including the Grand Canyon, Zion, Death Valley, Acadia, and Olympic. The law's key provision states:

> That the President of the United States is hereby authorized, in his discretion, to declare by public proclamation historic landmarks, historic and prehistoric structures, and other objects of historic or scientific interest that are situated

upon the lands owned or controlled by the Government of the United States to be national monuments...

The two monuments of relevance here are Grand Staircase Escalante, proclaimed by President Clinton in 1996, and Bears Ears proclaimed by President Obama in 2016. They are shown in Figures 9.2 and 9.3. President Trump reduced Bears Ears National Monument, from 1.3 million acres to 228,000 acres. The administration cut Grand Staircase-Escalante National Monument nearly in half, from 1.9 million acres to about one million acres. As one recent study shows, Trump took this and other actions despite the fact that of 2.8 million comments submitted to the DOI in 2017 on his administration's review of the status of National Monuments, 99 percent opposed the review and 80 percent of respondents polled in western states supported keeping monument designations intact.[21] As expected, a number of lawsuits followed that challenge the power of a president to *reduce* the size of a monument proclaimed by a prior president.

FIGURE 9.2 ■ Grand Staircase Escalante

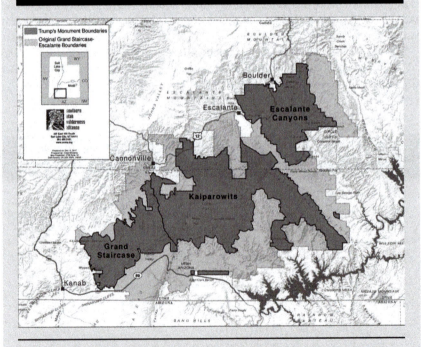

© Creed Murdock/Southern Utah Wilderness Alliance.

FIGURE 9.3 ■ Bears Ears National Monuments

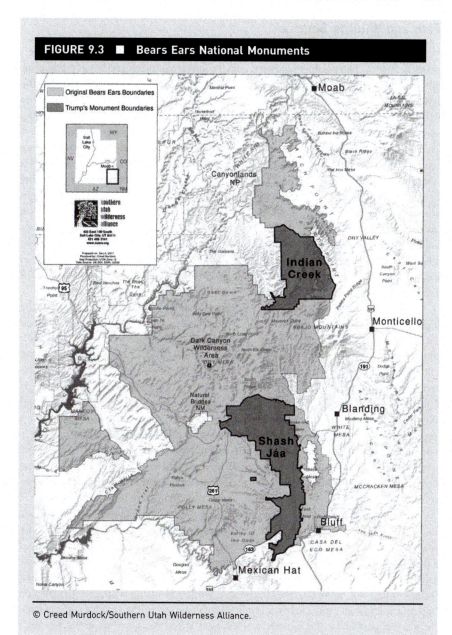

© Creed Murdock/Southern Utah Wilderness Alliance.

Fish and Wildlife Service

The USFWS is responsible for roughly 89 million acres in 568 wildlife refuges, many in Alaska. The product of several congressional resolutions and some predecessor agencies, the modern FWS dates to a 1970 reorganization. One summary describes the FWS as having "the most chaotic organizational history" of all public lands agencies that leads to a "weak position in the federal bureaucracy."[22] The agency lacks a strong, specific mandate, but is expected to manage the nation's fish and wildlife species. Having fish and animals as constituents does not help the agency's status with other federal entities such as Congress. As the saying goes, bears don't vote. Thus, the FWS is notoriously underfunded and understaffed. That situation has become even more dire in recent years. Specifically, the overall Refuge System budget, never large to begin with, dropped by 18 percent between 2010 and 2019. The number of refuge officers dropped by 65 percent between 1990 and 2015 to only 318.[23] Wildlife refuges are thus vulnerable to criticism and even hostile takeovers from antigovernment groups such as what happened in 2014 when a property rights militia occupied the Malheur Refuge in eastern Oregon.[24] Illustrative of the political manipulation of the agency, in 2018 President Trump pardoned the ranchers whose prosecution inspired the takeover.

At the heart of the mission for the FWS is enforcement of the ESA. In large part, the legislation is controversial because it calls for the listing and protection of species based on "biological grounds alone." Thus, the ESA has been the target of critics who claim it impedes economic development and infringes on property rights. Controversies over endangered species have intensified in recent years because listed species warrant protection and for some, such as polar bears, that would require addressing climate change. Not surprisingly then, Republicans in Congress have attempted to revise or even repeal this legislation numerous times in the last five decades. Consistent with the gridlock argument discussed earlier, Democrats have stopped Republican efforts and continued to fund the program, albeit at varying levels depending on the party of the presidential administration. In addition, the number of species listed for protection varies dramatically by the party of the president. The Clinton administration listed 522 species; the subsequent Bush administration listed only 62, while 268 were listed under Obama.[25]

Very few species have been listed under the ESA during the Trump years. Instead, in 2019, despite a just-released United Nations report claiming a million species at risk worldwide including many in the United States, the Trump administration announced a plan to revise implementation of the law. Specifically, for the first time ever, economic impacts of listing species would be considered in determining protections. Further, listings would not factor in climate change. While the lineup of political parties on opposite sides of the proposal was predictable, so too were the reactions of environmental groups such as the Center for Biological Diversity as well as seventeen states who sued to stop the rule change.[26]

While the FWS has often been characterized as struggling to protect species, they have had some success stories. Indeed, some 39 species, from bald eagles to American alligators, have been delisted due to recovery rates. Critics cite this low number as evidence of failure, but in fact over 1,600 species have been protected under the ESA and only 11 have gone extinct.[27] For many Americans, the ultimate success story of public agencies, specifically the NPS and FWS, fulfilling their conservation mandate is that of the Yellowstone wolves (Box 9.2).

BOX 9.2 CASE STUDY OF THE YELLOWSTONE WOLVES

Seeing or even hearing wolves in the wild is a truly remarkable experience. Yet, for decades, it was nearly impossible to do so in the continental United States. In the western United States, livestock interests despised gray wolves as the enemies of cattle and sheep and encouraged the federal government as well as ranchers and trappers to virtually eliminate gray wolves from their natural habitat in the northern Rocky Mountains. The loss of this predator caused significant changes in the ecosystem. In particular, elk and coyote populations increased dramatically. This in turn resulted in devastation of vegetation, erosion of stream beds, and declines in the populations of other animals such as beavers.

Thus, in the 1960s and 1970s, biologists from the FWS and NPS, as well as some environmental groups, began advocating for the reintroduction of wolves to Yellowstone National Park and central Idaho. Those calls inspired intense opposition from livestock interests in Wyoming and Montana and their allies in Congress. In the late 1970s and early 1980s, those supporting restoration expanded the sphere of conflict over the issue to involve the American public who, by overwhelming majorities, supported reintroduction.[28]

In 1994 and 1995, wildlife managers released 31 wolves into the park. They thrived and reproduced at rapid rates, the wolf population in the park growing to over 100 in just a few years. The wolves also reduced the elk and coyote populations and thus increased the vegetation and helped other species.[29] Economically, the wolves brought substantially increased tourism and tourist dollars to the park and nearby areas. While the entire costs of managing the reintroduction into Yellowstone (and neighboring Idaho) ran about $1.5 million annually, economic studies estimated total benefits to the area of over $20 million per year.[30] The wolves also, again predictably, did occasionally wander outside the park and kill cattle and sheep. Estimated kills were roughly about 100 animals per year, a fraction of fatalities from other causes, but still costs for ranchers. One environmental group, Defenders of Wildlife, attempted to defray the costs by reimbursing ranchers from a trust fund, but opposition to the wolves remained deep-seated among some local residents.

Not surprisingly then, the wolf program has been and continues to be subject to political decisions. In the late 1990s, a lawsuit based on the difficulty of differentiating reintroduced wolves (experimental status under the ESA) from wolves who migrate to the park on their own (and are thus fully protected) had the potential to abort the entire program until rejected by a federal judge. In 2007 and 2008, the George W. Bush administration used higher population numbers to attempt to delist the wolves from federal protection in neighboring states, only to be rejected by the courts after environmental groups sued. In 2011, Congress delisted gray wolves in the Northern Rockies in spite of objections from some scientists and environmental groups. In 2019, the Trump FWS proposed to strip all wolves from federal protection in the lower 48 states. Defenders of Wildlife and others moved immediately to try to stop the proposal.

The Forest Service

The USFS governs nearly 190 million acres of national forests under a multiple-use mandate. Prior to creation of the agency, many forests were cut down to build houses and roads with no plan for sustainable, long-term use. Thus, in response to demands from economic as well as environmental interests, some policymakers began setting aside large forest reserves in 1891. Defending the program in economic terms, President Theodore Roosevelt said, "The preservation of our forests is an imperative business necessity."[31] The FS was created in 1905 to manage national forests for perpetuity. The agency's mission is to balance multiple uses: sustainable timber production, watershed management, mineral extraction, grazing, and recreation. The agency also became the center of firefighting, largely as a result of heroic efforts to battle massive fires in the early twentieth century.[32]

As one of the most visible products of the Progressive movement's emphasis on scientific management, the FS was traditionally portrayed as a model of administrative efficiency.[33] From its beginning, however, the FS has been affected by politics. Early on, the agency developed a reputation for prioritizing timber production, in no small measure due to organizational placement in the Department of Agriculture.[34]

More recently, timber production from the national forests has varied, often corresponding again to the party of the president. In recent decades, high yields during the Reagan and George H. W. Bush years were followed by dramatic reductions under Clinton only to be revived again under George W. Bush, and then reduced again under Obama.[35] At the level of individual forests, each supervisor has to promulgate a plan for the amount of timber to be produced and those plans have always been subject to political pressures. The most controversial of these plans involved management of old-growth forests in the Pacific Northwest and the impacts of logging on species such as the northern spotted owl.[36] In addition, the FS has been frequently criticized by environmentalists as well as economists for subsidizing logging companies by building roads into forests, encouraging clear-cutting, and selling public timber at low prices.[37]

One recent manifestation of the impact of politics on forest management has been the inconsistent effort to set aside certain areas in national forests for purposes other than timber production. One controversy involved roadless areas in national forests, when in 1999, President Clinton proposed a ban on roads in 58 million acres. Of over 2 million comments on the proposal, 95 percent supported it, but lawsuits from the timber industry put the rule on hold. The subsequent George W. Bush administration then allowed revisions to the roadless rule, including for the 9-million acre Tongass National Forest in Alaska and in some cases allowing states to develop collaboratively, state-specific rules.[38] Later, a federal district judge reinstated the roadless rule but the controversy has not ended. Recently, Senator Lisa Murkowski (R-AK) introduced legislation to end roadless protections in the state of Alaska, including the Tongass. In response, Democratic Senators Patty Murray and Maria Cantwell of Washington state introduced the Roadless Conservation Act of 2018. The fate of these lands is not yet clear, although the Trump administration called for increased logging in the Tongass in 2020.

Summary of Policy Impacts Through Public Agencies

Presidents can significantly impact natural resource policies through bureaucratic actions but those impacts are limited by other institutions, as discussed in Chapter 4. Interest groups challenge presidential actions in the courts. State and local governments shape implementation of presidential directives at the subnational level. Finally, unilateral actions lack permanence. Subsequent presidents, especially if they are from the other party, can reverse or at least minimize the impacts from their predecessors. One needs, therefore, to consider the broad trends that affect these policies across administrations.

TRENDS IN NATURAL RESOURCE POLICIES

Some trends in natural resource policies transcend administrations or even crises. Indeed, while the coronavirus pandemic of 2019–2020 impacted all aspects of American life, including for example the closing of national parks, controversies in the issues discussed below continued. We discuss three broad topics that will be part of any deliberations on natural resource policies in the future.

Devolution and Privatization

The federally managed public lands of the United States have been the focus of historically long arguments about whether they are best managed by federal government, the states, or the private sector. If states were to manage these lands, they would have to be "transferred" by Congress to the states to manage per the Constitution, which gives Congress the power to manage federal property. Discussion usually centers on the BLM or USFS lands, not national parks or wildlife refuges. Transfer discussions began in 1912 and have continued off and on until the present day.[39] Indeed, as Chapter 2 describes, these arguments involve numerous policy areas in addition to public lands. The premise behind most of the arguments of those who favor transfer of national public lands is that the states can do a "better job" managing them. The term "better job" is ill defined. What is most likely is that the term is a code word for producing more revenue for the states by accelerating the production of timber, oil and gas, and other minerals, and less regulation of other activities such as grazing. Opponents argue that resource production and deregulation would increase harm to biological and other qualities of the public lands. The COVID-19 pandemic prompted intense discussions and protests over the responsibilities and relative roles of state and federal governments in all policy areas, as discussed in the Rabe chapter on federalism. Those debates in future years will certainly involve the management of public lands.

Arguments about the virtues of privatizing public lands have even deeper roots. Indeed, in response to such proposals in 1776 Adam Smith wrote, "Lands for the purpose of pleasure and magnificence—parks, gardens, public walks, etc…seem to be the only lands which, in a great and civilized monarchy, ought to belong to the crown."[40] Politicians and property rights advocates have frequently renewed a push to privatize public lands, whether that is selling federal land to corporations and others or maximizing possessory interests of concessioners. Proponents argue, "Markets offer a more effective, economical, and egalitarian alternative to the political management of federal lands."[41] Many conservationists and environmental groups have resisted these efforts with numerous counterarguments.[42] Markets need prices, they note, but putting a price on preservation is difficult. Establishing private ownership of public lands violates the principle of public goods that they be nonexcludable in that they belong to all people. Finally, public lands like parks are protected for future generations but how will their interests be accounted for in markets focused on bottom lines?

Collaboration and Compromise

One recent trend in natural resource policies is increased reliance on collaborative processes. In large part, this results from the polarization of parties and different interests who are involved in these issues. However, with collaboration comes compromise and that raises other questions, ones that date to the origins of the environmental movement.

Collaborative ecosystem management involves efforts to find cooperative solutions to potentially conflictual resource issues. Participants utilize science-based findings and communication between diverse stakeholders to seek win-win outcomes that satisfy all parties, or at least reduce the likelihood of litigation.[43] One version of this that is often used for large ecosystems is called adaptive management wherein resource planners work with stakeholders to take experimental actions, scientifically monitor the results, and then adjust plans for future actions. Policymakers have pursued long-term adaptive management programs on such large and complex ecosystems as the Everglades and the Colorado River in the Grand Canyon.

While the overall trend is one of increasing usage of collaborative management, as with all natural resource policies there is some variance depending on which party is in power. The Obama administration supported such efforts, notably in large-scale projects such as the Everglades Restoration Program and the sage-grouse initiative described at the outset of the chapter.[44] The Trump administration, as the sage-grouse situation reflects, often abandoned or underfunded such programs.

One obvious requirement for collaboration is the willingness of all interested parties to compromise to find some common ground. While this may sound logical, compromise has been a controversial topic in environmental politics. Indeed, many date the modern American environmental movement to the refusal of the Sierra Club and other groups to compromise on possible dam-building in the Grand Canyon in the late 1960s.[45] The perceived costs of compromise fostered the writing of Edward Abbey, the creation of groups such as Earth First!, and the regrets of some prominent environmentalists such as David Brower who said, "Every time I've compromised, I've lost."[46] Over the last several decades, environmental groups have taken a less confrontational tone and attempted to work with those pursuing economic development of natural resources. In just the last few years, some critics have argued that this trend has gone too far. One author, for instance, describes eco-pragmatism as "a gentle negotiation over how much of nature to cede to industry."[47]

Climate Change and Energy

Conflicts over energy development and natural resources have occurred for decades. For example, the two parties have battled over possible development of the Arctic National Wildlife Refuge (ANWR) in Alaska since passage of the Alaska Lands Act in late 1979. More broadly, about half of the undeveloped fossil fuels in the country are located in federal lands or offshore waters.[48] By 2017, oil or gas drilling occurred on 77 units (about 14 percent) of the National Wildlife Refuge system.[49] A dozen units of the National Park System had oil or gas production operations and over two dozen more were at risk of future development.[50] As of 2011, oil production on federal lands contributed about one-tenth of total US oil production while production from federal offshore facilities provided about one-third.[51]

The frequency and intensity of conflicts over these developments have increased markedly due to the impact of energy policies on climate change. Congressional Republicans who deny the threat of climate change have been supportive of these developments, such as legislation to prevent national park designations be used to block power plants.[52] In response to concerns that Congress would repeal Rule 9B (the 9B rules refer to nonfederal oil and gas rights) standards for energy development in national parks, the Obama NPS updated those rules in late 2016 to strengthen the guidelines even while processing oil and gas leases on other federal lands.[53] Many of these operations inspire controversy and protests and contribute momentum to the Keep it in the Ground movement.[54] This movement, in short, involves hundreds of groups demanding an end to fossil fuel usage by literally keeping those fuels in the ground. In addition to fossil fuel development, proposals for renewable projects have inspired battles in the Mojave Desert, near Death Valley, and in Joshua Tree National Parks, as well as with wind farms and transmission lines on Western lands.[55]

Prompted by these controversies, some policymakers have sought compromise solutions. For instance, a bipartisan group introduced the Public Lands and Renewable Energy Development Act in 2019 to facilitate leasing for solar and wind projects on public lands with royalties going into conservation funds. As another example, several federal agencies along with numerous stakeholders developed the Desert Renewable Energy Conservation Plan to guide energy development in the Mojave and Sonoran Deserts while ensuring that important wildlife areas are protected.[56] Finally, federal officials, states, and stakeholders continued to seek resolution of conflicts such as energy development on lands with endangered species (such as the sage-grouse) through collaborative ecosystem management.

Any momentum toward seeking compromise solutions for energy vs. natural resource conflicts ended abruptly with the Trump administration. Immediately after taking office, the administration commenced efforts to increase fossil fuel development on federal lands. These moves included the aforementioned reduction of the size of national monuments and opening sage-grouse habitat.[57] Abandoning even pretense to collaborative ecosystem management or compromise legislation, the administration pursued these goals with unilateral actions.

Notably, in March 2017, Trump issued an executive order requiring federal agencies to end policies, such as the Park Service's Rule 9B, that protected park units from harmful oil and gas development. In just the first year in office, the administration proposed leasing just outside the borders of more than 18 western parks, including one within a mile of Zion, helping restart the "external threats to national parks" debate.[58] The administration also reversed Obama-era restrictions on offshore oil development and promised to open up most coastal waters to exploration even while damages from the 2010 Deepwater Horizon spill continue to affect the Gulf of Mexico ecosystem.

Reactions to the administration's actions were predictable. For the most part, congressional Republicans supported the administration's moves while Democrats opposed them. One publicized exception involved offshore drilling. The administration's plan to open most US coastlines prompted more than a dozen governors from coastal states, several of them Republicans, to object. Those objections intensified after Secretary of Interior Zinke exempted the state of Florida after its Republican governor complained, an interesting dilemma for an administration rhetorically supportive of states' rights. In late 2020, shortly before the election, the administration reversed itself on offshore oil policies, at least in some states.

Other specific parts of the administration's agenda prompted objections from many groups. Native American groups, for example, vigorously protested expanded fracking operations near Chaco Culture National Historic Park, prompting cancellation of expanded leasing in 2018.[59] A proposed oil refinery just three miles from the border of Theodore Roosevelt National Park in North Dakota inspired protests from local groups as well as national conservation organizations such as the National Parks and Conservation Association.[60]

At present, Trump administration efforts to expand energy development on public lands continue despite opposition. The administration proposed conducting future ANWR lease sales, but, consistent with the themes in this chapter, without explicit congressional authorization. On many lands, the Trump efforts are further complicated by the fact that some states have proposed their own regulations and tax systems governing fracking operations on private or state-held lands.[61] Indeed, while many states opposed the Obama administration's proposed regulations, many others have resisted Trump's efforts to reverse them. Native American and conservation groups continue to battle Trump efforts to develop pipelines in court. Democrats, having retaken the House of Representatives in 2018, have also attempted to block some administration actions. Collaborative resolutions are rarely even discussed at the federal level, whereas some states, notably Colorado, have valiantly attempted to find common ground among different interests in their energy development oversight.

The coronavirus pandemic has impacted climate and energy issues in important ways. One involves oil prices and development. Stay-at-home orders and the reluctance of people to travel reduced oil consumption throughout the world to the point where oil was in such heavy supply in early 2020 that the price of a barrel dropped below $20, a level not seen in decades. While this has slowed oil development in some places, including plans for drilling in ANWR, long-term trends are unclear. Another impact involves concern over climate change. Although shelter-in-place behavior precluded the mass demonstrations planned for the fiftieth anniversary of Earth Day in 2020, scientists and activists did use the occasion to point out parallels between the necessary responses to both the pandemic and to climate change, notably the need to pay attention to scientific assessments and warnings.[62] In addition, the improvement in environmental indicators such as air pollution could give impetus to those seeking action on climate. If people do take those comparisons and indicators seriously, the long-term impacts on energy use and development could be profound.

CONCLUSIONS

Politics impacts natural resource policies in the United States. First, the two major political parties have very different priorities and goals. The two parties thus push different actions and those differences typically lead to gridlock in Congress. Second, presidents frequently use unilateral actions, typically through natural resource agencies, to pursue their party's agendas. Those actions motivate pushback from interest groups and some state governments through litigation and other actions. This pushback has, to some extent, limited what the Trump administration has been able to accomplish in their agenda regarding natural resources. Nevertheless, the Trump administration has had significant impacts on nearly all natural resource policies.

Third, these conflicting political directives force public agencies to respond to short-term pressures when their mandates call for long-term management of natural resources. Criticisms of those public agencies foster ongoing calls for alternative institutional arrangements, specifically privatizing public lands, using market-based mechanisms, or transferring control of federal properties to the states. Fourth, advocates and policymakers continue to pursue collaborative resolutions to conflicts over natural resources. Finally, tensions over natural resources are heightened by disagreements over energy and climate change.

Political conflicts over natural resources driven by short-term partisan goals are only likely to continue and even intensify in coming years. Indeed, even the immense crisis of the COVID-19 pandemic did not eliminate partisan divisiveness in the United States. These divisions will continue to impact natural resource policies in the future even though protections of natural resources such as lands and species depend upon long-term strategies and consistent implementation. Without those things, the United States continues to put at risk its most precious resources.

SUGGESTED WEBSITES

Center for Biological Diversity (www.biologicaldiversity.org) This environmental group is often at the forefront of conflicts over endangered species issues.

Competitive Enterprise Institute (www.cei.org) The CEI is a nonprofit organization that emphasizes free enterprise and individual liberty in making policy decisions.

Defenders of Wildlife (www.defenders.org) This environmental group publishes a monthly magazine and maintains a webpage that has much information about endangered species.

Heritage Foundation (www.heritage.org) The Heritage Foundation is a think tank in Washington DC, that promotes policies based on free enterprise and limited government.

High Country News (www.hcn.org) This is a weekly online newspaper that provides thorough analyses of natural resources, public lands, and species in the American West.

Mountain States Legal Foundation (www.mslegal.org) MSLF is a non-profit law firm that focuses on protecting property rights and economic liberty.

National Parks Conservation Association (www.npca.org) This environmental group uses information and advocacy for the further protection of America's national parks.

National Parks Traveler (www.nationalparkstraveler.org) This weekly online newsletter contains timely, informative stories about America's national parks.

Property and Environment Research Center (www.perc.org) PERC is a research institute located in Bozeman, Montana, dedicated to free-market environmentalism.

NOTES

1. Paul J. Culhane, *Public Lands Politics* (Baltimore: Johns Hopkins University Press, 1981).
2. Michael J. Graetz, *The End of Energy* (Cambridge, MA: MIT Press, 2011), 21.
3. William R. Lowry, "State Parks Found to be Source of Innovation," *Public Administration Times* 19 (1996): 1–13.
4. See, for example, William R. Lowry, "The Exceptionalism of the Open Space Issue in American Politics," *Social Science Quarterly* 99, no. 4 (2018): 1363–76.
5. Carl Segerstrom, "New Bill Leaves Lands Protected, Lawmaking Neglected," *High Country News*, February 20, 2019.
6. Matthew Daly, "Senate Passes Conservation Bill that Boosts Parks, Trails," *St. Louis Post-Dispatch*, February 13, 2019, A9.
7. Jacob Carter, Emily Berman, Anita Desikan, Charise Johnson, and Gretchen Goldman, "The State of Science in the Trump Era," Union of Concerned Scientists, 2019, www.ucsusa.org/ScienceUnderTrump.
8. Jeanne Nienaber Clarke and Daniel C. McCool, *Staking Out the Terrain*, 2nd ed. (Albany, NY: SUNY, 1996), 162; and Philip Foss, *Politics and Grass* (Seattle, WA: University of Washington Press, 1960).
9. For a recent scathing expose, see Christopher Ketcham, *This Land* (New York, NY: Viking Press, 2019).
10. John Freemuth, *Islands under Siege* (Lawrence, KS: University of Kansas Press, 1991); William R. Lowry, *The Capacity for Wonder* (Washington, DC: The Brookings Institution, 1994); and Alfred Runte, *National Parks: The American Experience* (Lincoln, NE: University of Nebraska Press, 1979).
11. U.S. Government Accountability Office, "Revenues from Fees and Donations Increased, but Some Enhancements are Needed to Continue this Trend" (Washington, DC: U.S. GAO, 2015).
12. Karen Chavez and Trevor Hughes, "Ordered to the Border," *USA Today*, December 2, 2019, 1A.
13. Statistics for visitation are found at www.nps.gov/aboutus/visitation-numbers.htm.
14. Julie Turkewitz, "National Parks Struggle with a Mounting Crisis: Too Many Visitors," *The New York Times*, September 27, 2017.
15. William Lowry, *Repairing Paradise* (Washington, DC: The Brookings Institution, 2009), Chapter 3.

16. Michael J. Yochim, *Yellowstone and the Snowmobile* (Lawrence, KS: University Press of Kansas, 2009).
17. Michael Grunwald, *The Swamp* (New York, NY: Simon and Schuster, 2007).
18. Nate Hegyi, "Yellowstone National Park Superintendent: 'I'm No Longer Wanted.'" www.npr.org/2018/06/07/617936487.
19. Karen Chavez and Trevor Hughes, "Ordered to the Border," *USA Today*, December 2, 2019, 1A.
20. John Freemuth, "In The Pandemic's Wake, Let's Create More Park Spaces," *National Parks Traveler*, https://www.nationalparkstraveler.org/2020/04/op-ed-pandemics-wake-lets-create-more-park-spaces, accessed April 21, 2020.
21. Christopher Ketcham, *This Land* (New York, NY: Viking Press, 2019), 152.
22. Clarke and McCool, 1996, 107.
23. Helen Santoro, "Wildlife Refuges Suffer under Budget Cuts and Staff Shortages," *High Country News*, November 20, 2019.
24. James Pogue, *Chosen Country* (New York, NY: Henry Holt and Company, 2018).
25. Center for Biological Diversity, "Study: Endangered Species Protection Taking Six Times Longer than Law Allows," press release dated August 8, 2016.
26. Timothy Puko, "Trump Administration Eases Endangered Species Act Protections," *The Wall Street Journal*, August 12, 2019.
27. Seth Borenstein, "Butterfly on a Bomb Range," *St. Louis Post-Dispatch*, December 8, 2019, A17.
28. Douglas W. Smith and Gary Ferguson, *Decade of the Wolf* (Guilford, CT: Lyons Press, 2005); and Lowry, 2009: Ch. 2.
29. For a recent review of these changes, see Peter Dockrill, "It Started with Just 14 Wolves," 2018, www.sciencealert.com/how-31-wolvestransformed-yellowstone-in-ways-nobody-could-ever-have-predicted-national-park-wolf-reintroduction-trophic-cascade.
30. J. W. Duffield, C. J. Neher, and D. A. Patterson, "Wolf Recovery in Yellowstone," *Yellowstone Science* 16, no. 1 (2008): 22.
31. Roosevelt quoted in H. W. Brands, *Dreams of El Dorado* (New York, NY: Basic Books, 2019), 473.
32. Timothy Egan, *The Big Burn* (New York, NY: Mariner Books, 2010).
33. Herbert Kaufman, *The Forest Ranger* (Baltimore, MD: Johns Hopkins Press, 1960); and Clarke and McCool, 1996: 50–3.
34. Pinchot quoted in Steve Olson, *Eruption* (New York, NY: W. W. Norton, 2016).
35. Congressional Research Service, "Timber Harvesting on Federal Lands," report R45688, dated April 12, 2019, p. 8.
36. For a thorough review of the case, see Judith A. Layzer, *The Environmental Case*, 3rd ed. (Washington, DC: CQ Press, 2012), Chapter 7.
37. Christopher Ketcham, *This Land* (New York, NY: Viking Press, 2019); and Randal O'Toole, *Reforming the Forest Service* (Washington, DC: Island Press, 1988).
38. https://www.fs.usda.gov/Internet/FSE_DOCUMENTS/stelprdb5053193.pdf, accessed January 8, 2020.
39. John Freemuth and Esther Babcock, "Governance, Science and the Public Lands," in *Environmental Politics and Policy in the West Third Edition*, eds. Zachary Smith and John Freemuth (Boulder, CO: University of Colorado Press, 2016).
40. Adam Smith, 1776, *The Wealth of Nations*, Vol. 2, 1960 edition, London: Dent, p. 306.
41. Reed Watson, "A Moral Case for Markets," 2017, PERC Reports, Vol. 36, No. 1: 4.
42. Steven Davis, *In Defense of Public Lands* (Philadelphia, PA: Temple University Press, 2018).
43. Julia M. Wondollek and Steven L. Yaffee, *Making Collaboration Work* (Washington, DC: Island Press, 2000); and Mark Lubell, "Collaborative Watershed Management," *Policy Studies Journal* 32, no. 3 (2004): 341–61.

44. Lubell and Segee, 2013: 198–200.

45. For example, see Roderick Frazier Nash, *Wilderness and the American Mind*, 4th ed. (New Haven, CT: Yale University Press, 2001), 235–37; and Marc Reisner, *Cadillac Desert* (New York, NY: Penguin Books, 1987), 295.

46. Quoted in Ketcham, 2019, 324.

47. Ketcham, 2019, 325.

48. Ben Adler, "Fossil Fuel Extraction on Public Lands is the Next Climate Fight," *High Country News*, September 29, 2015, 2.

49. U.S. General Accounting Office, "Wildlife Refuge Oil and Gas Activity," GAO-02-64R, October 31, 2001.

50. Kurt Repanshek, "Oil and Gas Production and the National Parks," *National Parks Traveler*, May 5, 2010.

51. Marc Humphries, "U.S. Crude Oil Production in Federal and Non-Federal Areas," *Congressional Research Service Report for Congress*, R42432, March 20, 2012.

52. Kurt Repanshek, "Republicans, Democrats Differ over Whether National Park Designations Should Block Energy Projects," *National Parks Traveler*, May 11, 2010.

53. Nicholas Lund, "The Facts on Oil and Gas Drilling in National Parks," *NPCA blog post*, 2/10/2017, www.npca.org/articles/1471/the-facts-on-oil-and-gas-drilling-in-national-parks; and Elizabeth Shogren, "Park Service May Strengthen its Oil and Gas Regulations," *High Country News*, October 27, 2015.

54. Kirk Nielsen, "Bad Vibes in Big Cypress," *Sierra*, July/August 2016: 50–51; Kate Schimel, "How the Keep it in the Ground Movement Came to Be," *High Country News*, July 19, 2016; and Joshua Zaffros, "Protests against Drilling on Public Lands are Escalating," *High Country News*, May 13, 2016.

55. Amy Harder and Cassandra Sweet, "Solar Push Stirs Concern," *The Wall Street Journal*, October 26, 2016, 6; and Elizabeth Shogren, "At the BLM, A Mixed Record for Renewables on Public Lands," *High Country News*, December 4, 2015.

56. Defenders of Wildlife, "Getting it Right," *Defenders*, Winter 2015, p. 16.

57. Rebecca Leber, "The Federal Land at Stake in Trump's Rush for More Drilling," *Mother Jones*, December 7, 2017; Brady McCombs, "Lands Stripped from Monuments Open to Claims," *Arizona Republic*, 2/4/2018: 11E; and Elizabeth Shogren, "Interior Moves Swiftly After Trump's Climate Order," *High Country News*, March 31, 2017.

58. Kurt Repanshek, "Change in Administrations Puts Zion National Park in Bind over Drilling Proposal," *National Parks Traveler*, February 5, 2017; John Miles, "Mixing Energy Development and National Parks," *National Parks Traveler*, July 15, 2018; and Mike Wirth, "Drilling in National Parks?" *National Parks*, Fall 2018: 10–1.

59. Jonathan Thompson, "Resistance to Drilling Grows on the Navajo Nation," *High Country News*, March 2, 2018.

60. Blake Nicholson, "Firm Denies Skirting Law to Build Refinery Near Park," *St. Louis Post-Dispatch*, December 24, 2017, A23.

61. Charles E. Davis, "Shaping State Fracking Policies in the United States," *State and Local Government Review* 49, no. 2 (2017): 140–50; Barry G. Rabe and Rachel L. Hampton, "Taxing Fracking," *Review of Policy Research* 32, no. 4 (2015): 389–412; and Rachel L. Hampton and Barry G. Rabe, "Leaving Money on the Table," *Commonwealth* 19, no. 1: 5–32.

62. See, for example, Daniel C. Esty, "Covid-19 Lessons for the 50th Anniversary of Earth Day," *The Hill*, April 22, 2020.

10 APPLYING MARKET PRINCIPLES TO ENVIRONMENTAL POLICY

Sheila M. Olmstead

Each day, you make decisions that require trade-offs. Should you walk to work or drive? Walking takes more time; driving costs money for gasoline and parking. You might also consider the benefits of exercise if you walk, or the costs to the environment of the emissions if you drive. In considering this question, you need to determine how to allocate important scarce resources—your time and money—to achieve a particular goal.

Economics is the study of the allocation of scarce resources, and economists typically apply two simple concepts, efficiency and cost-effectiveness, for systematically making decisions. Let's take a concrete environmental policy example. The Snake River in the Pacific Northwest provides water for drinking, agricultural irrigation, transportation, industrial production, and hydroelectricity generation. It also supports rapidly dwindling populations of endangered salmon species. If there is not enough water to provide each of these services and to satisfy everyone, we must trade off one good thing for another.

Some scientific evidence indicates that removing hydroelectric dams on the upper Snake River may assist in the recovery of salmon populations. Salmon declines may also be caused by too little water in the river, which might be addressed by reducing agricultural or urban water withdrawals. Each of these measures could be implemented at some cost. Benefit–cost analysis would compare the benefits of each measure (the expected increase in salmon populations) to its costs. An *efficient* policy choice would maximize net benefits; we would choose the policy that offered the greatest difference between benefits and costs.

To continue this example, what if the Endangered Species Act requires that a specific level of salmon recovery be achieved? In this case, the benefits of salmon recovery may never be monetized. But economics can still play a role in choosing policies to achieve salmon recovery. Cost-effectiveness analysis would compare the costs of each potential policy intervention that could achieve the mandated salmon recovery goal. Decision-makers would then choose the least costly or most *cost-effective* policy option.

This discussion is highly simplified. Explaining the causes of Snake River salmon decline and forecasting the impact of policy changes on salmon populations are complex scientific tasks, and different experts have different models that produce different results.[1] The trade-offs can also be multidimensional. Removing dams may sound like a great environmental idea, but hydroelectric power is an important source of clean energy in the Pacific Northwest. Would the dams' hydroelectricity be replaced by coal- or gas-fired power plants? What would be the impacts of the increased emissions of local and global air pollutants? In this chapter, we discuss some simple economic tools for examining such trade-offs.

ECONOMIC CONCEPTS AND ENVIRONMENTAL POLICY

Economic Efficiency and Benefit–Cost Analysis of Environmental Policy

Many countries regulate emissions of sulfur dioxide (SO_2), an air pollutant that can damage human health and also causes acid rain, which harms forests and aquatic ecosystems. In the United States, power plants are a major source of SO_2 emissions, regulated under the Clean Air Act (CAA). As evidence accumulated in the 1980s regarding the damages from acid rain in the northeastern United States, Congress considered updating the CAA so that it would cover many old power plants not regulated by the original legislation. This process culminated in the 1990 CAA Amendments, which set a new goal for SO_2 emissions reductions from older power plants. Assume that it is 1989, and you have been asked to tell the US Congress, from an economic perspective, how much SO_2 emissions should be reduced.

First, consider the costs of reducing SO_2 emissions. Economic costs are *opportunity costs*—what we must give up by abating each ton of emissions rather than spending that money on other important things. Emissions abatement can be achieved by removing SO_2 emissions from power plant smokestack gases using a "scrubber," which requires an up-front investment, as well as labor and materials for routine operation. Power plants can also change the fuels they use to generate electricity, switching from high-sulfur to more expensive low-sulfur coal or from coal to natural gas. Spending this money on pollution control leaves less to spend to improve a plant's operations or increase output. These costs are passed on by the firm to its employees (in the form of reduced wages), stockholders (in terms of lower share prices), consumers (in the form of higher prices), and other stakeholders.

If required to reduce emissions, firms will accomplish the cheapest abatement first, and resort to more and more expensive options as the amount of required abatement increases. The cost of abating each ton of pollution tends to rise slowly at first, as we abate the first tons of SO_2 emissions, and then more quickly. This typical pattern of costs is represented by the lower, convex curve in Figure 10.1, labeled total costs, or $C(Q)$.

The value of reducing emissions declines as we abate more and more tons of SO_2. At high levels of SO_2 emissions (low abatement), this pollutant causes acid rain as well as respiratory and cardiovascular ailments in populated areas. But as the air gets cleaner, low SO_2 concentrations cause fewer problems. Thus, while the total benefits of reducing SO_2 may always increase as we reduce emissions, the benefit of each additional ton of abatement will go down. This typical pattern of benefits is represented by the upper, concave curve in Figure 10.1, labeled total benefits, or $B(Q)$.[2]

Economic efficiency requires that we find the policy that will give us the greatest net benefits—the biggest difference between total benefits and total costs. In Figure 10.1, the efficient amount of SO_2 emissions abatement is marked as Q^*, where

FIGURE 10.1 ■ Comparing the Total Benefits and Costs of Pollution Abatement

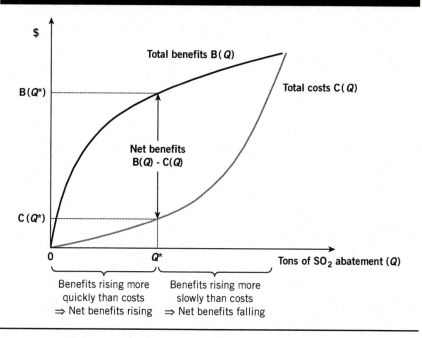

Benefits rising more quickly than costs ⇒ Net benefits rising   Benefits rising more slowly than costs ⇒ Net benefits falling

the vertical distance between the benefit and cost curves is biggest. Why would it be inefficient to abate more or less than Q^* tons of SO_2 emissions? To the right of Q^*, the total benefits of reducing pollution are still positive and still rising. But costs are rising faster than benefits. So for every dollar in benefit we gain by eliminating a ton of emissions, we incur greater costs. To the left of Q^*, the benefits of each ton of abatement are rising more quickly than costs, so if we move to the left, we will also reduce the policy's net benefits.

Note that we have emphasized the costs and benefits of reducing each individual ton of pollution. Where total benefits are steep, at low levels of abatement, the benefit of an additional ton is very high. Where the total benefits curve is flat, the benefit of an additional ton is low. This concept of the decreasing "benefit of an additional ton" defines the economic concept of *marginal benefit*. On the cost side, where total costs are almost flat, at low levels of abatement, the cost of adding an additional ton is very low. As total costs get very steep, the cost of abating an additional ton is high. This concept of the increasing "cost of an additional ton" defines the concept of *marginal cost*. The efficient quantity of pollution abatement is the number of tons at which the marginal benefit of abating an additional ton is exactly equal to the marginal cost.[3]

What we have just done is a benefit–cost analysis of a potential SO_2 emissions reduction policy. If you had completed the analysis to advise Congress, the information amassed on benefits, costs, and the efficient quantity of pollution to abate would illuminate the trade-offs involved in improving air quality. When Congress passed the CAA Amendments of 1990, it eventually required 10 million tons of SO_2 emissions abatement, roughly a 50 percent reduction in power plant emissions of this pollutant. Was this the efficient level of pollution control? Subsequent analysis (particularly of the human health benefits of avoided SO_2 emissions) suggests that the efficient amount of SO_2 abatement would have been higher than the 10-million-ton goal.[4] But economic efficiency is one of many criteria considered in the making of environmental policy, some others of which are detailed elsewhere in this book. An excellent summary of how economists see the role of benefit–cost analysis in public decision-making is offered by Nobel Laureate Kenneth Arrow and coauthors:

> Although formal benefit-cost analysis should not be viewed as either necessary or sufficient for designing sensible public policy, it can provide an exceptionally useful framework for consistently organizing disparate information, and in this way, it can greatly improve the process and, hence, the outcome of policy analysis.[5]

There are many critiques of benefit–cost analysis.[6] A common critique is that basing environmental policy decisions on whether benefits outweigh costs ignores important political and ethical considerations. As is clear from the preceding quotation, even when citizens and their governments design policy based on concerns other than efficiency, collecting information about benefits and costs can be extremely useful. Some critics of benefit–cost analysis object to placing a dollar value on environmental goods and services, suggesting that these "priceless" resources are devalued when treated in monetary terms.[7] But benefit–cost analysis simply makes explicit the trade-offs represented by a policy choice—it does not create the trade-offs. When environmental policy is made, we establish how much we are willing to spend to protect endangered species or avoid the human health impacts of pollution exposure. Whether we estimate the value of such things in advance and use these numbers to guide policy or set policy first based on other criteria and then back out our implied values for such things, we have still made the same trade-off. No economic argument can suggest whether explicit or implicit consideration of benefits and costs is *ethically* preferable. But the choice does not affect the outcome that a trade-off has been made. Used as one of many inputs to the consideration of policy choices, benefit–cost analysis is a powerful and illuminating tool.

The Measurement of Environmental Benefits and Costs

Thus far, we have discussed benefits and costs abstractly. In an actual economic analysis, benefits and costs would be estimated, so that the horizontal and vertical axes of Figure 10.1 would take on specific numerical units. Quantifying the costs of environmental policies can require rough approximations. For example, one study has estimated the costs of protecting California condors, designated an endangered species following their near extinction and their later reintroduction into the wild from a captive breeding program.[8] Figure 10.2 describes the costs of each potential step that policymakers might take to protect the condor population; when the number of condors saved per year is graphed against the cost of each potential step taken to save

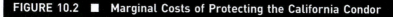

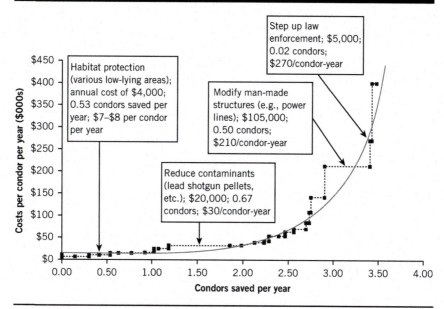

FIGURE 10.2 ■ Marginal Costs of Protecting the California Condor

Source: From *Markets and the Environment* (2nd ed.), by Nathaniel O. Keohane and Sheila M. Olmstead. Published by Island Press. Copyright © 2016 by the authors.

them, a marginal cost curve results that is upward sloping like those we discussed for pollution abatement.

Economists measure the benefits of an environmental policy as the sum of individuals' willingness to pay for the changes it may induce. This notion is clearly anthropocentric—the changes induced by an environmental policy are economically beneficial only to the extent that human beings value them. This does not suggest that improvements in ecosystem function or other "nonhuman" effects of a policy have no value. Many people value open space, endangered species preservation, and biodiversity and have shown through their donations to environmental advocacy groups, campaigning for local referenda and contributions to political candidates that they are willing to sacrifice much for these causes. The economic value of an environmental amenity (such as clean air or water or open space) comprises the value that people experience from using it, and the so-called nonuse value. Nonuse value captures the value people have for simply knowing that an endangered species (like the grizzly bear) or a pristine area (like the Arctic National Wildlife Refuge) exists, even if they never plan to see or use such resources. To carry out a benefit–cost analysis, however, we must know more than that people have some value for a policy's goal; we must estimate that value to compare it to the policy's costs.

An in-depth discussion of environmental benefit estimation methods is beyond the scope of this chapter.[9] But we can sketch out the basic intuition behind the major approaches. Some benefits of environmental policies can be measured straightforwardly through their impacts on actual markets. For example, if we are considering a policy to reduce water pollution that may increase commercial fish populations, estimates of the increased market value of the total catch would be included in an estimate of the policy's total benefits.

For most environmental goods and services, however, estimating benefits is much trickier. The values that people have for using environmental amenities can often be measured indirectly through their behavior in markets. For example, many people spend money on wilderness vacations. While they are not purchasing wilderness, per se, when they do this, economists can estimate the recreational value of wilderness sites from travelers' expenditures. Travel cost models are a class of statistical methods that economists use for this purpose.

Another method, the hedonic property model, is based on the idea that what people are willing to pay for a home reflects, in part, the environmental attributes of its neighborhood. Economists use statistical techniques to estimate what portion of a home's price is determined by environmental attributes, such as the surrounding air quality, controlling for other price determinants, for example, the home's physical characteristics, school district quality, and proximity to jobs and transportation. To estimate the value of human health impacts of an environmental policy, economists primarily use hedonic wage studies. These models estimate people's willingness to pay for small decreases in risks to life and health by examining the differences in wages for jobs with different levels of risk. As in the hedonic property model, statistical methods must be used to control for the many other determinants of wages (for example, how skilled or educated a worker must be to take a particular job).

Travel cost and hedonic models are *revealed preference* models—they estimate how people value a particular aspect of an environmental amenity from their actual behavior, revealed in markets. But if we were to stop with values expressed in markets, the benefits we could estimate for things like wilderness areas and species preservation would be incomplete. Nonuse value leaves no footprint in any market. Thus a *stated preference* approach must be used to quantify nonuse values. Economists design carefully structured surveys to ask people how much they are willing to pay for a specific improvement in environmental quality or a natural resource amenity, and they sum across individuals to assess aggregate willingness to pay.

The Environmental Protection Agency's (EPA's) 1985 benefit–cost analysis of reducing the lead content of gasoline offers an example of what each side of such an analysis might include. The analysis quantified the main benefits from phasing out leaded gas: reduced human health damages from lead exposure (retardation of children's cognitive and physiological development and exacerbation of high blood pressure in adult males), reduction in other local air pollutants from vehicle emissions (since leaded gas destroyed catalytic converters, which reduce emissions), and lower costs of engine maintenance and related increases in fuel economy. The costs were primarily the installation of new refinery equipment and the production of alternative fuel additives. The study found that the lead phasedown policy had projected annual net benefits of $7 billion (in 1983 dollars), even though only a portion of benefits were actually monetized. The health benefits of the regulation that the EPA estimated included the avoided costs of medical care and of remedial education for affected children. Americans, if surveyed, would likely have had significant willingness to pay to avoid the lasting health and cognitive impacts of childhood lead exposure, but these benefits were never monetized. Even with these gaps, acknowledged by the study's authors, this analysis helped to "sell" the regulation; a few years earlier, the EPA had decided on a much weaker rule, citing potential costs to refineries.[10]

The fact that the EPA did not monetize some benefits of the US lead phasedown brings us to an important point. In some cases, existing estimation methods may be sufficient to evaluate the benefits of an environmental policy but are too complex and expensive to implement. In the lead case, this was immaterial. The benefits of the policy exceeded the costs by a large margin, even excluding the unmonetized benefits, so the eventual policy decision was not affected by this choice. In other cases, when benefits are hard to monetize, doing so may matter for the ultimate policy outcome.

In some cases, economic tools simply prove insufficient to estimate the benefits (or avoided damages) from environmental policy. For example, climate science suggests that sudden, catastrophic events (like the reversal of thermohaline circulations or sudden collapse of the Greenland or West Antarctic ice sheets) are possible outcomes of the current warming trend. The probabilities of such disastrous events may be very small. When these low probabilities are combined with the fact that important climate change impacts may occur in the distant future, estimating the benefits of current climate change policy becomes a challenging and controversial task.[11] Some analysts have attempted, incorrectly, to estimate the benefits of avoiding the elimination of vital ecosystem services, such as pollination and nutrient cycling, using economic benefit estimation tools.[12] Used correctly, these tools measure our collective willingness to pay for small changes in the status quo. The elimination of Earth's vital ecosystem services would cause dramatic shifts in human and market activity of all kinds. While the benefit estimation techniques we have discussed are well suited to assessing the net effect of specific policies, like reducing air pollutant concentrations or setting aside land to preserve open space, they are inadequate to the task of measuring the value of drastic changes in global ecosystems—efforts to use them for this purpose have resulted in, as one economist quipped, a "serious underestimate of infinity."[13] Estimation of the benefits from environmental policy is the subject of a great deal of economic research, and much progress has been made. But, in some situations, the limits of these tools remain a significant challenge to comprehensive benefit–cost analysis.

Cost-Effective Environmental Policy

The economists' goal of maximizing net benefits is one of many competing goals in the policy process. Even when an environmental standard is inefficient (too stringent or not stringent enough), economic analysis can still help to select the particular policy instruments used to achieve that goal. Earlier we defined the concept of cost-effectiveness as choosing the policy that can achieve a given environmental standard at least cost. Let's return to our SO_2 example to see how this works in practice.

Imagine that you are a policy analyst at the EPA, given the job of figuring out how US power plants will meet the 10-million-ton reduction in SO_2 emissions required under the 1990 CAA Amendments. One important issue to consider is how much the policy will cost. All else equal, you would like to attain the new standard as cheaply as possible. We can reduce this problem to a simple case to demonstrate how an

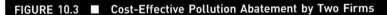

FIGURE 10.3 ■ Cost-Effective Pollution Abatement by Two Firms

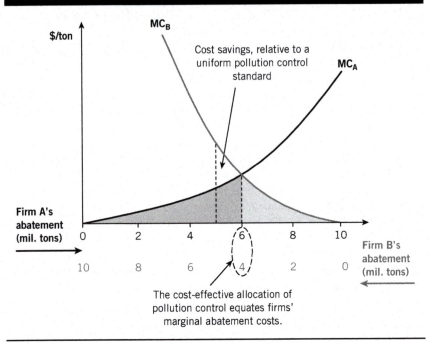

economist would answer this question. Assume that the entire 10-million-ton reduction will be achieved by two power plants, firms A and B. Each has a set of SO_2 abatement technologies, and the sequence of technologies for each firm and their associated costs determine the marginal cost curves in Figure 10.3, labeled MC_A and MC_B. Notice that abatement increases from left to right for firm A, and from right to left for firm B. At any point along the horizontal axis, the sum of the two firms' emissions reductions will always equal 10 million tons, as the CAA Amendments require.

Let's begin with one simple solution that seems like a fair approach: divide the total required reduction in half and ask each firm to abate 5 million tons of SO_2. This allocation of pollution control is represented by the left-most dotted vertical line in Figure 10.3 and is often referred to as a "uniform pollution control standard" because the abatement requirement is uniform across firms. Is this the cheapest way to reduce pollution by 10 million tons? Suppose we require firm A to reduce one extra ton and require firm B to reduce one ton less. We would still achieve a 10-million-ton reduction, but that last ton would cost less than it did before. Firm B's cost curve lies above A's at the uniform standard, so when we shift responsibility for abating that ton from B to A, we reduce the total cost of achieving the new standard. How long can we move to the right along the horizontal axis and continue to lower total costs? Until the marginal costs of abatement for the two firms are exactly equal: where the two curves intersect, when firm A abates 6 million tons and B abates 4 million tons.

The cost savings from allocating abatement in this way rather than using the uniform standard is equivalent to the difference in costs between firm A and firm B for the last million tons of abatement. On Figure 10.3, that is equal to the area between the two firms' marginal cost curves (the cost to firm B minus the cost to firm A), bounded by the dotted lines marking the uniform standard and the cost-effective allocation. In Figure 10.3, we demonstrate how this principle works for two firms. But the rule of thumb for a cost-effective environmental policy instrument is the same for a large number of firms as it is in this simple example: a pollution control policy minimizes costs if the marginal abatement costs of all firms reducing pollution under the policy are equal.

You may have noticed that, in order to identify the cost-effective pollution control allocation in this simple case, we needed a lot of information about each firm's abatement costs—we used their marginal cost curves to accomplish the task. As an EPA policy analyst, you would not likely have this information, and having it would be even less likely if you were to move from considering two power plants to looking at the thousands eventually covered by the CAA Amendments of 1990. In a market economy, the structure of firms' costs is proprietary information that they will not easily share with regulators. So how can a cost-effective pollution control policy be designed?

One class of environmental policy instruments, often called "market-based" or "incentive-based" approaches, does not require regulators to have specific information about individual firms' marginal abatement costs in order to attain a particular pollution control standard at least cost. In the case of the SO_2 emissions reduction required by the CAA Amendments of 1990, regulators chose one of these approaches, a system of tradable pollution permits, to achieve this goal.

A market for tradable pollution permits works quite simply. Return to the world of two firms and assume that you have advised the EPA administrator to allocate responsibility for 5 million tons of SO_2 emissions abatement to each of the two firms. You add, however, the provision that the firms should be able to trade their allocations, so long as the aggregate reduction of 10 million tons is achieved. If we begin at the (5 million, 5 million) allocation in Figure 10.3, we can imagine the incentives for the two firms under this trading policy.

The last ton abated by firm B costs much more than the last ton abated by A if each sticks to the 5-million-ton abatement requirement. Firm B would be willing to pay A to increase its own abatement, so that B could abate fewer tons of SO_2. In fact, B would pay any price that lies below its own marginal abatement cost. Firm A would be willing to make such a deal, so long as B paid more than A's own marginal abatement cost. So the vertical distance between the two cost curves at (5 million, 5 million) represents the potential gain from trading for that ton of abatement. The same is true of the next ton, and the next, all the way to the point at which B is abating 4 million tons and A is abating 6 million. Notice that the firms will trade permits until exactly the point at which the total costs of reducing 10 million tons of SO_2 are minimized.[14] The bigger the difference in abatement costs across regulated firms, the larger the potential cost savings from trading.

How much of a difference did this make in the SO_2 emissions abatement regulation we have been discussing? While it operated, the US SO_2 trading program produced cost savings of about $1.8 billion annually, compared with the most likely alternative policy considered during deliberations over the 1990 CAA Amendments (which would have required each firm to install the same technology to reduce emissions).[15] In fact, the Environmental Defense Fund, an environmental advocacy organization, agreed to endorse and help write the legislation proposed by the George H. W. Bush administration to amend the CAA if the administration would increase the required emissions reduction (from 8 million tons, which it was proposing) to the eventual 10 million tons of SO_2, based on the potential cost savings of the tradable permit approach. In this case, the tradable permit policy was not only cost-effective, but it allowed political actors to take a step closer to the efficient level of abatement, with significant benefits for human and ecosystem health.[16]

PRINCIPLES OF MARKET-BASED ENVIRONMENTAL POLICY

You may have noticed that the notion that firms A and B are inherently doing something wrong when they produce SO_2 emissions along with electricity has been lacking from our discussion. In economic terms, pollution is the result of a set of incentives facing firms and consumers that is "stacked against" environmental protection. The consumption and production decisions that result can be changed if we alter the relevant incentives. The social problem of pollution results from what economists call market failure. The three types of market failure most relevant to environmental policy are externalities, public goods, and the "tragedy of the commons."[17]

Let's consider externalities first. If you have driven a car, checked e-mail, or turned on a light today, you have contributed to the problem of global climate change. With near unanimity, the scientific community agrees that the accumulation of carbon dioxide (CO_2) and other heat-trapping gases in Earth's upper atmosphere has increased global mean surface temperatures by about 1 degree Fahrenheit since the start of the twentieth century, with consequences including sea level rise, regional changes in precipitation, increased frequency of extreme weather events, species migration and extinction, and spatial shifts in the prevalence of disease. Ask an economist, however, and he or she will suggest that the roots of the problem are not only in the complex dynamics of Earth's atmosphere but also in the incentives facing individuals and firms when they choose to consume and produce energy. Each such decision imposes a small cost in terms of its contribution to future atmospheric carbon concentrations. However, the individuals making these decisions do not bear these costs. Your electricity bill does include the cost of producing electricity and moving it from the plant to your home, but (in most locations) it does not include the cost of the carbon emissions from electricity production. Carbon emissions are an *externality*—their costs are external to the transaction between the buyer and the seller of electricity. The market for energy is incomplete, since its price does not reflect the full cost of its provision.[18]

You also may have noticed that bringing nations together to negotiate a solution to the problem of global climate change seems to be difficult. Clean air and a stable global climate are *public goods*: everyone benefits from their provision, whether or not they have contributed, and we can all enjoy these goods without interfering with the ability of others to enjoy them. Other public goods include national defense, weather forecasting, and public parks. If you have ever listened to public radio or watched public television without contributing money to these institutions, then you have been a free rider. Free riding is a rational response to the incentives created by a public good: many beneficiaries will pay nothing for its provision, and those who do pay will generally pay less than what, in their heart of hearts, they would be willing to pay. Markets for public goods are incomplete. Left to their own devices, markets will underprovide these valuable goods and services.

A third category of market failure relevant here is the tragedy of the commons.[19] A group of individuals sharing access to a common resource (a pasture for grazing cattle, a fishery, or a busy highway) will tend to overexploit it. The "tragedy" is that, if individuals could self-regulate and reduce their collective use of the resource, the productivity of the resource would increase to everyone's benefit. But the actions of single individuals are not enough to make a difference. Individuals restricting their own use only bear the costs of their restraint, with no benefits. The resulting spiral of overexploitation can destroy the resource entirely. A prominent example of the tragedy of the commons is the collapse of many deep-sea fisheries. Climate change is another example. The global upper atmosphere is a resource shared by everyone and owned by no one. The incentive for individual citizens to reduce carbon emissions (their exploitation of this resource) is small; thus the resource is overexploited.

When markets fail in these three ways and environmental damages result, government intervention may be required to fix the situation. Governments can correct externalities, provide public goods, and avert the tragedy of the commons. From an economic perspective, market principles should be used to correct these market failures.

Using Market Principles to Solve Environmental Problems

Like the global damages from carbon emissions, the local and regional damages from SO_2 emissions (for example, human health problems and acid rain) are external to power plants' production decisions and their consumers' decisions about how much energy to use. The tradable permit program described earlier is an example of using market principles to correct market failures. The government distributed permits to power plants and allowed them to trade. Plants made these trades by deciding how to minimize the costs of producing power—for each ton of SO_2 that they produced before the regulation was passed, they now faced a choice. Either they could continue to emit that ton, and use one of their permits, or they could spend money to abate that ton, freeing up a permit to sell to another power plant (and earning the permit price). The result was an active market for SO_2 emissions permits. In June 2008 alone, 250,000 tons of emissions were traded in this market, at an end-of-month price of $325 per ton. By putting a price on pollution, the government internalized its cost, represented by the price of a permit to emit one ton of SO_2, a cost that firms took into account when they decided how much electricity to produce. Unfortunately, regulatory and judicial actions in 2008–2010 effectively dismantled the US SO_2 allowance market, one of the most significant experiments with a market-based approach to pollution control.[20] When governments create environmental markets, even well-functioning markets that dramatically reduce the costs of pollution control, they can also dismantle them.

Another way to use market principles to reduce pollution is to impose a tax. Rather than imposing a cap on the quantity of pollution, and allowing regulated firms to trade emissions permits to establish a market price for pollution, a tax imposes a specific price on pollution and allows firms to decide how much to pollute in response. A tax has an effect on firms' decisions that resembles the effect of the permit price created by a cap-and-trade policy; polluters decide, for each ton of emissions, whether to abate that ton (incurring the resulting abatement costs) or to pay the tax and continue to emit that ton.

Some important differences exist between taxes and tradable permits. First, a cap-and-trade system pins down a total quantity of allowable pollution, and subsequent permit trading establishes the permit price. Before any trading has taken place, we know exactly how much pollution the policy is going to allow, but we are uncertain how high the costs will be. In setting a tax, regulators pin down the price of pollution instead. As under the permit policy, firms make private choices about how much pollution to emit, comparing their abatement costs to the tax. But the total quantity of pollution that will result is uncertain. To be certain about how much pollution will result after the tax is imposed, regulators need to have good information about the cost of reducing pollution in the regulated industry. Both the total quantity of allowable pollution and the total cost of pollution reduction are important pieces of information to consider in designing environmental policy. Each of the two primary market-based approaches to internalizing the cost of pollution, taxes and permits, offers certainty over one, but not both, of these important variables.[21]

Taxes and tradable permits can also differ in their costs to regulated firms. Under a pollution tax, a firm must pay the tax for every ton of pollution it emits. Under a permit system, permits are typically given to firms for free, at least initially. So a permit system requires a firm to pay only for pollution in excess of its permit allocation. For this reason, complying with a tax can be more expensive for firms than complying with a permit system.[22] If taxes and tradable permits are fundamentally equivalent pollution regulations from an economic perspective, what about those extra compliance costs for firms? Taxes create revenues for the government agency that collects them, exactly equal to the aggregate tax bill for all firms, whereas tradable permits distributed for free do not. The net impact of this difference between the two policies depends on how tax revenues are spent. From the standpoint of efficiency, the best thing to do with environmental tax revenues may be to use them to reduce other taxes in the economy that tend to distort consumers' and firms' decisions—taxes on income, sales, and capital gains, for example.[23] The economy may benefit to a smaller degree if governments rebate tax revenues to citizens, or use revenues to provide additional goods and services.

Pollution taxes may also be imposed indirectly. For example, many countries tax gasoline. In the United States, the revenues from most state gasoline taxes are used to maintain and expand transportation infrastructure; US gas taxes are not explicit pollution control policies. However, economists have estimated the optimal US gasoline tax, taking into account the most significant externalities: emissions of local pollutants (particulate matter and nitrous oxides), CO_2 emissions (which contribute to global climate change), traffic congestion, and the costs of accidents not borne by drivers.[24] In their estimation, the efficient tax would be $0.83 per gallon. A more recent study suggests potential regional heterogeneity in the optimal gas tax; California's efficient tax may be about $1.37 per gallon.[25] However, most of these externalities depend on the number of miles driven, not the amount of gasoline consumed (only a proxy for miles driven); taxing miles driven rather than gasoline would be better from an economic perspective. This highlights the important fact that the choice of what to tax may be as important as the level of a tax.

Notice that as we have discussed market-based approaches to environmental policy, all of the options we have mentioned require some role for government intervention, setting a tax, for example, or enforcing a cap on pollution. This is a critical point. Market-based approaches should not be conflated with voluntary (nonregulatory) environmental policies.

MARKET-BASED ENVIRONMENTAL POLICY INSTRUMENTS IN PRACTICE

Reducing Air and Water Pollution

Taxes on air and water pollution are common, though existing environmental taxes tend to be lower than efficient levels.[26] The Canadian province of British Columbia established a carbon tax in 2008, currently set at about $30 (US) per ton. In 2019, Canada implemented a nationwide carbon pricing policy, which sets a minimum carbon price (ramping up over time) and allows provinces to achieve it by various means, including cap-and-trade (like the system already in place in Quebec) and a carbon tax (such as that in British Columbia). The Canadian government imposes the national minimum price on provinces without a system of their own. Australia levied a carbon tax beginning in 2012 but repealed it in July 2014, after an election in which debate over the carbon tax played an important role. Among developing countries, China's pollution levy system appears to have reduced both air and water pollution.[27] Also in China, the Lake Tai Basin "pay for permit" policy reduced water pollution emissions from regulated industrial plants by about 40 percent in its first two years.[28]

We have already discussed the US SO_2 trading program as a classic cap-and-trade example, and there are many others. For example, in the 1980s, the EPA implemented a lead trading policy to enforce a regulation reducing the allowable lead content of US gasoline by 90 percent. Earlier in this chapter, we discussed the benefit–cost analysis of this policy, which suggested that the benefits of eliminating lead in gasoline exceeded its costs. The policy the EPA chose to implement the lead phasedown had something to do with this. Refiners producing gasoline with a lower lead content than was required earned credits that could be traded and banked. This policy successfully met its environmental goal, at a cost savings of approximately $250 million per year, relative to prescriptive approaches, until the phasedown was completed in 1987.[29]

Many cap-and-trade policies focus on mitigating the greenhouse gas (GHG) emissions that are changing the global climate.[30] Among industrialized countries that took on emissions reduction targets under the 1997 Kyoto Protocol, the countries of the EU opted to use an emissions trading system (ETS), established in 2005, to meet their emissions reduction targets, and this program remains an important component of the EU's approach to meeting its obligations under the 2015 Paris Agreement. The ETS sets a cap on CO_2 emissions for the EU as a whole, allocated by the EU to member countries. Covered industries include: electric power generation; refineries; iron and steel; cement, glass, and ceramics; pulp and paper; and air travel. The EU ETS is the world's largest ETS, covering almost 12,000 facilities in 27 countries in 2020, and accounting for about 45 percent of EU CO_2 emissions.[31] Determining the impact of the EU ETS on emissions is difficult, in part because the binding of caps beginning in 2008 coincided with the Great Recession (resulting in falling CO_2 emissions). However, studies suggest that the ETS has decreased emissions by 2–5 percent below "business as usual" in the pilot phase, by about 8 percent in 2008–2009, and by almost 4 percent from 2008 to 2016.[32] The cap in the EU system in the pilot phase was a small reduction below expected emissions in the absence of the policy, and in retrospect permits in the pilot phase were overallocated. Multiple reductions in the cap from the pilot through phase 3, partial postponement of permit auctioning from the early years of phase 3 toward the later years, and the 2019 introduction of a Market Stability Reserve (which responds to surpluses and shortages automatically by adjusting available permits) have put upward pressure on prices since the program's early years.[33]

China's emerging cap-and-trade system for CO_2 emissions, launched in 2017, could eventually become the world's largest, surpassing the EU ETS. It is too early to assess this system's effectiveness, however.

The Obama administration committed the United States to reduce GHG emissions by about 25 percent over 2005 levels by 2025 when it signed the 2015 Paris Agreement, largely relying on prescriptive regulations (discussed further in Chapters 7 and 12) to meet this goal.[34] Many of these regulations have been repealed or are in the process of repeal by the Trump administration, which announced in 2017 its intention to withdraw from the Paris Agreement (effective November 2020). However, support for action on climate change has led some US states to enact cap-and-trade policies. California's Global Warming Solutions Act of 2006 (AB 32) committed to reducing GHG emissions to 1990 levels by 2020, with further commitments to reduce by 40 percent below 1990 by 2030. A cap-and-trade policy, which began in 2013, is an important component of California's approach. California's market covers 85 percent of GHG emissions (from power plants, large industrial sources, natural gas distribution, and the transportation sector). The cap declines annually, and the program linked with Quebec's permit market in 2013 (an aspect challenged in court by the Trump administration in 2019). In 2020, permits trade around $15 per ton, a level many suggest is too low to make a substantial dent in emissions, due to a relatively lenient cap thus far.

Another US market-based initiative is the Regional Greenhouse Gas Initiative (RGGI), a cap-and-trade system among electricity generators in nine northeastern states, covering about 18 percent of regional emissions in 2020. The RGGI began in 2009, so similar to the EU ETS, this timing makes it difficult to disentangle Great Recession–related emissions decreases from those due to RGGI. Careful work accounting for this suggests that RGGI has reduced CO_2 emissions within participating states by about 9 million tons per year (2004–2012), counterbalanced by an emissions increase in surrounding states equal to about one-half of that effect.[35]

Individual Tradable Quotas for Fishing

Thus far, we have talked about cap-and-trade policies as if they applied only to pollution problems. But a common application is to fisheries management, to avert the tragedy of the commons. The world's largest market for tradable individual fishing quota (IFQ), created in 1986, is in New Zealand.[36] By 2004, it covered seventy different fish species, and the government of New Zealand had divided coastal waters into "species-regions," generating 275 separate markets that covered more than 85 percent of the commercial catch in New Zealand's territorial waters. In the United States, Pacific halibut and sablefish off the coast of Alaska, mid-Atlantic surf clams and ocean quahogs, South Atlantic wreckfish, and red snapper in the Gulf of Mexico are all regulated using IFQ markets. Iceland manages stocks of twenty fish and shellfish species using IFQ markets, in a system established in 1990.

A market for fishing quota works similarly to a market for pollution permits. The government establishes a total allowable catch (TAC), distributing shares to individual fishers. Fishers can trade their assigned quota, which represents a percentage of the TAC for a particular species-region. An analysis of catch statistics from 11,315 global fisheries between 1950 and 2003 provides the first large-scale empirical evidence for the effectiveness of these approaches in halting, and even reversing, the global trend toward fisheries collapse. The authors empirically estimate the relative advantage of IFQ fisheries over non-IFQ fisheries in terms of a lower probability of collapse, and estimate that, had all non-IFQ global fisheries switched to management through tradable quota in 1970, the percentage of collapsed global fisheries by 2003 could have been reduced from more than 25 percent to about 9 percent.[37]

Waste Management Policies

Market-based approaches have also been used to manage solid waste. Some waste products have high recycling value. If you live in a community with curbside recycling, you may have seen low-income residents of the community picking aluminum cans out of recycling bins at the curb; they do this because it is much less costly to produce aluminum from scrap metal than from virgin ore, and as a result, those cans are quite valuable. But most household waste ends up as trash, disposed of legally in landfills or incinerators or illegally dumped. The marginal cost of public garbage collection and disposal for an American household has been estimated at $1.03 per trash bag, but until recently, the marginal cost of disposal borne by households was approximately zero.[38] An increasingly common waste management policy is the "pay-as-you-throw" system, a volume-based waste disposal charge often assessed as a requirement for the purchase of official garbage bags, stickers to attach to bags of specific volume, periodic disposal charges for official city trash cans of particular sizes, and (rarely) charges based on the measured weight of curbside trash. These systems function like an environmental tax, internalizing the costs of disposing of household waste. In 2006, more than 7,000 US communities had some form of pay-as-you-throw disposal.[39] Studies suggest that these policies increase the sorted waste ratio by 25–35 percent, and reduce municipal solid waste disposal by 15–74 percent.[40] For example, an 80-cent tax per bag in Charlottesville, Virginia, reduced the number of bags households threw out by about 37 percent.[41] However, the effect was offset by two factors: (1) the reduction in the total weight of trash thrown away was much smaller (about 14 percent), since consumers compacted their trash in order to reduce the number of bags they used; and (2) illegal disposal increased. A study of two hundred towns in New Hampshire suggests that the introduction of such policies reduces municipal solid waste generation very significantly, with an additional marginal effect of increasing the disposal cost per bag.[42]

Habitat and Land Management Policies

Tradable development rights (TDRs) have been applied to solve problems as diverse as deforestation in the Brazilian Amazon and the development of former farmland in the Maryland suburbs of Washington, DC.[43] About 140 US communities have implemented TDRs, with many other potential programs in the pipeline.[44] The program in Calvert County, Maryland, preserved an estimated thirteen thousand acres of farmland between 1978 and 2005. In Brazil since 1998, TDRs have been used to slow the conversion of ecologically valuable lands to agriculture; each parcel of private property that is developed must be offset by preserving a forested parcel elsewhere (within the same ecosystem and with land of greater or equal ecological value). Landowners can develop the most profitable land and preserve less profitable land. But this raises two key issues. First, how can regulators ensure that a preserved parcel is really additional—that it would not have remained in forest without the developer's efforts? Second, are the traded land parcels really ecologically equivalent? In the case of carbon emissions, each ton of emissions has essentially the same impact on atmospheric carbon concentrations, no matter where it is emitted. In contrast, the effects of land preservation are spatially heterogeneous.

A related policy, wetlands mitigation banking, holds similar promise and faces similar challenges. Wetlands are classic public goods. They provide a rich set of ecosystem services, for which there are no markets and from which everyone benefits, regardless of who pays for their preservation. Wetlands have been depleted rapidly in the United States and other parts of the world by conversion to agricultural and urban use. The externalities to wetlands conversion include increased flood risk; loss of habitat for birds, fish, and mammals; and reduced groundwater recharge. Since the early 1990s, the United States has experimented with mitigation banking, a policy under which land developers must compensate for any lost wetlands by preserving, expanding, or creating wetlands elsewhere.[45] Wetland banks serve as central brokers, allowing developers to purchase credits and to fulfill credits through the physical process of wetlands preservation, creation, and management. In 2020, there were more than 3,700 US wetland banks listed in the US Army Corps of Engineers' database that tracks these institutions.

As in the case of TDRs, wetlands mitigation banking reduces the costs of preserving wetlands acreage but faces significant challenges. A wetland in a particular location provides a specific portfolio of biophysical services. For example, coastal wetlands support shellfish nurseries and may reduce damages from storm-related flooding. Inland wetlands may filter contaminants and provide islands of habitat for migratory bird species in overland flight. If development pressures in coastal cities create incentives for landowners to develop wetlands in these locations, and to pay for wetlands creation inland, the net effect of these kinds of trades must be considered. This is a significant change from the simple ton-for-ton trading that occurs for some air pollutants.

This section offered a handful of examples of the many applications of market principles to environmental policy.[46] We emphasized the strong arguments in favor of taxes, tradable permits, and other market-based approaches, especially given that these policies can achieve environmental policy goals at less cost than more prescriptive approaches. But they are not appropriate solutions to all environmental problems. The issues we raised in our discussion of market-based land management policies arise in other contexts, as well. Market-based policies can be designed for situations in which the location of pollution emissions or natural resource amenities matters for the benefits of pollution control or resource management. But they are not workable in extreme cases. For example, the impacts of a toxic waste dump are highly localized. Economists would not advise setting a national limit on toxic waste disposal, allowing firms to trade disposal permits, and letting the waste end up where it may. Environmental problems at this end of the spectrum—the opposite end from a problem like carbon emissions, which can be reduced anywhere with essentially the same net effect—may be better addressed through prescriptive approaches.

CONCLUSION

Economics offers a powerful pair of tools—efficiency and cost-effectiveness—with which to evaluate environmental policy trade-offs. Efficiency has to do with the setting of environmental policy goals: how much pollution should we reduce, or how many acres of wetlands should be preserved? An efficient pollution control policy equalizes the monetized benefits and costs of the last ton of pollution eliminated. The process used to determine whether an environmental policy is efficient is benefit–cost analysis. If a strict benefit–cost test were applied to the decision of how much SO_2 pollution to eliminate from power plant smokestacks in the 1990s, additional reductions would have been required. Benefit–cost analysis would also have suggested that lead be eliminated from US gasoline sooner than it was. In other cases, applying the rule of efficiency suggests that environmental standards should be weakened.

However, efficiency is not the only potential input to good environmental policy. The treatment of benefit–cost analysis in the major US environmental statutes is a good indication of our ambivalence toward analyzing environmental trade-offs in this systematic way; the statutes alternately "forbid, inhibit, tolerate, allow, invite, or require the use of economic analysis in environmental decision making."[47] For example, the CAA forbids the consideration of costs in setting the National Ambient Air Quality Standards, and the US Safe Drinking Water Act requires benefit–cost analysis of all new drinking water contaminant standards. Environmental regulatory agencies are not economic agencies. While laying out the trade-offs involved in setting environmental standards is critically important from an economic perspective, political, social, and ethical concerns may hold more influence than benefit–cost analysis in this process.

Even when environmental standards are inefficient, however, policymakers can choose policies to achieve those standards at least cost. The billions of dollars saved by policies like SO_2 trading, paired with their proven environmental effectiveness, have made market-based approaches such as tradable permits and taxes appealing policy instruments for solving environmental problems. Ironically, as the theoretical and empirical evidence of the cost savings and environmental effectiveness of these policies has stacked up, US political support for them has not only eroded but also been thoroughly demonized, and especially so by Republicans in Congress,[48] in direct conflict with their support for markets in most other contexts. At the same time, Democratic politicians have shifted away from discussion of carbon taxes and cap-and-trade toward ambitious but prescriptive climate proposals such as the Green New Deal. These political shifts may make it much harder in the future to implement market-based policies, unnecessarily raising the cost of environmental regulation. Market-based environmental policy approaches are not appropriate for all situations. But where market incentives create environmental problems, market principles should be harnessed to solve them.

SUGGESTED WEBSITES

Center for Effective Government (www.foreffectivegov.org) This center follows budgets and regulatory policies.

Climate Leadership Council (www.clcouncil.org) This nonprofit, bipartisan institute advocates for a specific US carbon tax proposal of its own design, endorsed by thousands of economists from across the political spectrum.

EPA National Center for Environmental Economics (www.epa.gov/environmental-economics) This page provides access to research reports, regulatory impact analyses, and other EPA publications.

Resources for the Future (www.rff.org) This nonprofit research organization is devoted to environmental and resource economics and policy.

NOTES

1. For competing scientific opinions about Snake River salmon decline, see Charles C. Mann and Mark L. Plummer, "Can Science Rescue Salmon?" *Science* 298 (2000): 716–19, with letters and responses. See also David L. Halsing and Michael R. Moore, "Cost-Effective Management Alternatives for Snake River Chinook Salmon: A Biological-Economic Synthesis," *Conservation Biology* 22, no. 2 (2000): 338–50.
2. In reality, each ton of emissions is not equivalent. SO_2 emissions in urban areas or those upwind of critical ecosystems may cause relatively greater harm. In addition, costs and benefits may not follow the smooth, continuous functions we depict in Figure 10.1. The simple curves help us to develop intuition that carries through even in more complex situations.
3. Since marginal benefits and costs represent the rate at which benefits and costs change when we add an additional ton of emissions reduction, they are also measured by the slope of the total benefit and cost curves; notice that net benefits are largest in Figure 10.1 (at Q^*) when the slopes of B(Q) and C(Q) are equal.
4. Dallas Burtraw, Alan Krupnick, Erin Mansur, David Austin, and Deirdre Farrell, "The Costs and Benefits of Reducing Air Pollutants Related to Acid Rain," *Contemporary Economic Policy* 16 (1998): 379–400.
5. Kenneth J. Arrow, Maureen L. Cropper, George C. Eads, Robert W. Hahn, Lester B. Lave, Roger G. Noll, Paul R. Portney, Milton Russell, Richard Schmalensee, V. Kerry Smith, and Robert N. Stavins, "Is There a Role for Benefit-Cost Analysis in Environmental, Health, and Safety Regulation?" *Science* (April 12, 1996): 221–22.

6. Stephen Kelman, "Cost-Benefit Analysis: An Ethical Critique," with replies, *AEI Journal on Government and Social Regulation* 5, no. 1 (January/February 1981): 33–40.

7. Frank Ackerman and Lisa Heinzerling, *Priceless: On Knowing the Price of Everything and the Value of Nothing* (New York, NY: New Press, 2004).

8. This discussion is based on Nathaniel O. Keohane, Benjamin Van Roy, and Richard J. Zeckhauser, "Managing the Quality of a Resource with Stock and Flow Controls," *Journal of Public Economics* 91 (2007): 541–69.

9. See A. Myrick Freeman III, Joseph A. Herriges, and Catherine L. Kling, *The Measurement of Environmental and Resource Values*, 3rd ed. (Washington, DC: Resources for the Future, 2014).

10. Albert L. Nichols, "Lead in Gasoline," in *Economic Analyses at EPA: Assessing Regulatory Impact*, ed. Richard D. Morgenstern (Washington, DC: Resources for the Future, 1997), 49–86.

11. See Martin L. Weitzman, "A Review of the *Stern Review on the Economics of Climate Change*," *Journal of Economic Literature* 45, no. 3 (2007): 703–24.

12. Robert Costanza, Ralph d'Arge, Rudolf de Groot, Stephen Farber, Monica Grasso, Bruce Hannon, Karin Limburg, Shahid Naeem, Robert V. O'Neill, Jose Paruelo, Robert G. Raskin, Paul Sutton, and Marjan van den Belt, "The Value of the World's Ecosystem Services and Natural Capital," *Nature* 387 (1997): 253–60.

13. Michael Toman, "Why Not to Calculate the Value of the World's Ecosystem Services and Natural Capital," *Ecological Economics* 25 (1998): 57–60.

14. We emphasize the cost-effectiveness of market-based approaches to environmental policy in the short run, a critical concept and one that is relatively easy to develop at an intuitive level. However, the greatest potential cost savings from these types of environmental policies may be achieved in the long run. Because they require firms to pay to pollute, market-based policies provide strong incentives for regulated firms to invest in technologies that reduce pollution abatement costs over time, either developing these technologies themselves or adopting cheaper pollution control technologies developed elsewhere.

15. Nathaniel O. Keohane, "Cost Savings from Allowance Trading in the 1990 Clean Air Act," in *Moving to Markets in Environmental Regulation: Lessons from Twenty Years of Experience*, eds. Charles E. Kolstad and Jody Freeman (New York, NY: Oxford University Press, 2007).

16. The definitive overview of the SO_2 permit-trading program is found in A. Denny Ellerman, Richard Schmalensee, Elizabeth M. Bailey, Paul L. Joskow, and Juan-Pablo Montero, *Markets for Clean Air: The U.S. Acid Rain Program* (New York, NY: Cambridge University Press, 2000).

17. These concepts are described in much greater detail in Nathaniel O. Keohane and Sheila M. Olmstead, *Markets and the Environment*, 2nd ed. (Washington, DC: Island Press, 2007), Chapter 5.

18. Pollution is a negative externality, but externalities can also be positive. For example, a child vaccinated against measles benefits because she is unlikely to contract that disease. But a vaccinated child in turn benefits her family, neighbors, and schoolmates, as she is less likely to expose them to disease.

19. Garrett Hardin, "The Tragedy of the Commons," *Science* 162 (1968): 1243–48.

20. Richard Schmalensee and Robert N. Stavins, "The SO_2 Allowance Trading System: The Ironic History of a Grand Policy Experiment," *Journal of Economic Perspectives* 27 (2013): 103–22.

21. This difference between the two approaches—taxes and permits—can cause one approach to be more efficient than the other; see Martin L. Weitzman, "Prices v. Quantities," *Review of Economic Studies* 41 (1974): 477–91.

22. This distinction disappears if permits are auctioned rather than given away. But auctioned permit systems are less common than free allocation.

23. This decision of how best to use the revenues from an environmental tax is complicated because environmental taxes can exacerbate the distortions introduced by other taxes. For a straightforward discussion of this problem and for other comparisons between taxes and permits, see Lawrence H. Goulder and Ian W. H. Parry, "Instrument Choice in Environmental Policy," *Review of Environmental Economics and Policy* 2, no. 2 (2007): 152–74.

24. Ian Parry and Kenneth Small, "Does Britain or the United States Have the Right Gasoline Tax?" *American Economic Review* 95 (2005): 1276–89.

25. C.-Y. Cynthia Lin and Lea Prince, "The Optimal Gas Tax for California," *Energy Policy* 37, no. 12 (2009): 5173–83.

26. Robert N. Stavins, "Experience with Market-Based Environmental Policy Instruments," in *Handbook of Environmental Economics*, vol. 1, eds. Karl-Göran Mäler and Jeffrey Vincent (Amsterdam: Elsevier Science, 2003), 355–435.

27. Hua Wang, "Pollution Regulation and Abatement Efforts: Evidence from China," *Ecological Economics* 41, no. 1 (2002): 85–94; and Avraham Ebenstein, "The Consequences of Industrialization: Evidence from Water Pollution and Digestive Cancers in China," *Review of Economics and Statistics* 94, no. 1 (2012): 186–201.

28. Pan He and Zhang Bing, "Environmental Tax, Polluting Plants' Strategies and Effectiveness: Evidence from China," *Journal of Policy Analysis and Management* 37, no. 3 (2018): 493–520.

29. EPA, Office of Policy Analysis, *Costs and Benefits of Reducing Lead in Gasoline, Final Regulatory Impact Analysis* (Washington, DC: EPA, 1985).

30. The World Bank surveys global carbon pricing policies every two years. See: World Bank Group, *State and Trends of Carbon Pricing 2020*, Washington, DC, available at: https://openknowledge.worldbank.org/handle/10986/33809.

31. See A. Denny Ellerman and Barbara K. Buchner, "The European Union Emissions Trading Scheme: Origins, Allocation, and Early Results," *Review of Environmental Economics and Policy* 1, no. 1 (2007): 66–87; A. Denny Ellerman and Paul L. Joskow, *The European Union's Emissions Trading System in Perspective* (Washington, DC: Pew Center on Global Climate Change, 2008); Frank J. Convery and Luke Redmond, "Market and Price Developments in the European Union Emissions Trading Scheme," *Review of Environmental Economics and Policy* 1, no. 1 (2007): 88–111; and European Commission, *The EU Emissions Trading System (EU ETS)*, http://ec.europa.eu/clima/policies/ets/index_en.htm.

32. Christian Egenhofer, Monica Alessi, Anton Georgiev, and Noriko Fujiwara, *The EU Emissions Trading System and Climate Policy towards 2050: Real Incentives to Reduce Emissions and Drive Innovation?* CEPS Special Report (Brussels: Centre for European Policy Studies, 2011); and Patrick Bayer and Michaël Aklin, "The European Union Emissions Trading System Reduced CO_2 Emissions Despite Low Prices," *Proceedings of the National Academy of Sciences* 117 (2020): 8804–12.

33. See A. Denny Ellerman and Barbara K. Buchner, "The European Union Emissions Trading Scheme: Origins, Allocation, and Early Results," *Review of Environmental Economics and Policy* 1, no. 1 (2007): 66–87; A. Denny Ellerman and Paul L. Joskow, *The European Union's Emissions Trading System in Perspective* (Washington, DC: Pew Center on Global Climate Change, 2008); Frank J. Convery and Luke Redmond, "Market and Price Developments in the European Union Emissions Trading Scheme," *Review of Environmental Economics and Policy* 1, no. 1 (2007): 88–111; and Sascha Kollenberg and Luca Taschini, "Dynamic Supply Adjustment and Banking Under Uncertainty in an

Emission Trading Scheme: The Market Stability Reserve," *European Economic Review* 118 (2019): 213–26.

34. In June 2009, the US federal government appeared to be taking significant steps toward setting up an economy-wide cap-and-trade system to reduce CO_2 emissions, with the passage in the House of Representatives of the American Clean Energy and Security Act, also known as the Waxman-Markey bill. In July 2010, the US Senate abandoned its effort to draft companion legislation (see Chapter 5).

35. Harrison Fell and Peter Maniloff, "Leakage in Regional Environmental Policy: The Case of the Regional Greenhouse Gas Initiative," *Journal of Environmental Economics and Management* 87 (2018): 1–23.

36. See Suzanne Iudicello, Michael Weber, and Robert Wieland, *Fish, Markets and Fishermen: The Economics of Overfishing* (Washington, DC: Island Press, 1999). For assessments of New Zealand's policy, see John H. Annala, "New Zealand's ITQ System: Have the First Eight Years Been a Success or a Failure?" *Reviews in Fish Biology and Fisheries* 6 (1996): 43–62; and Richard G. Newell, James N. Sanchirico, and Suzi Kerr, "Fishing Quota Markets," *Journal of Environmental Economics and Management* 49, no. 3 (2005): 437–62.

37. Christopher Costello, Stephen D. Gaines, and John Lynham, "Can Catch Shares Prevent Fisheries Collapse?" *Science* 321 (2008): 1678–81.

38. See Robert Repetto, Roger C. Dower, Robin Jenkins, and Jacqueline Geoghegan, *Green Fees: How a Tax Shift Can Work for the Environment and the Economy* (Washington, DC: World Resources Institute, 1992).

39. Skumatz Economic Research Associates, Inc., "Pay as You Throw (PAYT) in the U.S.: 2006 Update and Analyses," Final report to the EPA Office of Solid Waste, December 30, 2006, Superior, CO. See https://archive.epa.gov/wastes/conserve/tools/payt/web/pdf/sera06.pdf.

40. Ju-Chin Huang, John M. Halstead, and Shanna B. Saunders, "Managing Municipal Solid Waste with Unit-Based Pricing: Policy Effects and Responsiveness to Pricing," *Land Economics* 87, no. 4 (2011): 645–60; and Alessandro Bucciol, Natalia Montinari, and Marco Piovesan, "Do Not Trash the Incentive! Monetary Incentives and Waste Sorting," *Scandinavian Journal of Economics* 117, no. 4 (2015): 1204–29.

41. Don Fullerton and Thomas C. Kinnaman, "Household Responses to Pricing Garbage by the Bag," *American Economic Review* 86, no. 4 (1996): 971–84.

42. Ju-Chin Huang, John M. Halstead, and Shanna B. Saunders, "Managing Municipal Solid Waste with Unit-Based Pricing: Policy Effects and Responsiveness to Pricing," *Land Economics* 87, no. 4 (2011): 645–60.

43. On Brazil, see Kenneth M. Chomitz, "Transferable Development Rights and Forest Protection: An Exploratory Analysis," *International Regional Science Review* 27, no. 3 (2004): 348–73. On Calvert County, Maryland, see Virginia McConnell, Margaret Walls, and Elizabeth Kopits, "Zoning, Transferable Development Rights and the Density of Development," *Journal of Urban Economics* 59 (2006): 440–57.

44. Virginia McConnell and Margaret Walls, "U.S. Experience with Transferable Development Rights," *Review of Environmental Economics and Policy* 3, no. 2 (2009): 288–303.

45. National Research Council, *Compensating for Wetland Losses under the Clean Water Act* (Washington, DC: National Academies Press, 2001); and David Salvesen, Lindell L. Marsh, and Douglas R. Porter, eds., *Mitigation Banking: Theory and Practice* (Washington, DC: Island Press, 1996).

46. For surveys of these approaches, see Stavins, "Experience with Market-Based Environmental Policy Instruments"; Thomas Sterner and Jessica Coria, *Policy Instruments for Environmental and Natural Resource Management*, 2nd ed. (Washington, DC: Resources

for the Future, 2011); and Theodore Panayotou, *Instruments of Change: Motivating and Financing Sustainable Development* (London: Earthscan, 1998).

47. Richard D. Morgenstern, "Decision Making at EPA: Economics, Incentives and Efficiency," draft conference paper in *EPA at Thirty: Evaluating and Improving the Environmental Protection Agency* (Durham, NC: Duke University Press, 2000), 36–8.

48. Richard Schmalensee and Robert N. Stavins, "The SO₂ Allowance Trading System: The Ironic History of a Grand Policy Experiment," *Journal of Economic Perspectives* 27 (2013): 103–22.

11

SUSTAINABILITY AND RESILIENCE IN CITIES
What Cities Are Doing

Kent E. Portney and Bryce Hannibal

If the "sustainability epoch" of environmental policy described by Daniel Mazmanian and Michael Kraft seems completely foreign to national policymakers in Washington, DC, it has nonetheless taken hold in many of the world's cities, including in numerous cities in the United States.[1] This chapter takes a close look at what cities in the United States are doing in order to try to become more sustainable and resilient. The focus here is on local public policies and programs that are designed to make progress toward protecting and improving cities' biophysical environments while still seeking to grow municipal economies. Over the last twenty years or more, many US cities have made significant commitments to achieving these goals. They have enacted and implemented many different types of programs and policies and have sought to do this with an eye toward becoming more livable, sustainable, and resilient places.

Of course, cities vary in how seriously they pursue these goals, and they vary in what kinds of programs and policies they create. After briefly discussing the idea and origins of city sustainability, this chapter contrasts sustainability with the emerging idea of resilience, and then reviews the wide array of local policies and programs that major US cities have adopted and implemented in the pursuit of these goals, providing examples from a range of cities that have made significant commitments. The chapter ends with a look at some of the challenges that cities face as they try to do more and then suggests some strategies to facilitate these efforts.

THE IDEA OF SUSTAINABILITY IN CITIES

Sustainability is a very broad, and sometimes misunderstood, concept that has evolved over the last thirty years or so.[2] The core of the concept of sustainability, as articulated in the 1987 report of the United Nations' World Commission on Environment and Development, refers to economic development activity that "meets the needs of the present without compromising the ability of future generations to meet their own needs."[3] This means that sustainability does not readily accept a trade-off between protecting the environment and growing the economy; rather, the two go hand in hand to support and improve human well-being. Sustainability has become an important principle in a variety of contexts, including the private sector (as discussed by Olmstead in Chapter 10 and Fiorino in Chapter 14), state government (as discussed by Rabe in Chapter 2), and others.

Over the more than twenty years since US cities started developing sustainability programs, experience and research have painted a fairly clear picture of what cities can do if they wish to try to become more sustainable places. One aspect of sustainability that seems to be constant across nearly all definitions is the fact that the environment is at the core of what it means to be, or become more, sustainable. Today, as will be elaborated below, cities' efforts to try to become more sustainable have included policies and programs on mass transit and transit-oriented development, smart growth, energy efficiency, housing densification, water conservation and protection, climate protection (climate mitigation and adaptation), carbon footprint reduction, urban agriculture and food systems, and many other programs. Although early research argued that such sustainability programs only were found in wealthy places, places with a well-educated population, or cities on the West Coast, subsequent analysis has firmly established that communities of all sorts have successfully pursued sustainability policies. The fact is, many US cities—indeed, cities around the world—have come to understand that to grow and thrive, they need to pay close attention to the quality of life offered to current and prospective residents. Cities may embark on sustainability policies because they wish to be responsive to demands placed by residents through local environmental and other nonprofit groups, but they quickly learn that these policies represent a new and effective way to engage in economic development.

BUILDING RESILIENCE IN CITIES

Resilience has risen in prominence and popularity in urban planning and among city managers in the recent years. Many cities and several nonprofit organizations have dedicated significant resources to building resilience. Like the concept of sustainability, resilience is broadly defined and applicable to a wide array of social, psychological, physical, and biological phenomena. Building resilience can include numerous policy strategies and programs.

Defining resilience has been a challenge for many stakeholders and scholars. Cities use resilience as a concept that reflects the ability to "bounce back" from some external shocks or stressors such as natural disasters. In the narrowest definition, cities seek to prepare for various potential shocks by improving their respective disaster and emergency management functions. Other, more expansive, definitions try to capture the idea that instead of bringing cities back to the condition before the shock, they should strike to "bounce forward" so that the city is better prepared to resist and recover from future shocks. Such broader definitions include aspects of adaptation to changing climate conditions, such as moving people out of harm's way and reducing social vulnerabilities. Building city resilience, then, can include a variety of strategies that assist in facilitating bouncing back, recovering quickly, and adapting.

Cities undertake many initiatives to build resilience such as hardening physical infrastructure, creating redundant city service systems, poverty reduction, equitable access to healthy food and clean water, and others. Numerous policy areas have been used by cities to try to build resilience. Specifically, some cities have used land use policies, regulated through zoning and building codes, transportation infrastructure, climate mitigation through pollution and carbon reduction programs, enhanced public health, and others. Building resilience does not look the same in all cities. Resilience building efforts in coastal cities, such as Fort Lauderdale or San Diego, look dramatically different than resilience efforts in noncoastal Tulsa or Dallas. Pragmatically, building resilience in a city needs to take into account what areas of that city are the most vulnerable. This also influences what policies and programs carry the most impact to the population. Regardless of the specific policies and programs, specific projects and upgrades to build resilience, such a renewed or replaced infrastructure, cost money, often in the form of capital expenses. Cities often need to create innovative financing strategies in order to fund some of the more capital-intensive projects.

Perhaps the most obvious need in many cities today is to build this kind of resilience in anticipation of climate change that could well pose major risks to infrastructure and public well-being, particularly to coastal cities and others that are prone to flooding from increased rainfall or to droughts from insufficient precipitation. As the evidence of climate change mounts (see Chapter 12), cities and states may well find themselves searching for effective ways to respond. Similarly, the coronavirus pandemic in 2020 demonstrated significant risks to cities and neighborhoods where high density housing and concentrated retail stores and restaurants make it difficult for residents to maintain a safe distance when circumstances warrant doing so. Impacts from both climate change and pandemics also point to the need to rethink city and state planning and budgets to ensure adequate resources are available when needed.

CITY SUSTAINABILITY POLICIES AND PROGRAMS

What can a city do to try to become more sustainable? In fact, based on the efforts of cities all around the United States, cities can do a lot. What have cities done as a matter of local public policy in order to try to become more sustainable? By 2016, at least forty-nine of the fifty-five largest cities in the United States had created significant sustainability programs, and only five seem to have made no effort to try explicitly to become more sustainable as a matter of public policy. By 2020, nearly every major city in the nation had created some program oriented toward building or becoming more sustainable.[4] Sustainability programs take many different forms and include many different programmatic elements. In Seattle, the heart of the program is found in the city's comprehensive plan *Toward a Sustainable Seattle*. In New York City, sustainability represents the core of its *OneNYC*. In Philadelphia, the program is encapsulated in its *Greenworks Philadelphia* program. In Denver, the sustainability program got its start in the *Greenprint Denver* program. After more than twenty years of experience, cities have tended to settle on several dozen specific types of programs and policies. Of course, not all cities are the same. Some take a much more comprehensive approach to their policies than others. After this chapter reviews an array of municipal sustainability policies and programs, variations across the fifty-five largest US cities will become evident.

The content of cities' sustainability efforts spans a wide range of programs and environmental results. The policies and programs include general statements of sustainability policy as reflected in city comprehensive plans, planning documents, and resolutions of chief executives and legislative bodies. They also include specific programs designed to address climate protection (climate change mitigation and adaptation), energy efficiency, smart growth, and the reduction of metropolitan sprawl. Many other initiatives are pursued as well, such as developing transit-oriented housing, housing densification, and the protection of environmentally sensitive lands. We can think of thirty-eight categories of these kinds of actions, and they focus on programs and policies that include the following: household and industrial recycling, hazardous waste recycling, brownfield redevelopment, hazardous waste site remediation, tax incentives for environmentally friendly development, alternatively fueled city vehicles, car pool lanes, the operation of public transportation, eco-industrial projects, the use of zoning to protect environmentally vulnerable land, air emissions (greenhouse gas) reduction and climate mitigation and adaptation, bicycle ridership, asbestos and lead abatement, green building, urban infill and transit-oriented housing development, energy efficiency, renewable energy for city and general residential customers, water conservation and water quality protection, recycling wastewater and reduction in stormwater runoff, creating a citywide comprehensive sustainability plan, having a single city agency responsible for managing sustainability, green city purchasing, and initiating a sustainability indicators or performance management program.

Table 11.1 shows the results of an assessment designed to determine how many different sustainability program elements each of the largest US cities had adopted and

City	2012 Sustainability Score* (Number of Sustainability Programs Adopted and Implemented Out of a Possible Total of 38)	2019 Resilience Score** (Number of Resilience Programs Adopted and Implemented Out of a Possible Total of 109)
Portland	35	49
San Francisco	35	96
Seattle	35	83
Denver	33	70
Albuquerque	32	55
Charlotte	32	58
Oakland	32	78
Chicago	31	79
Columbus	31	65
Minneapolis	31	75
Philadelphia	31	72

TABLE 11.1 ■ 2012 Sustainability and 2019 Resilience Scores for the Fifty-Five Largest US Cities

City	2012 Sustainability Score* (Number of Sustainability Programs Adopted and Implemented Out of a Possible Total of 38)	2019 Resilience Score** (Number of Resilience Programs Adopted and Implemented Out of a Possible Total of 109)
TABLE 11.1 ■ 2012 Sustainability and 2019 Resilience Scores for the Fifty-Five Largest US Cities—contd.		
Phoenix	31	77
Sacramento	31	66
New York City	30	97
San Diego	30	70
San Jose	30	70
Austin	29	87
Dallas	29	71
Fort Worth	29	61
Nashville	29	59
Tucson	29	58
Washington, DC	29	82
Boston	28	66
Kansas City	28	75
Los Angeles	28	81
Indianapolis	27	63
Fresno	26	54
Las Vegas	26	42
Louisville	26	69
Miami	26	53
Raleigh	26	60
San Antonio	26	82
Baltimore	25	69
El Paso	25	61

Continued

TABLE 11.1 ■ 2012 Sustainability and 2019 Resilience Scores for the Fifty-Five Largest US Cities

City	2012 Sustainability Score* (Number of Sustainability Programs Adopted and Implemented Out of a Possible Total of 38)	2019 Resilience Score** (Number of Resilience Programs Adopted and Implemented Out of a Possible Total of 109)
Cleveland	24	56
Milwaukee	24	59
Atlanta	23	66
Jacksonville	23	65
Honolulu	22	49
Houston	22	72
Long Beach	22	60
Mesa	22	48
Arlington, TX	20	50
Memphis	20	49
Tampa	20	65
Tulsa	20	54
Colorado Springs	19	44
Omaha	19	45
St. Louis	19	65
Oklahoma City	18	46
Detroit	17	50
Virginia Beach	17	51
Pittsburgh	16	63
Santa Ana	16	47
Wichita	7	36

Sources: *Drawn from Kent E. Portney, *Taking Sustainable Cities Seriously: Economic Development, the Environment, and Quality of Life in American Cities*, 2nd ed. (Cambridge, MA: MIT Press, 2013), 23–24, updated. **Derived from the 2019 *Resilient Cities Policies and Programs Project Online*.

implemented as of July 2012, though a similar assessment in 2020 would likely show similar results. This assessment is based on a comprehensive effort to examine cities' policies and programs by considering materials available on each city's respective website, by consulting independent surveys of city officials, and by talking to selected city administrators. In order for a city's program to be counted, there must be tangible evidence that it has actually been implemented in some way. A program that exists only on paper would not be counted in this assessment. The evidence to support the assessments is found in the context of city sustainability and comprehensive plans, and from the programs that are operated by a variety of different city departments.

The results show that three cities have adopted and implemented thirty-five of the thirty-eight different policies and programs—Portland, Oregon; San Francisco, California; and Seattle, Washington. Denver, Colorado, has adopted thirty-three programs. Oakland, California; Charlotte, North Carolina; and Albuquerque, New Mexico, have adopted thirty-two; and six cities have adopted thirty-one each. At the lower end, Virginia Beach, Virginia, and Detroit, Michigan, have adopted seventeen; Santa Ana, California, and Pittsburgh, Pennsylvania, have adopted sixteen each; and Wichita, Kansas, has adopted only seven. The exact content of these efforts may well vary considerably. Although Portland, San Francisco, and Seattle reside at the top of the list, the character of any one particular type of program could very well differ. All three cities have energy efficiency programs, yet the programs differ in how they have gone about trying to achieve greater efficiency. These latter differences are not captured in the sustainability scores presented in Table 11.1.

It should also be noted that this chapter focuses on cities' policies and programs—the programs that they have adopted and implemented in their efforts to try to become more sustainable. It is perhaps natural to want to know what these programs have achieved—what their results are. Yet assessing the "outcomes" of these programs is no small task, and indeed, even as of 2020 there are no comprehensive data available to allow such assessments to be conducted. As cities push forward with their programs, our understanding of the efficacy of these programs will improve. At this moment, however, it is not possible to know with any degree of accuracy whether these programs have improved the quality of the air, the water, and the environment broadly.

CITY SUSTAINABILITY PROGRAMS: A CLOSER LOOK

One way to get a sense of the character of cities' sustainability programs is to look at them as individual cases and to describe in detail what they are doing. This chapter provides at least a glimpse into what a number of US cities are doing in their efforts to try to become more sustainable and resilient. We will first look at sustainability in four large cities—New York City, Los Angeles, Chicago, and Philadelphia. These cities are very similar in terms of the numbers of specific programs they have implemented, having adopted and implemented between twenty-eight and thirty-one of the programs. After discussing these four cities, our attention will turn to examining resilience efforts and policy in four major cities—San Francisco, Austin, Seattle, and Washington, DC. These cities are also similar in the number of policies they have enacted in efforts to build resilience. What these cities also tend to have in common is that they have been able to pursue sustainability and grow their local economies by protecting the environment. We think the efforts these cities have made over the past decade and more are broadly representative of larger US cities. But even medium to smaller cities have engaged in comparable efforts even though they often have fewer resources to pursue sustainability and resilience policies.

Sustainability in Four Large Cities

Although four of the five largest cities in the United States, New York City, Los Angeles, Chicago, and Philadelphia, came to the pursuit of sustainability somewhat later than many other cities, they nonetheless have gotten very serious about enacting and implementing programs. Because of their population sizes, and perhaps their geographic sizes as well, these four cities face a somewhat more formidable set of challenges than do other cities, and that is perhaps why these cities did not stand out as early adopters of sustainability programs and policies.

New York City, NY

New York City's sustainability effort is summarized in its *OneNYC* plan, an updated version of the earlier *PlaNYC*, a comprehensive plan first released in 2007 during the administration of Mayor Michael Bloomberg. *PlaNYC* addressed ten different sustainability-related policy areas and involved at least twenty-five major city departments and agencies,[5] all geared toward the city's effort to create jobs while protecting the environment and becoming more resilient in the face of rising sea levels and increasingly severe storms. The 2011 update of *PlaNYC* contained 132 specific program initiatives with over four hundred "milestones" or goals to be achieved between 2013 and 2030, and it addressed sustainability issues related to the city's housing and neighborhoods, parks and public spaces, brownfield redevelopment, waterways and water resources, transportation, energy, air quality and climate action, and solid waste, as well as "crosscutting" issues such as public health, food, natural systems, green building, waterfront development, economic development, and public engagement. The 2011 plan update provided a comprehensive implementation plan and assessment, showing each milestone or goal, its timeline, what had been accomplished, who was responsible for accomplishing it, where the funding was supposed to come from, and what progress had been made to date.[6] It also presented the city's sustainability indicators report, with annual progress report assessments of accomplishments in twenty-nine broad indicators. The explicit inclusion and treatment of public health as part of its sustainability plan stands out, as very few other cities make this connection.[7]

OneNYC, released in 2015 under the administration of Mayor Bill de Blasio, represents an expansion of *PlaNYC*, adding goals related to making the city more resilient and increasing emphasis on issues of environmental equity. It is organized around four broad "visions": to promote economic growth and development, to achieve greater equity in public services and public well-being, to become more sustainable and aggressive in fighting climate change, and to adapt to the realities of climate change by becoming more resilient.

What does *OneNYC* prescribe for programs and goals related to the environment and environmental sustainability? Most of the goals associated with the biophysical environment are incorporated in Vision 3, referred to as "Our Sustainability City," and address climate protection, energy, air quality, water, hazardous and solid waste, and many other areas. The city's climate plan contains dozens of different initiatives with associated milestones, including conducting and releasing an annual inventory of greenhouse gas emissions, assessing opportunities to further reduce greenhouse gas emissions by 80 percent by 2050, regularly assessing climate change projections, partnering with the Federal Emergency Management Agency to update flood insurance rate maps, developing tools to measure the city's current and future climate exposure, updating regulations to increase the resilience of buildings, working with the insurance industry to develop strategies to encourage the use of flood protection in buildings, protecting New York City's critical infrastructure, identifying and evaluating citywide coastal protective measures, mitigating the urban heat island effect, and enhancing the city's understanding of the impacts of climate change on public health.

Ostensibly, all of the programs and projects outlined in the sustainability vision, along with many of those associated with the fourth vision, "Our Resilient City," represent a full statement of the city's policy and planning around the environment. A few of the pertinent policies outlined in Vision 4 include the effort to integrate climate change projections into emergency management and preparedness, working with communities and neighborhoods to increase their climate resilience, and improving the city's critical infrastructure.[8]

OneNYC is designed to be used, in part, as an implementation tool, as it specifies performance metrics or measures, along with goals and milestones for each action area. Even so, some of the action areas and milestones do not seem terribly well tailored to serve this purpose, particularly when the milestone is just a restatement of the action area. *OneNYC* provides assessments of progress on each milestone, although, unlike many other sustainability plans, it has no clear-cut interventions prescribed when milestones are not met. Presumably, the designation of funding sources, while typically not very specific, does provide some sense that internal funding decisions may well be tied to these action areas, milestones, and goals. And, of course, there is always the question as to whether any of the specific programs, when implemented, will have their intended effects on the environment or resiliency. For example, although there are no serious assessments available, Brian Paul argued that the NYC approach to transit-oriented development would actually increase the city's greenhouse gas emissions by putting many more motor vehicles on the streets.[9]

In April 2019, Mayor de Blasio announced *OneNYC 2050*, a new expansion of the plan that set forth a vision for the future of New York City and a strategy focused on building a "strong and fair city."[10] The 2019 report shows the significant progress made by the City in working towards goals established in the 2015 plan[11] and details a new strategy comprised of eight goals and thirty initiatives to reach the vision for 2050. The plan consists of volumes that each detail a different goal, including a vibrant democracy, thriving neighborhoods, equity and excellence in education, efficient mobility, an inclusive economy, healthy lives, a livable climate, and modern infrastructure. Through these volumes, the plan sets more aggressive goals than the original *OneNYC* plan, including a commitment to carbon neutrality by 2050. The actions laid out in the plan, along with previous actions taken by the Mayor de Blasio administration and former Mayor Bloomberg administration, set the city up to reach a 40 percent reduction in emissions from 2005 levels by 2030 and to reach the carbon neutrality goal by 2050.[12] A 2020 report of the city's progress on their goals and initiatives is forthcoming.[13]

Los Angeles, CA

Although Los Angeles had developed elements of a sustainability plan earlier, its sustainability policy became official in 2007 when the then mayor Antonio Villaraigosa issued Executive Directive 10, instructing all city departments to create sustainability plans. The city's Department of Environmental Affairs had primary responsibility for managing and coordinating these plans. Much of this sustainability effort focused on the environment. Programs such as the "Mayor's Green Agenda" and green infrastructure, air quality and climate change, green building, energy, solid and hazardous waste, brownfield redevelopment, water, and urban habitats initiatives were all placed under the auspices of the Environmental Affairs Office. Los Angeles has not worked from a single comprehensive sustainability plan; instead, it defined specific programs, which, taken together, make up its initiative. Sustainability plans in many cities change when the political leadership changes, and this is true in Los Angeles. After the election of Eric Garcetti as mayor in 2013, questions arose about his commitment to sustainability, particularly after his early elimination of the Department of Environmental Affairs, whose functions have been distributed to numerous other agencies, and in his message of "back to basics," which appeared to downplay earlier emphases on sustainability. However, these concerns declined during later stages of his tenure.

Over the last decade, one of the key components of the Los Angeles sustainability program has been the city's Green LA climate change initiative. The program is complemented by a Clean Air Action Plan, which calls for the city to reduce its greenhouse gas emissions 35 percent below 1990 levels by 2030 and to increase the city's use of renewable energy to 40 percent by 2020. The plan includes about fifty specific initiatives designed to produce the targeted carbon reduction levels, such as the increased use of energy-efficient lighting, extensive plans for green building, and the hallmark creation of an extensive wind turbine farm. In 2009, Los Angeles completed the construction of the Pine Tree Wind Power Project, consisting of eighty municipally owned wind turbines producing about 120 MW of electricity to serve about 56,000 households in the city. The Los Angeles Department of Water and Power operates this facility, along with numerous other energy efficiency and alternative energy programs, and claims to be on track to generate 20 percent of the city's energy from renewable sources by 2020.

The Los Angeles sustainability program also includes efforts to convert the city's fleet vehicles to ones using alternative fuels, a bicycle ridership program, a significant brownfield redevelopment effort, and extensive tree and native planting efforts. Two specific initiatives deserve special mention. The first, a major initiative on water conservation, is important because of the limited supplies available to the city. The second, a commitment to public transit, focuses its attention on the city's relatively new subway system. Operated by the Los Angeles County Metropolitan Transportation Authority

(Metro), the regional transit authority, the subway system is much smaller than those in other major cities, including New York, Boston, Philadelphia, San Francisco, and Washington, DC. About 150,000 people ride the subway daily, with estimated ridership per mile of track making it the ninth busiest subway in the United States. The system has received extensive criticism, primarily for its cost and, because the population of Los Angeles is so widely dispersed, for "not going anywhere" that people want to go. Yet one of the purposes of the system was to influence the patterns of new development, encouraging more transit-oriented development and nudging people back toward the "downtown" of the city. Whether and to what extent this has started to happen is still unclear.

In April 2019, Mayor Garcetti launched Los Angeles' *Green New Deal: Sustainable City pLAn*. The 2019 plan serves as an update to his 2015 sustainability plan and addresses a number of topics relative to sustainability in LA.[14] Among these are some of the same areas addressed by sustainability initiatives in the past, including renewable energy, urban ecosystems, local water efforts, mobility and public transit, among others. The plan also sets more ambitious goals and targets than previously in place, including reducing greenhouse gas emissions by 50 percent below 1990 levels by 2025, carbon neutrality by 2050,[15] 55 percent renewable energy supplied by 2025 and 100 percent by 2045.[16] A key chapter included in the 2015 and 2019 versions of the plan is that of "Lead by Example," which commits the city government as a leader in sustainability initiatives by setting more aggressive targets for government-owned properties and city services than for the city as a whole.[17]

Chicago, IL

By the time Mayor Richard M. Daley declared that Chicago would become the greenest city in America, he had already made a commitment to improving the biophysical environment of the city. Particularly over the last five or so years of his twenty-two-year tenure as mayor, Daley and his administration made significant commitments to sustainability. Although making substantial changes to the way the people operate in a city is fraught with challenges and obstacles, the city created a Department of Environment, headed by a commissioner, with four offices each headed by a deputy commissioner. These offices, Energy and Sustainable Business, Permitting and Enforcement, Natural Resources and Water Quality, and Urban Management and Brownfield Redevelopment, combine a number of traditional city functions, such as permitting, with a broader environmental mandate. Additionally, the city's Department of Planning and Development has a division of Sustainable Development with responsibility for applying sustainable design to all of the city's economic and housing development projects. Together, these made up the heart of the administrative apparatus responsible for sustainability until 2012 when the city released a comprehensive sustainability plan. Under Mayor Rahm Emanuel, who served from 2011 through 2019, a Sustainability Council, composed of numerous department heads, was established, as was the office of "Chief Sustainability Officer."

Unlike New York City or Philadelphia, Chicago did not have a unified, comprehensive sustainability plan until 2012. Previously, "greening" efforts were defined fairly piecemeal, with many different departments developing a part of the overall effort. For example, the Sustainable Development division worked with a guide for encouraging, and sometimes requiring, green urban design in new construction. This plan, *Adding Green to Urban Design*, was probably the closest thing to a full sustainability plan in the city until the release of the 2015 Sustainable Chicago Action Plan.[18] The primary purpose of the earlier document was to provide all city departments, the city council, and the city Plan Commission with some guidance as to what constitutes sustainable development, so they could apply sustainability principles when it was time to review and approve specific development projects. The guide provided a rationale for why each proposed project needed to take into consideration a wide array of environmental impacts on the land, the air, the water, and ultimately the quality of life in the city. Presumably, the intent behind this effort was to encourage the disapproval of projects that did not meet the implied standards of what constitutes sustainable development.

But certainly not all of the programs and policies that make up Chicago's sustainability activities are contained in this, or any other, single document. For example, the Sustainable Development division also has responsibility for the *Eat Local, Live Healthy* program, outlined in a separate document.[19] This program seeks to coordinate Chicago's sustainable food system in order to support local and regional agriculture while improving human nutrition and health. Another program not represented in any comprehensive plan is the Department of Aviation's Sustainable Airport Initiatives, designed to encourage the airports in the city to adopt a variety of practices that are consistent with sustainability in construction, renovation, and daily operations.[20] The city's Department of Transportation has responsibility for traffic management, including a bicycle ridership program, but the mass transit system is managed by the Chicago Transit Authority, an independent agency. Climate action has been tackled by a task force appointed by the mayor and operating as a nonprofit organization called the Chicago Climate Action Plan. This group, which works with the Chicago Department of the Environment and Sustainability and many other city agencies, has responsibility for the city's greenhouse gas inventory, and for the plan to reduce the city's carbon footprint.[21] The city still operates under the 2015 Action Plan, which incorporates and consolidates the various elements developed previously.[22]

In 2018, the Mayor's office created Sustain Chicago, a project that provides resources and guidance to residents on climate change and sustainability.[23] Sustain Chicago has a number of goals established with the focus of reducing emissions and improving air quality, as well as preparedness for the impacts of climate change. Just before that, in late 2017, former Chicago mayor Emanuel signed a Chicago Climate Charter, committing the city to achieve carbon emissions reductions in line with the Paris Agreement on Climate Change.[24]

Philadelphia, PA

Although there were rumblings of interest in sustainability during the latter part of the term of former mayor John F. Street, it wasn't until the beginning of Mayor Michael Nutter's administration in 2008 that the city seemingly started to get interested in sustainability issues. Nutter campaigned, in part, on a platform of engaging the city in a major sustainability effort, and, upon taking office, he created the Mayor's Office of Sustainability, now a permanent part of the city's administrative structure.

The city's comprehensive plan is founded on the "five *E*s" of sustainability: energy, environment, equity, economy, and engagement. It has outlined three target areas for energy: to lower city government energy use by 30 percent between 2008 and 2015; to reduce citywide building energy consumption by 10 percent over this same period; and to retrofit 15 percent of the city's housing stock with insulation, air sealing, and cool roofs. The three environmental target areas include reducing greenhouse gas emissions by 20 percent, improving overall air quality to attain federal standards, and the diversion of 70 percent of solid waste from landfills. The plan's four equity targets are to manage stormwater to meet federal standards, to provide parks and recreation resources within ten minutes for 75 percent of residents, to bring local food within ten minutes of 75 percent of residents, and to increase tree coverage toward 30 percent in all neighborhoods by 2025. The three economy targets focus on reducing vehicle miles traveled by 10 percent, increasing the state of repair of the city's resilient infrastructure, and doubling the number of low- and high-skill green jobs. The single "engagement target" is to engage residents in the definition, planning, and evaluation of the results of the initiatives used to achieve the plan's targets.

Philadelphia's Local Action Plan for Climate Change is a significant part of the responsibility of Greenworks Philadelphia. Unlike many other cities, Philadelphia conceives of climate action as an explicit part of sustainability. Its greenhouse gas emission reduction goals are embedded in its environmental targets.[25] Its greenhouse gas inventory, efforts to reduce emissions, and progress toward achieving those reductions are reported in the Greenworks Philadelphia annual report.[26] Additionally, the city has worked closely with its regional transit authority, the Southeastern Pennsylvania Transit Authority (SEPTA), to coordinate emissions reduction strategies.

Although the *Greenworks Philadelphia* progress report itemizes dozens of specific activities, projects, and programs designed to achieve results in the fifteen target areas, including the installation of biodiesel fueling stations for city vehicles, the installation of porous pavement on city streets, and partnerships with local businesses, including an arrangement with the Philadelphia Eagles professional football team to make its stadium energy independent, three specific efforts at implementing its programs deserve additional discussion here. First, as one of *Greenworks Philadelphia*'s equity targets, the Mayor's Office of Sustainability worked with the US Forest Service to launch a local carbon offset market. This simple idea makes a personal carbon footprint calculator available to residents and provides an opportunity for residents to purchase carbon offsets; collected funds go to the Fairmount Park Conservancy to support its extensive tree-planting program.[27] Second, in order to implement its energy conservation and climate action programs, the city developed numerous initiatives to help building owners retrofit their facilities. Although many cities have created similar programs, Philadelphia has explicitly made this an equity target with the intent of making sure that the benefits of energy savings are broadly distributed across neighborhoods within the city. And third, the city took the unusual step of creating an Office of Watersheds within the Water Department to manage its water resources and wastewater with an eye toward explicitly protecting the seven watershed areas that service the city. Although all cities are serviced by watersheds and all cities have administrative agencies responsible for managing water and wastewater, Philadelphia is perhaps the first to organize and manage these administrative functions while paying such unequivocal attention to watersheds. This, of course, requires a high degree of collaboration and cooperation with other municipal water agencies.[28]

CITY RESILIENCE POLICIES AND PROGRAMS

The idea of city resilience has emerged at least in part in response to growing recognition that cities face a variety of new challenges as a result of climate change. There is no dominant understanding of what constitutes resilience, and as noted earlier, there is a wide range of possible definitions. Some conceptions of resilience look a lot like sustainability, and others are far more narrowly drawn. Some discussions of city resilience advocate using this terminology instead of sustainability because it is thought to generate less political debate and opposition. Other discussions suggest that resilience is part of sustainability, but only a part. A city cannot be sustainable if it is not resilient, so the argument goes, but a city could pursue resilience policies and programs independent of any effort to try to become more sustainable.

A project conducted at Texas A&M University during 2019 studied city resilience policies and programs much like the earlier project studied city sustainability policies.[29] The resilient cities project created a list of some 109 different specific policies and programs that have been put forth as important pieces of resilience. These include programs to mitigate various natural and human-made hazards, pollution reduction and prevention, hardening critical infrastructure, renewable energy and energy efficiency, and many others. A full list of these programs is available from the authors. The project then conducted extensive analysis of city websites to determine whether each of the largest fifty-five cities had adopted and implemented each program. The results of this analysis are also reported in Table 11.1. The number reported in the right-most column represents the total number of the 109 policies and programs each city had adopted and implemented. As a general rule, cities that have done a lot to pursue sustainability have also done a lot to try to become more resilient. The correlation between these two scores is +.625, reflecting this tendency.

CITY RESILIENCE PROGRAMS: A CLOSER LOOK

The tendency for cities to enact and implement both sustainability and resilience policies hides some important differences between the two. A comparison of the scores in Table 11.1 shows that there are some cities with high sustainability scores but relatively low resilience scores, and some cities with high resilience scores but low sustainability scores. Portland, OR, for example, is highly ranked for its sustainability policies and programs, but is ranked relatively low on resilience. A number of cities have relatively low sustainability scores but high resilience scores, many of which are in Texas. Presumably, recent experiences with flooding, especially from hurricanes, has prompted those cities to take resilience seriously. The details of what cities' resilience initiatives look like are apparent in the four cities profiled below.

Resilience in Four Major Cities

Like sustainability, major cities have taken building resilience seriously in the recent years. This may largely be due to the increasing costs of impacts from natural and man-made hazards and disasters and the effect that preventative measures can have on the population and city budgets. Like sustainability, resilience is a multidimensional concept that includes numerous factors. As defined above, what a city can do to more easily bounce back from hazards and disasters is largely dependent on what type of hazards and disasters are common in the area. Though not all hazards and disasters are predictable, a city can examine its geographic history with natural disasters, assess industry risks and hazards and make preparations accordingly. As a result, building resilience may look quite differently depending on the city, though there are some commonalities, such as refraining from promoting development in vulnerable or risky areas. What may vary, however, is how a particular city chooses to define vulnerability or risk.

A range of strategies and policies have been considered, enacted, and implemented in "building resilience." For example, building resilience to natural disasters may involve hardening the city's water, electricity, and gas delivery infrastructure systems, or providing redundant systems for distribution and delivery of goods and services. Building social resilience often involves pursuing equity programs such as ensuring sufficient affordable housing, creating small business support programs such as recovery plans, job training or workforce development programs, among others. Cities may also promote individual or personal resilience through wellness and exercise opportunities available to the public such as creating or expanding bike paths, providing asbestos or lead paint and water pipe removal assistance in older homes, or expanding HOV lanes to reduce air pollution. Below, we highlight the efforts that four cities are making or planning in their resilience building policies.

San Francisco, CA

San Francisco has had a long history with devastating fires and earthquakes. Its booming technology, finance, and tourist industries require secure infrastructure to continue growth and economic development. To further complicate matters, San Francisco's geography as a coastal city near the San Andreas Fault requires that serious thought, planning, and strategy be dedicated toward making San Francisco resilient to acute and prolonged hazards. As a response, the city has developed and implemented several policy goals, strategies, and programs to help build resilience to various hazards and disaster that potentially threaten the area.

In April 2016, San Francisco released its resilience strategy titled *Resilient San Francisco—Stronger Today, Stronger Tomorrow*.[30] Resilient San Francisco boasts four individual resilience goals: (1) Plan and Prepare for Tomorrow; (2) Retrofit, Mitigate and Adapt; (3) Ensure Housing for San Francisco Today and After a Disaster; and (4) Empower Neighborhoods Through Improved Connections. As of 2020, the San Francisco Resilience website lists their ongoing and completed projects.

A significantly serious and daunting task for a major coastal city is preparing for sea level rise. San Francisco's Sea Level Rise Action Plan was released in March 2016. Its purpose was to define goals and a guiding overarching action plan to prepare for and mitigate potential sea level rise risks through four strategic tasks that address planning vulnerability and assessment risks, address data gaps, and provide a foundation for a city-wide sea level rise plan. Out of the Sea Level Rise Action plan came the Sea Level Rise Vulnerability and Consequences Assessment. This assessment identifies public infrastructure that fall within the sea level rise vulnerability zone as well as provides assessment and potential consequences for people, the economy, and the environment as a result of sea level rise.

Another innovative goal of San Francisco's resilience plan is the Better Roofs initiative. As of 2017, San Francisco became the first city to mandate living and solar roofs on most new construction projects. It was determined that 30 percent of the city's total land are rooftops. This unused space carried the potential to create or increase economic, social, and benefits. The initial cost–benefit analysis yielded largely positive results of the living roof. The largest cost of the roof seems to be the initial setup or installation of the roof. This is offset partially by avoiding the stormwater management equipment that would be incurred with a nonliving roof. Other potential benefits come from increased real estate values, added insulation which can reduce utility fees, and community benefit for neighbors and residents.

The final project to be discussed here is *Bay Area Resilient by Design*.[31] *Bay Area Resilient by Design* provides comprehensive and innovative strategies to reduce vulnerability that business, people, and infrastructure may experience as a result of earthquakes or other climate change–related issues. As stated on their website, this acts as a call to residents, organizations, public officials, and local, national, and international experts in efforts to strengthen resilience to hazards and disasters. There are numerous subprojects within the larger project that will use efforts such as elevating vulnerable areas, widening, deepening, and clearing sediment from channels and creeks, creating more public green spaces where there is a lack, as well as redesigning underutilized industrial areas to be more conducive to ecological and modern industrial development. The *Bay Area Resilient by Design* purports to take on large-scale issues exacerbated by potential threats of climate change. Its undertaking is massive, and efforts to realize the proposed changes must be equal in size. The results, however, would set precedent among large US cities of what is possible when communities, public officials, and industry unite to take on threats to progress.

Austin, TX

Geographically, Austin may not appear to resemble San Francisco much at all. Yet its dedication to building resilience is notable. Austin is the capital of Texas and one of the largest cities in the United States. Like many Texas cities, Austin is confronted by seemingly opposing threats, including intense heat which brings threat of drought and water scarcity, and flooding. Central Texas experiences numerous storms which can often bring about flash flooding. In fact, some metropolitan areas in Texas experience nearly annual flooding and serious storms and hurricanes have become expected in the hurricane season. The rising costs of recovery from these storms has brought focus to what cities need to do to become more resilient to these ongoing threats.

In 2013, numerous research teams began to identify potential environmental, economic, and social impacts resulting from climate change. The final report, "Toward a Climate Resilient Austin," provides details on goals and recommendations from the project and helped guide strategies listed in the Climate Resilient Action Plan. At the forefront of Austin's environmentally centered focus is an aggressive carbon neutrality goal as well as a net-zero community-wide greenhouse emission goal. The Climate Resilience Action Plan identifies three city-owned critical assets for resilience: (1) utility infrastructure, (2) transportation infrastructure, and (3) community facilities such as recreation centers and libraries. It further identifies four climate hazards specific to Austin: extreme heat, drought, flooding, and wildfire. Finally, recommendations and strategies are provided to build resilience in these areas, which is what we focus on below. There are four resilience strategies outlined in the Action Plan: strengthen emergency response, expand staff safety plans, evaluate and upgrade existing facilities and infrastructure, and future-proof new facilities and infrastructure.

There are several ongoing efforts to strengthening emergency responses in Austin, some of which directly involve efforts from the Department of Homeland Security. It is clear, however, that upgrading and improving critical infrastructure systems has been given priority to some extent. Like many cities, Austin's infrastructure was in need of serious upgrades. As a result, City Council approved a $1.4 million capital renewal fund to address facilities maintenance. While this fund is a step in the right direction for building resilience, the Action Plans estimate a $70 million backlog in facilities maintenance.

Austin has also emphasized and made efforts to harden critical infrastructure. Beginning with hardening pavement with innovative pavement technology, Austin transportation has also assessed the impact of heat on access to community facilities. Improvements in this area will potentially allow for more access to community facilities for vulnerable populations. With increased flooding in central Texas, Austin has set out to make improvements to its flood protection and flood control measures by hardening roadways, using erosion-resistant materials. The Austin Convention Center is a specific area of emphasis and focus for these measures.

Creating redundant utilities system is a lofty, yet ideal strategy to build resilience. The redundant systems allow for distribution of public goods even after the initial system fails. Austin has laid out plans to create a redundant power supply which will provide critical facilities with at least one redundant source of electricity which will make it capable of operating if a primary source of electricity fails. A similar strategy is intended for the cooling system as the Austin Convention center. It should be noted that as of 2020, the financial impact of this redundant system has not been released, though several private companies have formed that can supply temporary sources of power.

Austin provides a good example of efforts cities can take to build resilience in the face of unique hazards. Its efforts also show that regardless of geographic realities, there are a variety of policies, plans, and programs that cities can take on in order to build resilience to a variety of hazards and disasters.

Seattle, WA

As seen above, Seattle stands out as a leader in sustainability. In the recent years, however, Seattle has made building resilience a central focus in their planning efforts. Like San Francisco, Seattle is a coastal city which draws a lot of tourism and industrial growth, primarily from the Internet and technology sectors. Due to its economic growth, Seattle has come to experience several transportation congestion issues as well as a lack of affordable housing options as income inequality grows.

In 2019, Seattle released its resilience roadmap.[32] The roadmap provides a comprehensive statement of the City's resilience policies and programs, and details 15 goals and 69 action points that fit into four overall themes: building opportunities, making Seattle more affordable for all, creating a city where everyone is welcome, and fostering generational investments. The resilience roadmap also fits in with other development and comprehensive plans developed previously.

According to the *100 Resilient Cities* project, Seattle's greatest environmental threat is earthquakes. The impact of an earthquake may be exacerbated by infrastructure issues in the aging masonry buildings, many of which are occupied by low-income residents. The City of Seattle counts an estimated 1,164 unreinforced, highly vulnerable, masonry buildings (URMs). The Office of Emergency Management and Seattle's Department of Construction and Inspections worked for years to explore and develop policy options for retrofitting the buildings to increase public safety. In the absence of a city resilience policy, these buildings create a serious threat to public safety. Building resilience requires that these buildings be made less vulnerable to earthquake damage.

Retrofitting public buildings has been under way in Seattle for some time. All fire stations and some water reservoirs have been seismically strengthened. Natural gas emergency shutoff valves have been installed in 35 critical city facilities to isolate and minimize damage from broken gas mains. Another upgrade Seattle is making is an early warning system that will "automatically warn infrastructure control systems and personnel about earthquakes" with enough advance that appropriate measures can be taken to ensure safety. These early warning systems, among other things, can slow trains or return elevators to the ground floor when seismic activity is detected.

Local policies designed to build resilience in Seattle also incorporate a variety of nonenvironmental characteristics and issues. Social issues, such as income inequality and poverty, can increase exposure of entire populations of people to hazards and disasters. Building resilience for all populations can accelerate recovery for the city as a whole and Seattle has taken this charge very seriously. At least nine of the fifteen goals outlined in the City's Resilience Roadmap address inequality, and some of them are dedicated solely to reducing inequality.

One strategy to assist in building resilience for all is to assure widespread access to affordable housing. Seattle has experienced significant economic opportunity and growth which often is followed by increasing home costs and property taxes, among other negative outcomes, all of which threaten to increase homelessness. Seattle has proposed and enacted numerous measures to help protect vulnerable populations. For example, the *Just Cause Eviction* provides specifications to landlords about what conditions must be present before an eviction can be pursued legally. Additionally, Mayor Durkan signed an executive order to develop and implement strategies to increase access to affordable benefits, including rental assistance for lower-income residents. Seattle plans to double its homeownership opportunities with the Affordable Middle-Income Housing Advisory Council. One goal of the Council is to find solutions that will help middle-income families stay in Seattle. Other housing efforts include support for homeless populations, develop land oriented toward affordable housing, prevent displacement, and increase the diversity of housing financing sources.

Out of concern that lower-income areas have less access to healthy food options, Seattle's resilience policies and programs include efforts to deal with access to healthy food regardless of income level. Fresh fruits and vegetables are important for building individual resilience and healthy younger populations. Seattle's *Fresh Bucks* program provides assistance to families in need to purchase fresh fruits and vegetables. There are over 60 locations in Seattle and King Counties which include farmer's markets, farm stands, and neighborhood grocers including all *Safeway* grocery stores. The City sponsors two different *Fresh Bucks* programs, the SNAP Market Match program and the Complete Eats program, that together provide fresh food assistance to people who might otherwise find such foods inaccessible (see www.freshbuckseattle.org).

Washington, DC

After over two years since the initial announcement, Mayor Bowser released Washington DC's (hereafter DC) resilience plan in early 2019, titled *Resilient DC*.[33] *Resilient DC* gives emphasis to 4 goals: inclusive growth, climate action, smarter DC, and safe and healthy Washingtonians. Unlike the other cities mentioned above, *Resilient DC* elaborates on how technological advancements can be used to build resilience. We spend the majority of focus on that aspect of building resilience in DC. There are three components of the Smarter DC goal: automation and the future of work, movement of people and goods, and increased cyber threats in an increasingly connected city.

Projections and employment automation trends suggest that many aspects of today's work environment will be automated in the future. This automation may range from ordering food via a touchscreen computer today to autonomous vehicles in the future. *Resilient DC* projects that fewer jobs will be available to individuals without advanced degrees or training. Similarly, there will be a shortage of technology workers that have the requisite skills for cybersecurity and software development. To prepare for this, DC set out to be the highest ranked city on the Economist Intelligence Unit Digital Security Index. This will require significant and strategic advancements and policy emphasis in digital security.

To meet the digital security objectives, *Resilient DC* lays out 4 initiatives with tasks. The first is to adopt cybersecurity best practices to improve the District Government cybersecurity posture. Increasing cybersecurity efforts in DC is incredibly important as DC information systems are likely the targets of numerous attacks. In fact, in 2018, there was a five-day review period in which over 330,000 potential threats to DC networks were mitigated. The second initiative is to launch a cybersecurity partnership to best practices. This voluntary partnership is intended to be up and running before 2023. The third initiative is to launch a cybersecurity corps program to train the next generation of cybersecurity professionals. This program will involve numerous universities in the area to train high school and university students in cybersecurity and defense techniques that will propel their careers. Finally, the fourth initiative is to ensure that all DC agencies plan for cyberthreats by 2023. To date, the plans to mitigate threats have been primarily focused on physical threats, such as floods. These increased efforts will ensure that they are prepared for cyber threats as well.

Resilient DC focuses on climate action as well. Heat waves, river tide, land subsistence, and flooding as a result of seal level rise are of particular concern to DC residents and agencies. Virtually all environmental threats have increased over the past century and they are projected to continue to pose increased threat to DC residents. Of notable concern is the rise of days experiencing extreme heat and the number of residents that do not have adequate cooling, which can result in dehydration and other heat-related health issues. A simple solution is to provide cooling for these residents. However, increased air condition units put added strain on the city's electrical grid and increases greenhouse gas emissions, potentially exacerbating the problem. To begin to address these potential issues, DC sets out to be carbon neutral by 2050. Some policies and programs to achieve this lofty goal are incorporating climate projections into land management and capital improvement project, retrofitting at-risk buildings, designing climate-ready neighborhoods including increasing resilience at the neighborhood and household level, and creating a tool for individuals to track and prepare for climate events and risks.

TAKING CITY SUSTAINABILITY AND RESILIENCE POLICIES AND PROGRAMS SERIOUSLY: A SUMMARY AND THOUGHTS FOR THE FUTURE

City sustainability and resilience policies span a wide array of specific programs designed to try to improve and protect the quality of the biophysical environment. These policies do this by addressing the quality of the air and by seeking to reduce greenhouse gas emissions, by trying to improve and protect access to high-quality drinking water, by managing wastewater and stormwater runoff, by managing how environmentally vulnerable or sensitive land is used, and by pursuing many other goals. When we look at efforts in these nine US cities, it is clear that there is substantial variation in the extent to which cities seem to take the goal of sustainability seriously. Some cities, notably Portland, Seattle, San Francisco, and Denver, stand out as having the greatest commitment to sustainability, as is seen in their policies and programs. But all of the cities profiled here have accomplished impressive policy changes in pursuit of their sustainability and resilience goals. The exact content of their sustainability and resilience policies and programs varies, perhaps having been tailored to the specific conditions these cities face. Yet they all seem to have found ways of defining these policies so as to promote sustainability and resilience to protect and improve the quality of the environment in ways never entertained years ago.

As these and other cities face the growing need to create or improve resilience and sustainability programs, they will discover that the challenge only gets more difficult. With two to three decades of experience, cities now face the challenge of developing a deeper understanding of what their policies and programs have accomplished. Have cities' climate protection programs really decreased their carbon and other greenhouse gas emissions? Have their energy conservation programs reduced energy consumption or altered the forms of energy their citizens consume? Has the quality of drinking water improved? Have cities' concerted efforts to build resilience actually allowed them to be protected from the consequences of disasters, and to bounce back or bounce forward after disasters strike? These are the kinds of questions that cities and researchers need to focus on in the future.

Despite the idea that they will be able to share their experiences and find "best practices" to use as models, cities will face new environmental challenges that will undoubtedly make their work more daunting. Problems surrounding climate change—rising sea levels and increased storm surges from more intense weather events in coastal communities, drought and water shortages, and rising carbon dioxide emissions—promise to be long-term challenges. At the same time, many cities find that their efforts to pursue sustainability meet with political resistance from those who neither understand the magnitude of the environmental problems nor appreciate the governmental actions required to address them. A number of state governments, for example, have established policies barring their cities from operating sustainability or related programs. Although the US federal government, particularly the US Department of Energy and the US Environmental Protection Agency, has provided modest support in the form of grants to help cities create sustainability programs, the future of these programs is anything but assured. Yet cities will likely continue to provide an important location for the pursuits of sustainability and resilience.

SUGGESTED WEBSITES

100 Resilient Cities (https://www.rockefellerfoundation.org/100-resilient-cities/) This nonprofit organization, started with support from the Rockefeller Foundation, provided guidance and assistance to cities around the world as they search for ways to cope with and prepare for the physical, economic, and social challenges associated with environmental changes. After declaring its efforts a success, it ceased operations in 2019.

Clinton Foundation—Climate Change (www.clintonfoundation.org/our-work/by-topic/climate-change) This international effort provides technical assistance and resources to major cities around the world in order to help them reduce their air pollution and carbon emissions.

ICLEI Local Governments for Sustainability USA (www.icleiusa.org) This is the American chapter of the international organization that was originally formed to help cities implement local Agenda 21 programs. It provides information and technical assistance to cities and towns that elect to become members.

International City/County Management Association—Resilient and Sustainable Communities (https://icma.org/topics/resilient-and-sustainable-communities) This section of the ICMA website provides members with extensive information about "best practices" and facilitates communications across cities on both resilience and sustainability.

Partnership for Sustainable Communities (www.sustainablecommunities.gov) This collaborative effort on the part of the US Environmental Protection Agency, the US Department of Housing and Urban Development, and the US Department of Transportation provides information and support to cities and towns.

Smart Growth America (www.smartgrowthamerica.org) The organization advocates for a wide range of local "smart growth" policies and programs, and it provides support and advice to communities to promote sustainability leadership and coalition building.

STAR Communities (www.starcommunities.org/about) This nonprofit organization based in Washington, DC, provides its member cities and towns with "Sustainability Tools for Assessing and Rating Communities." Now part of the US Green Building Council.

US Conference of Mayors Climate Protection Center (https://www.usmayors.org/programs/mayors-climate-protection-center/) The center provides mayors who enroll in its program with guidance, technical assistance, and "best practices" information focused on efforts to reduce carbon emissions.

NOTES

1. See Daniel A. Mazmanian and Michael E. Kraft, "The Three Epochs of the Environmental Movement," in *Toward Sustainable Communities: Transition and Transformations in Environmental Policy*, eds. Daniel A. Mazmanian and Michael E. Kraft (Cambridge, MA: MIT Press, 2009), 3–32.
2. Kent E. Portney, *Sustainability* (Cambridge, MA: MIT Press, 2015), 1–56.
3. World Commission on Environment and Development, *Our Common Future* (New York, NY: Oxford University Press, 1987), 39.
4. Kent E. Portney, *Taking Sustainable Cities Seriously: Economic Development, the Environment, and Quality of Life in American Cities*, 2nd ed. (Cambridge, MA: MIT Press, 2013), 23–24.
5. City of New York, *OneNYC* (New York, NY: City of New York, 2015), http://www.nyc.gov/html/onenyc/downloads/pdf/publications/OneNYC.pdf.
6. City of New York, *PlaNYC, 2011 Update* (New York, NY: City of New York, 2011), http://www.nyc.gov/html/planyc/downloads/pdf/publications/planyc_2011_planyc_full_report.pdf.
7. Jason Corburn, *Toward the Healthy City: People, Places, and the Politics of Urban Planning* (Cambridge, MA: MIT Press, 2009).
8. City of New York, *OneNYC*, 214–51.
9. Brian Paul, "How 'Transit-Oriented Development' Will Put More New Yorkers in Cars," *Gotham Gazette*, April 21, 2010, http://www.gothamgazette.com/article/Transportation/20100421/16/3247.
10. City of New York, *OneNYC 2050: Building a Strong and Fair City* (New York, NY: City of New York, 2019), http://1w3f31pzvdm485dou3dppkcq.wpengine.netdna-cdn.com/wp-content/uploads/2020/01/OneNYC-2050-Full-Report-1.3.pdf.
11. City of New York, *OneNYC2050: Building a Strong and Fair City*, 10.
12. New York City Press Office, "Action on Global Warming: NYC's Green New Deal," April 22, 2019, https://www1.nyc.gov/office-of-the-mayor/news/209-19/action-global-warming-nyc-s-green-new-deal#/0.
13. Daniel A. Zarrilli, "Memo to NYC Council Speaker Corey Johnson Re: OneNYC 2050 Progress Report," April 22, 2020, http://1w3f31pzvdm485dou3dppkcq.wpengine.netdna-cdn.com/wp-content/uploads/2020/04/OneNYC-Letter-20200422_FINAL.pdf.

14. Los Angeles, *L.A.'s Green New Deal: Sustainable City pLAn 2019* (Los Angeles, CA: Los Angeles Mayor Eric Garcetti, 2019), https://plan.lamayor.org/sites/default/files/pLAn_2019_final.pdf. For the 2015 plan, see https://d3n8a8pro7vhmx.cloudfront.net/mayorofla/pages/17002/attachments/original/1428470093/pLAn.pdf?1428470093.

15. Los Angeles, *L.A.'s Green New Deal: Sustainable City pLAn 2019*, 13.

16. Los Angeles, *L.A.'s Green New Deal: Sustainable City pLAn 2019*, 36.

17. Los Angeles, *L.A.'s Green New Deal: Sustainable City pLAn 2019*, 138.

18. City of Chicago, *Adding Green to Urban Design: A City for Us and Future Generations* (Chicago, IL: City Plan Commission, November 2008), https://www.chicago.gov/dam/city/depts/zlup/Sustainable_Development/Publications/Green_Urban_Design/GUD_booklet.pdf.

19. City of Chicago, *Eat Local, Live Healthy* (Chicago, IL: Department of Planning and Development, November 2006), https://www.chicago.gov/content/dam/city/depts/zlup/Sustainable_Development/Publications/Eat_Local_Live_Healthy_Brochure/Eat_Local_Live_Healthy.pdf.

20. Chicago Department of Aviation, *Sustainable Airport Manual* (Chicago, IL: City of Chicago, 2012), https://www.flychicago.com/community/environment/sam/Pages/default.aspx.

21. City of Chicago, *Climate Action Plan* (Chicago, IL: City of Chicago, 2008), https://www.chicago.gov/city/en/progs/env/climateaction.html.

22. City of Chicago, *2015 Sustainable Chicago: Action Agenda 2012–2105 Highlights and a Look Ahead* (Chicago, IL: City of Chicago, December 2015), https://www.chicago.gov/content/dam/city/progs/env/Sustainable_Chicago_2012-2015_Highlights.pdf.

23. Sustain Chicago: A Project by the City of Chicago, "About," https://sustainchicago.cityofchicago.org/; Rahm Emanuel, *The Nation City: Why Mayors are Now Running the World* (New York, NY: Knopf, 2020).

24. The pledge is available at: www.chicago.gov/city/en/depts/mayor/press_room/press_releases/2017/december/ChicagoClimateSummitCharter.html.

25. City of Philadelphia, *Greenworks Philadelphia: A Vision for a Sustainable Philadelphia* (Philadelphia, PA: City of Philadelphia Office of Sustainability, 2016), https://beta.phila.gov/media/20161101174249/2016-Greenworks-Vision_Office-of-Sustainability.pdf.

26. Richard Freeh and Sarah Wu, *Greenworks Philadelphia 2015 Progress Report* (Philadelphia, PA: City of Philadelphia Office of Sustainability, 2015), https://beta.phila.gov/media/20160419140539/2015-greenworks-progress-report.pdf.

27. See "Erase Your Trace," *Ms. Philly Organic*, August 24, 2009, http://msphillyorganic.wordpress.com/2009/08/24/erase-your-trace/.

28. See Philadelphia Water Department, Office of Watersheds, "What's on Tap," https://www.nrdc.org/sites/default/files/philadelphia.pdf.

29. See Sierra C. Woodruff, Ann Bowman, Richard Feiock, Bryce Hannibal, Ki eun Kang, Jeongmin Oh, and Garett Sansom, "Resilience in U.S. Cities: A Survey of Policies and Programs," Texas A & M University, 2020, https://oaktrust.library.tamu.edu/bitstream/handle/1969.1/189324/Final%20report%20T1%20101%20resilient%20cities%20survey.pdf?sequence=1&isAllowed=y.

30. See https://sfgsa.org/sites/default/files/Document/Resilient%20San%20Francisco.pdf.

31. See www.resilientbayarea.org/.

32. See https://durkan.seattle.gov/wp-content/uploads/sites/9/2019/08/Resilient-Seattle_ONLINE.pdf.

33. See https://app.box.com/s/d40hk5ltvcn9fqas1viaje0xbnbsfwga.

GLOBAL ISSUES AND CONTROVERSIES

PART

IV

12

GLOBAL CLIMATE CHANGE GOVERNANCE

Can the Promise of Paris Be Realized?

Henrik Selin and Stacy D. VanDeveer

Our planet is warming. The 5 years from 2014 and 2018 were the warmest on record. Starting in 1880, 18 of the 19 warmest years occurred since 2000.[1] In the Arctic, indigenous peoples struggle to sustain their economic and cultural lives, polar bears fight to stay alive, and melting sea ice opens up new shipping lanes for oil tankers, cargo vessels, and warships. In coastal cities from Boston to Amsterdam, Dhaka, and Guangzhou, and on low-lying islands from South Asia to the South Pacific and the Caribbean, people worry about the consequences of sea level rise and intensifying storm surges. Farmers from California to Cameroon grow more anxious about severe droughts and extreme weather events, while historic fires raged from Portugal to California and Australia. In response, people participate in climate change initiatives in churches, schools, boardrooms, and public offices from city governments to the United Nations (UN), where former UN secretary-general Ban Ki-moon declared climate change "the defining issue of our time."[2] They also organize and join protests and climate strikes around the globe. Despite strong political statements and the clear dangers of changing the climate system, international institutions and the states that drive decision-making have moved much too slowly over the last 30 years, failing to reverse global greenhouse gas (GHG) emissions.

Climate change politics and policymaking focus on both mitigation and adaptation issues. Mitigation efforts center on different ways to reduce GHG emissions. Many current mitigation policies focus on switching to less carbon-intensive energy sources (including wind, solar, and hydro power); improving energy efficiency for vehicles, buildings, and appliances; and supporting the development and deployment of technologies that help reduce GHG emissions. Adaptation efforts seek to improve the ability of human societies (broadly) and local communities (more specifically) to adjust to a changing climate (for example, to alter agricultural practices in response to seasonal and precipitation changes or to prepare for rising sea levels, localized flooding, intense heatwaves, and other severe weather occurrences). Complicating climate change policymaking is the fact that mitigation and adaptation issues are fraught with difficult political, moral, and ethical challenges, as some countries contribute much more to the causes of climate change than do others, and many people and societies struggle to adapt to dangerous climatic changes they largely did not cause. Our world is also a very unequal place, and that makes climate politics and policymaking both more difficult and more important.

This chapter explores climate change governance across global, regional, national, and local levels. As countries struggle to formulate meaningful mitigation and adaptation policies, more aggressive action by intergovernmental forums, firms, small communities, and individuals is necessary to meet the challenges posed by climate change causes and impacts. The next section discusses the history of climate change science and the Intergovernmental Panel on Climate Change (IPCC). This is followed by an outline of the global political framework on climate change that has developed in conjunction with the IPCC assessments, the 1992 UN Framework Convention on Climate Change (UNFCCC), the 1997 Kyoto Protocol, the 2009 Copenhagen Accord, and the 2015 Paris Agreement. Next, three important aspects of climate change politics are addressed: (1) European Union (EU) leadership and policy responses, (2) North American climate change policymaking, and (3) the challenges facing the developing world and the BRICS (Brazil, Russia, India, China, and South Africa). The chapter ends with a few remarks about the future of global climate change governance.

SCIENCE, GHG EMISSIONS, AND THE IPCC

Energy from the sun reaches Earth in the form of visible light. Most of this sunlight reaches the Earth's surface while some is reflected back into space by clouds before it can get through. The sun's energy is absorbed by land and water that heats up. As the Earth's surface warms, it gives up energy in the form of infrared radiation. Naturally occurring GHGs in the lower atmosphere trap some of this outgoing infrared radiation before it can go back into space, which makes the Earth warmer in what has been termed the greenhouse effect. GHGs have been present in the atmosphere for much of Earth's 4.5 billion-year history; without them, the planet would have average surface temperatures of approximately $-20°C$ ($0°F$). The amount of energy that remains trapped in the atmosphere by GHGs has important long-term effects on the climate, as human activities add both naturally occurring and human-made GHG to the atmosphere.[3]

Researchers in different academic fields for over 150 years have contributed to our current—and still developing—understanding of the global climate system. American scientist Eunice Foote and British researcher John Tyndall in the 1850s studied how carbon dioxide (CO_2) and other gases in the atmosphere influence the Earth's temperature, arguing that surface temperature levels are higher with CO_2 than without CO_2 (i.e., early work establishing the dynamics of the greenhouse effect).[4] In 1896, Swedish scientist Svante Arrhenius explored what could happen to the climate system if atmospheric CO_2 concentrations increased, but he did not predict actual, significant changes. In 1938, however, British engineer Guy Stewart Callendar proposed that human CO_2 emissions were changing the climate. Furthermore, in 1956, Gilbert Plass, an American scientist, calculated that adding CO_2 to the atmosphere would have significant heat-trapping effects, contributing to a warming climate.

Science advanced when Charles David Keeling at the Mauna Loa Observatory in Hawaii began measuring actual CO_2 concentrations in open air in 1960. Before industrialization, atmospheric CO_2 concentrations were approximately 280 parts per million by volume (ppmv). Ice core data show that historical concentrations were relatively stable for up to 800,000 years prior to the Industrial Revolution. Measured atmospheric CO_2 concentrations at the Mauna Loa Observatory exceeded 415 ppmv in 2019, then the highest ever throughout human history, and concentrations continue to grow.[5] While other GHGs besides CO_2 also add to the ongoing warming, other emissions (mostly sulfate aerosols and other particulates) have a cooling effect in that they repel incoming sunlight. The future relationship between GHGs and particulates on the overall warming trend and how changes in cloud formation may impact the Earth's climate system are major areas of contemporary climate change science.[6]

Current global climate changes are different from earlier alterations between warmer and cooler eras in that recent critical changes are driven by human behavior. Since the beginning of industrialization, human activities have dramatically altered the composition of GHGs in the lower atmosphere by adding to the volume of naturally occurring gases (for example, CO_2 and methane) as well as by releasing human-made GHGs (for example, hydrofluorocarbons, or HFCs). Human activities influence the amount of energy trapped by GHGs (largely by releasing CO_2 into the atmosphere through the burning of fossil fuels in manufacturing and transport, and GHGs stemming from agricultural production) and the amount of incoming energy absorbed by the Earth's surface (through land use changes including deforestation). Approximately 10 percent of global anthropogenic CO_2 emissions stem from deforestation and other land use changes.[7]

Emissions are commonly calculated as either CO_2 or CO_2 equivalents, where the global warming potential of other GHGs is converted into that of CO_2.[8] Many GHGs, including methane and HFCs, trap much more energy in the atmosphere than does the same amount of CO_2. There exist varying calculations of CO_2 emissions, but one estimate is that since the beginning of the Industrial Revolution, the United States has emitted 25 percent of all CO_2 emissions, followed by the EU (22 percent), China (13 percent), the Russian Federation (6 percent), and Japan (4 percent).[9] In 2017, China (27.5 percent), the United States (14.7 percent), and the EU (9.8 percent) were responsible for over 50 percent of global CO_2 emissions, but with vastly different emission trends (see Table 12.1). The latter is also true for other major country emitters of CO_2. These large differences result from multiple factors, including the scope and stringency of environmental policy, the profile of the national economy, country size and population density, public transportation infrastructure, trade patterns, consumption habits, varying access to renewable energy, and differences in energy policy.

There were also major differences in CO_2 emissions per capita across the world's countries in 2017 (see Table 12.1). The global average CO_2 emissions per person were 4.8 metric tons. Canadian and US per capita emissions were over 15 metric tons. Per capita emissions in Saudi Arabia were over 19 metric tons while the world's highest were found in Qatar at 49 metric tons (the latter not shown in the table). EU-28 emissions per person were under 9 metric tons, and some member states were much lower with, for example, Sweden at 4.2 metric tons. Per capita emissions in India remained under 2 metric tons, and many other developing countries were lower still. The average American in 2017 emitted as much CO_2 in 2.3 days as the average Malian or Nigerian did during the entire year.[10]

Today, most scientists agree that changes to Earth's climate system pose significant ecological, humanitarian, and economic risks. A 2016 study confirms that approximately 97 percent of published climate research agrees that humans are causing contemporary climate change.[11] Through the IPCC—established in 1988 by the

TABLE 12.1 ■ Top Ten Global Emitters of CO_2 in 2017			
Country/Region	Percentage Change in Emissions 1990–2017	Percentage Share of Global Emissions in 2017	Metric Tons CO_2 Emissions per Capita in 2017
1. China	+307	27.5	7.0
2. United States	+3	14.7	16.2
3. EU-28	−21	9.8	8.8[a]
4. India	+298	6.9	1.8
5. Russian Federation	−35	4.6	11.8
6. Japan	+3	3.3	9.5
7. Iran	+228	1.9	8.3
8. South Korea	+159	1.8	12.1
9. Saudi Arabia	+241	1.8	19.3
10. Canada	+24	1.6	15.6

[a]EU-28 per capita emissions for 2017 are for CO_2 equivalent based on data provided by Eurostat (https://ec.europa.eu/eurostat/databrowser/view/t2020_rd300/default/table?lang=en).

Source: Data on changes in country CO_2 emissions 1990–2017 and percentage of global CO_2 emissions in 2017 are from the Global Carbon Project (https://www.globalcarbonproject.org/carbonbudget/19/data.htm). Data on per capita CO_2 emissions in 2017, with the exception of the EU-28, are from Our World in Data (https://ourworldindata.org/co2-and-other-greenhouse-gas-emissions#per-capita-co2-emissions).

World Meteorological Organization and the UN Environment Programme—thousands of climate change scientists and experts from around the world work together, tasked with assessing, summarizing, and publishing the latest peer-reviewed data and analysis. The IPCC was created to inform policymaking, but not to formulate policy. Most IPCC work is divided into three working groups (WGs). WG I studies the physical science basis of the climate system and climate change. WG II focuses on the vulnerability of socioeconomic and natural systems to climate change, the negative and positive consequences of climate change, and adaptation options. WG III examines mitigating options through limiting or preventing GHG emissions and enhancing activities that remove them from the atmosphere.

The IPCC has produced five sets of assessments with WG I reports released in 1990, 1995, 2001, 2007, and 2013.[12] The first of these reports stated that, although much data indicated that human activity affected the variability of the climate system, the authors could not reach consensus. Signaling a higher degree of consensus, the 1995 report stated that the "balance of evidence" suggested "a discernible human influence on the climate." The subsequent three reports were critical in building a much higher degree of confidence within the scientific community. The 2013 report states that the human influence is clear, as the warming of the climate system (atmosphere and ocean) is "unequivocal" with atmospheric concentrations of CO_2, methane, and nitrous oxide having increased to levels unprecedented in at least the last 800,000 years. Additional GHG emissions will cause further warming and changes in all components of the climate system.

The 2014 WG II report outlines a litany of impacts of climate change, including altered precipitation patterns and amounts, ice loss, sea level rise, ocean acidification, and the frequency and increased intensity of extreme weather events. Such changes already impact ecological systems and human health and societies, raising the risk of conflict. WG III reports include emissions scenarios that project possible GHG emissions levels decades into the future, based on a different set of assumptions about future levels of economic growth and the choices made by governments and citizens that affect the generation of GHG emissions. Scenarios are designed to help decision-makers and planners think about how climate change may impact societies and how quickly various mitigation strategies could change future emissions. They also inform thinking about projects such as building new sewage treatment systems in a coastal area, new power plants, or the design of new water policies in drought-stricken regions. Together with reports by WG I and WG II, they make it clear that all countries face adaptation challenges, even if these vary tremendously across societies and regions.

The sixth set of IPCC reports are scheduled for publication in 2021 and 2022, but two other major reports came out recently. A 2018 special IPCC report on global warming of 1.5 degrees Celsius found that the 1.5 degrees Celsius target would likely be met sometime between 2030 and 2052; the world is already more than halfway there with an estimated likely range of a 0.8–1.2 degrees Celsius increase so far.[13] Achieving the 1.5 degrees goal is probably politically impossible, given that GHG emissions must be reduced by 45 percent from 2010 levels by 2030 and by 100 percent by 2050. A 2019 report released by the United Nations Environment Programme concluded that countries collectively failed to stop global GHG emissions growth, resulting in a need for much deeper and faster emission cuts.[14] Reports like these—and reactions to them—shape political debate because they are widely reviewed and cited by people in international organizations, local governments, large and small firms, environmental advocacy groups, journalism, and scientific research.

INTERNATIONAL LAW AND CLIMATE CHANGE NEGOTIATIONS

International law on climate change is shaped by a complex mix of evolving scientific consensus and the material interests and values of state, nongovernmental, and private sector actors. Over two decades of global negotiations have left many frustrated and disappointed with the results and dispirited about the prospects of addressing the problem.[15] Along with a growing number of policy responses at every level of the public, private, and civil society spheres, the world's countries have adopted four major multilateral agreements: the 1992 UNFCCC, the 1997 Kyoto Protocol, the 2009 Copenhagen Accord, and the 2015 Paris Agreement. The UNFCCC was negotiated between publication of the first IPCC report and the 1992 UN Conference on

Environment and Development in Rio de Janeiro, where it was adopted. It entered into force in 1994, and by 2020, 196 countries and the EU were parties to the treaty. As a framework convention, the UNFCCC sets out a broad strategy for countries to work jointly to address climate change.

Like other framework conventions, the UNFCCC defines the issue at hand, sets up an administrative secretariat to oversee treaty activities, and lays out a legal and political framework under which states cooperate over time. The UNFCCC contains shared commitments by states to continue to research climate change, to periodically report their findings and relevant domestic implementation activities, and to meet regularly to discuss common mitigation and adaptation issues at conferences of the parties (COPs). Usually, framework conventions do not include detailed regulatory commitments, leaving those issues to be addressed in subsequent agreements. Similar approaches using the framework convention–protocol model have been applied to issues such as protection of the stratospheric ozone layer, acid rain and related transboundary air pollution problems, and biodiversity loss.

The UNFCCC sets the long-term objective of "stabilization of greenhouse gas concentrations in the atmosphere at a level that would prevent dangerous anthropogenic interference with the climate system" (Article 2). The UNFCCC establishes that the world's countries have common but differentiated responsibilities in addressing climate change. This principle refers to the notion that all countries share an obligation to act, but industrialized countries and countries with economies in transition (then, former communist countries) have a particular responsibility to take the lead because of their relative wealth and disproportionate contribution to the problem through historical emissions. These forty countries, plus the EU, are listed in Annex I of the UNFCCC. They agreed to reduce their anthropogenic emissions to 1990 levels, but no clear deadline was set for this target. The UNFCCC did not assign developing countries any GHG reduction commitments.

Responding to mounting scientific evidence about human-induced climate change in the mid-1990s, much of it presented in the second IPCC report, and to growing concern about the negative economic and social effects of climate change among environmental advocates and policymakers, the UNFCCC parties negotiated the Kyoto Protocol between 1995 and 1997. The protocol came together when the EU and US negotiators struck a deal, helped along by last-minute compromises brokered, in part, by then vice president Al Gore. The Kyoto Protocol covered six GHGs: CO_2, methane, nitrous oxide, perfluorocarbons, HFCs, and sulfur hexafluoride. UNFCCC Annex I countries committed to reduce their GHG emissions collectively by 5.2 percent below 1990 levels by 2008–2012. Toward this goal, thirty-nine states had individual targets.[16] Some agreed to cut their emissions, while others merely consented to slow their emissions growth. For example, the EU-15 took on a collective target of an 8 percent reduction, while the United States and Canada committed to cuts of 7 percent and 6 percent, respectively. In contrast, Iceland committed to limit its emissions at 10 percent above 1990 levels.

Annex I countries could meet their targets through different measures, including the development of national policies that lower domestic GHG emissions. They could also calculate the benefits from domestic carbon sinks that soak up more carbon than they emit by, for example, preserving forests or adopting sustainable practices such as reforestation. In addition, they could participate in transnational emissions trading schemes with other Annex I parties, develop joint implementation (JI) programs with other Annex I parties and get credit for lowering GHG emissions in those countries, or design a partnership venture with a non-Annex I country through what is known as the Clean Development Mechanism (CDM) and get credit for lowering GHG emissions in the partner country. These three options were intended to provide flexibility and reduce mitigation costs by allowing parties to cut emissions wherever it was most efficient, including in other countries.

Developing countries, lacking individual GHG reduction requirements based on the principle of common but differentiated responsibility, strongly supported the Kyoto Protocol, which entered into force in 2005. Among industrialized countries, a stark transatlantic difference emerged.[17] The EU took an early leadership role in defense of the Kyoto Protocol, succeeding in reducing GHG emissions beyond its target (see also EU section below). In contrast, the United States refused to ratify the Kyoto Protocol (and tried to convince other countries to do the same). Canada ratified the Kyoto Protocol in 2002 but announced in 2011 its official withdrawal from the agreement, after failing to curb its emissions. Both US and Canadian GHG emissions increased significantly after 1990 (see Table 12.1). Australia, Japan, New Zealand, and Russia all became parties but displayed varying levels of support.

In Bali in 2007, the UNFCCC parties launched a process to negotiate a follow-up agreement to the Kyoto Protocol, to be adopted at the 2009 COP in Copenhagen.[18] However, national leaders were unable to agree to a legally binding agreement. Instead, they settled on a Copenhagen Accord, under which countries adopted their own voluntary and widely divergent GHG targets for 2020. The Copenhagen Accord also set the goal that average global temperature increases should remain below 2°C (a target related to the UNFCCC goal to prevent dangerous anthropogenic interference with the climate system). In addition, the Copenhagen Accord noted that industrialized countries would try to mobilize $30 billion from 2010 to 2020, with the goal of reaching $100 billion a year by 2020, to support mitigation and adaptation projects in developing countries. Parties at the 2010 COP in Cancun created the Green Climate Fund as a central financial mechanism to operate alongside other bodies, including the Global Environment Facility, the World Bank, and regional development banks.

At the 2011 COP in Durban, the parties adopted the Durban Platform for Enhanced Action, launching a new round of negotiations to develop "a protocol, another legal instrument or an agreed outcome with legal force," to be concluded in 2015. The 2012 COP in Doha formulated a second Kyoto commitment period wherein Annex I countries agreed to reduce their overall GHG emissions by at least 18 percent (collectively) below 1990 levels by 2020. However, some Annex I countries—Canada, Japan, New Zealand, Russia, and the United States—refused to take on a second, formal round of Kyoto targets for 2020. Parties also added nitrogen trifluoride (NF_3) to the list of regulated GHGs, bringing the total to seven, and they extended the CDM and JI mechanisms. The 2013 COP in Warsaw and the 2014 COP in Lima laid the foundation for a global agreement in Paris in 2015, as both industrialized and developing countries for the first time moved forward with a system of GHG reduction plans for the post-2020 period.[19] A bilateral climate change agreement between the United States and China in November 2014 helped pave the way for the Paris

Agreement, including by specifying more significant emission targets for the two countries than had been previously announced.[20]

The Paris Agreement entered into force in November 2016, and 188 countries and the EU were parties by early 2020. The United States became a party in early 2016. This was made possible by the Obama administration classifying it as an executive agreement that did not need Senate approval because it required no changes to US federal law. The only UNFCCC parties not joining the Paris Agreement by 2020 were Eritrea, Iran, Iraq, Libya, South Sudan, Turkey, and Yemen. The conclusion of the Paris Agreement returns to the adoption of international legally binding agreements, but it also builds on language in the Copenhagen Accord, setting the goal of holding global average temperature increases to well below 2°C above preindustrial levels and pursuing efforts to limit it to 1.5°C. To this end, the parties aim to reach a global peak in GHG emissions as soon as possible, as well as a balance between GHGs from anthropogenic sources and a corresponding removal by sinks between 2,050 and 2,100 (i.e., emissions would come down to "net zero"). Recognizing the principles of equity and common but differentiated responsibilities and respective capabilities in the light of different national circumstances, developed countries must continue to lead by undertaking economy-wide absolute GHG emission reduction targets, while developing countries are encouraged to do the same over time.

Continuing the approach from the Copenhagen Accord, the Paris Agreement allows countries to determine their own commitments. All parties voluntarily formulate their own nationally determined contributions (NDCs) under a "pledge and review" system.[21] Importantly, NDCs are not part of the Paris Agreement but are submitted separately by each party—done to satisfy those countries rejecting legally binding GHG commitments. Total emission cuts by the initial NDCs fall well short of those required to meet the agreed-upon temperature goals. Even if all countries were to fulfill their individual NDC pledges—which is uncertain—global average temperatures are expected to increase by between 2.6 and 3.1°C by 2100 (with some estimates higher than that).[22] The parties to the Paris Agreement agreed to start the process of submitting updated NDCs in 2020, to be repeated every 5 years. However, the 26th COP for the UNFCCC, scheduled for November of 2020, was postponed because of the COVID-19 crisis. It is estimated that countries must increase their ambitions in their initial NDCs threefold for realizing the 2°C target and by more than fivefold for meeting the 1.5°C target.[23] Parties are expected to make finance flows consistent with a pathway toward low GHG emissions, and developed countries shall provide financial resources to assist developing countries. However, developed countries did not take on any individual finance targets but determine themselves how much they want to contribute.

Because many climatic changes are already under way and significant amounts of additional GHGs will be emitted over the coming decades, the Paris Agreement recognizes the importance of adaptation and climate-resilient development, mainly as a result from demands by developing countries. Developing countries also secured the inclusion of language on "loss and damage" as separate from adaptation. The idea is that countries contributing very little to the problem should be compensated for impacts related to climate change that cause extensive or irreversible damage, including, for example, loss of lives and property during extreme weather or the disappearance of coastal areas and small islands due to sea level rise. At the insistence of industrialized countries, however, a conference decision attached to the Paris Agreement states that such language does not constitute a legal basis for any liability or compensation. Wealthier countries secured this decision in order to avoid any clear financial or other responsibilities to compensate poorer ones for losses induced by climate change.

The Paris Agreement is widely hailed as a significant step forward by policymakers and environmental activists from many parts of the globe. Global cooperation on nearly every issue moves quite slowly, so calling Paris Agreement the "end of the beginning" for global climate politics or "good enough governance" qualifies as endorsement in international politics.[24] But it is far from perfect, and its success rests on what states and subnational and private sector actors do next. Global climate change governance is increasingly "polycentric" as it happens at multiple levels of authority and in many types of forums beyond the UNFCCC.[25] Some decisions under the Montreal Protocol to protect the stratospheric ozone layer on substances such as HFCs introduced as substitutes for other substances that depleted the stratospheric ozone layer but that are potent GHGs can have climate benefits. Other examples include decisions taken within the International Civil Aviation Organization, the International Maritime Organization, the World Bank, and regional development banks—to name only a few. All of these bodies will need to do more to help rapidly curb GHG emissions if the 2°C goal is to be met, as countries must continue to act more aggressively to reduce emissions. For example, as some countries dramatically scale back coal usage, others burn more, as coal interests fight curbs on CO_2 emissions.[26]

EU LEADERSHIP AND POLICY RESPONSES

Since the 1990s, the EU, representing the majority of the world's industrial countries, has emerged (comparatively) as a global leader in climate change politics and policymaking.[27] Even as the EU grew from fifteen members in the mid-1990s to the current twenty-seven members (following the UK's "Brexit" in 2020), EU bodies such as the European Commission (the administrative bureaucracy), the Council of the European Union (comprising member state government officials), and the European Parliament (of representatives elected by member state citizens) have worked together and collaborated with civil society and private sector actors to enact and implement a set of pan-European climate change and energy-related goals and policies. With a population of roughly four hundred and fifty million people, the EU remains a major contributor to climate change and a major player in climate change politics and policymaking, including in the implementation of the Paris Agreement.[28]

The desire to meet the Kyoto target served as an important early impetus for EU policymakers to develop a growing number of joint policies and initiatives. The then 15 EU member states' collective Kyoto target (8 percent below 1990 GHG emission levels by 2012) was divided among members under a 1998 burden-sharing agreement. To facilitate intra-EU policymaking and implementation, each member state took on a differentiated target under this burden-sharing approach.[29] Several more economically developed member states took on relatively far-reaching commitments to reduce national GHG emissions, while less wealthy member states could increase their GHG emissions in the period up to 2012, as part of these countries' efforts to expand industrial production and accelerate economic growth.

In 2009, the EU adopted a major climate and energy package in support of the so-called 20-20-20 by 2020 goals. These goals, all with a 2020 deadline, refer to a 20 percent reduction in GHGs below 1990 levels, 20 percent of the total energy consumption coming from renewable sources, and a 20 percent reduction in primary energy use compared with projected trends. The 20 percent GHG reduction goal by 2020 was submitted as the EU goal under the Copenhagen Accord and was also included in the Kyoto Protocol's second commitment period. In its NDC for the Paris Agreement, the EU raised its goals to 40 percent for GHG reduction and 27 percent for both renewable energy generation and improved energy savings, all to be achieved by 2030. The suite of policies toward the 2020 and 2030 goals also uses a burden-sharing approach in which each member state is allotted national targets.

Receiving much attention as the world's first international GHG trading scheme, the EU Emissions Trading System (ETS) serves as a main policy instrument. Ironically, the EU was opposed to GHG emissions trading during the Kyoto negotiations—an issue championed by the United States in part because of its domestic experience with emissions trading for sulfur dioxide and nitrogen oxide. The EU attempted to enact a carbon tax in the late 1990s, but this effort failed when member states could not agree on a common tax. In the face of this policy failure and the EU's need to meet its Kyoto target, EU officials developed the ETS.[30] The ETS launched with a first phase (2005 and 2007), a second trading period (2008–2012), and a third starting in 2013. As the fourth phase of the ETS starts in 2021, the scheme has been significantly amended and expanded since 2005, and it includes all EU members as well as Iceland, Lichtenstein, and Norway (while the United Kingdom plans to leave the ETS in 2021).

Currently, the ETS sets a regional cap for three GHGs—CO_2, nitrous oxide, and perfluorocarbons—from over eleven thousand major point sources in power generation and manufacturing as well as airlines operating between member states, collectively covering 45 percent of all EU GHGs.[31] The cap was designed to shrink annually so that emissions from covered sectors in 2020 would be 21 percent lower than in 2005. Every year, the EU allots emissions allowances to each participating country, which, in turn, allocates these to domestic firms. Allowances are increasingly auctioned off, rather than distributed for free, across different economic sectors. Firms can also use some emission credits generated under the CDM and JI mechanisms to meet their obligations. Those without enough allowances to cover their emissions by the end of each year are fined. This is part of an overall strategy to increase the stringency and effectiveness of the ETS over time (see also Chapter 8).

As the EU looks, and sometimes struggles, to maintain a prominent leadership role both regionally and globally, EU data show that the EU-15 cut collective GHG emissions by 12.2 percent from 1990 to the 2008–2012 period, thus exceeding the Kyoto Protocol target.[32] The EU is on track to exceed its 2020 GHG reduction goal of 20 percent, having achieved over a 23 percent reduction by 2018. However, region-wide GHG emissions are not on a strong enough downward trajectory to reach the 2030 target of a 40 percent reduction without further policy measures. Based on current numbers and trends, the EU seems likely to meet its 20 percent renewable energy goal by 2020, having achieved 18 percent in 2018. Again, it appears likely that additional policy actions will be needed to meet the 2030 renewables goal. Additionally, the EU may not meet either the 2020 or 2030 energy efficiency goals based on existing measures, but further initiatives are required.[33]

NORTH AMERICAN AND US CLIMATE CHANGE POLICY AND POLITICS

In contrast to EU bodies and many EU member states, North American federal governments have been slow to act, and they produced a much less regionally integrated response despite sharing a common market under the North American Free Trade Agreement (NAFTA).[34] The United States and Canada have been laggards among industrialized countries for more than 20 years, starting in the mid-1990s. National CO_2 emissions increased in all North American countries between 1990 and 2017—United States +3 percent, Canada +23 percent, and Mexico +55 percent.

While the United States ratified the UNFCCC rather quickly, US skepticism of global climate change policy dates back to the Kyoto Protocol era.[35] Vice President Al Gore signed the Kyoto Protocol on behalf of the Clinton administration (1993–2001), although President Clinton never pushed for its ratification because of strong opposition in the US Senate. President George W. Bush (2001–2009) made rejection of the Kyoto Protocol official US foreign policy. The Canadian federal government, led by Prime Minister Jean Chrétien (1993–2003), ratified the Kyoto Protocol in 2002 but failed to enact any meaningful implementation policies. In the face of growing GHG emissions, the subsequent Stephen Harper governments (2006–2015) took the very rare step of formally withdrawing Canada from the Kyoto Protocol in 2011, before the agreement's commitment period came to an end. Mexico joined the Kyoto Protocol in 2000.

At the 2009 Copenhagen COP, both the Obama administration (2009–2017) and the Harper administration opposed the idea of a legally binding agreement mandating national GHG reductions. The United States and Canada submitted identical voluntary targets under the Copenhagen Accord: a 17 percent reduction below 2005 levels by 2020. Mexico stated a goal to reduce GHG emissions by 30 percent by 2020, as compared to emission levels in a business-as-usual scenario, but only conditionally upon receiving adequate financial and technological support from developed countries. During the Durban Platform negotiations, the United States and Canada stressed that all major GHG emitters must be included in an agreement. The Obama administration, with support from Canada's new Justin Trudeau government (2015–) as well as from the EU, China, and other countries, worked to build momentum for a more significant agreement in Paris.[36]

All three North American countries submitted NDCs in the lead-up to the Paris conference. The United States stated a goal to reduce GHG emissions by 26–28 percent below its 2005 level by 2025 and to make best efforts to reduce its emissions by 28 percent (stemming from the 2014 joint US-China climate change deal that helped secure American and Chinese support for the Paris Agreement).[37] Canada, where the new government had campaigned on promises to take climate change policy more seriously, committed to reducing national GHG emissions by 30 percent below 2005 levels by 2030. Mexico intends to reduce GHG emissions by 25 percent relative to a business-as-usual baseline by 2030, which would translate into a net peaking of national emissions by 2026. Mexico also declared that this reduction target could be increased to 40 percent if it were provided external assistance at a scale commensurate to the challenge of global climate change. The leaders of Canada, the United States, and Mexico also announced the North American Climate, Clean Energy, and Environment Partnership in June 2016, which, among other things, set the goal to reduce methane emissions from the oil and gas sector by 40–45 percent by 2025.[38]

North American climate change politics have changed significantly at national and subnational levels. President George W. Bush opposed mandatory national GHG reductions throughout his presidency.[39] US federal policy under the Bush administration focused instead mainly on voluntary programs and the funding of scientific research and technology development. The Obama administration expressed early support for regulating GHG emissions and appointed many climate change scientists and policy advocates to government. In 2009 the Environmental Protection Agency (EPA), based on a 2007 US Supreme Court ruling that CO_2 can be classified as a pollutant under the Clean Air Act, issued an endangerment finding stating that current and projected atmospheric GHG concentrations threatened the public health and welfare of current and future generations.

Obama's first term saw little regulatory action beyond increases in automobile fuel efficiency standards, as the US Congress failed to pass climate change legislation. It was not until 2015 that the EPA issued a regulatory plan containing a series of measures. Chief among these was the proposal to cut national CO_2 emissions from large power plants by 30 percent below 2005 levels by 2030. The plan also supported expanded investments in renewable energy production and measures to improve energy efficiency. In addition, it stressed the need to prepare the United States for the impacts of climate change. The proposed EPA rules came under immediate scrutiny, as opponents took legal action to block their implementation. The Obama administration also raised fuel efficiency requirements for heavy trucks, proposed a set of regulations to curb methane emissions from the natural gas industry, launched a series of energy efficiency increases for appliances, denied permits to the Keystone XL pipeline, and worked with climate and energy policy leaders in various US states to facilitate state-level policy development. Climate change, energy efficiency, and renewables also became priority issues in US national security assessment and planning.

The 2017 inauguration of Donald Trump as US president resulted in a sharp reversal of federal efforts to address climate change mitigation and adaptation (see Chapter 4).[40] As part of a wide-ranging environmental rollback agenda, the Trump administration quickly launched initiatives to end or review Obama-era regulations aimed at reducing GHG emissions and other pollutants from power plants, the natural gas industry, and cars and heavy trucks.[41] The administration installed climate science skeptics in leadership positions in a host of federal agencies, including the EPA; proposed deep cuts in programs related to climate change and in federal funding for climate change science and public health initiatives; and reversed the Obama decisions regarding the Keystone XL pipeline and his increases in vehicle fuel efficiency standards. In June 2017, President Trump announced that he would withdraw the United States from the Paris Agreement. Formal notice of withdrawal can be given 3 years after the date of the agreement's entry into force, which the Trump administration did on November 4th, 2019. The withdrawal came into effect 1 year after that. The Trump administration also ceased all implementation of the nonbinding parts of the Paris Agreement, including the goals and actions outlined in the US NDC, and any further contributions to the Green Climate Fund. The incoming Joe Biden administration has pledged to rejoin the Paris Agreement and its related initiatives.

Similar to climate change policy in the United States under George W. Bush, Canadian federal GHG policy under the Harper government did not mandate or incentivize GHG reductions. The 2015 election of Trudeau, together with growing climate policy action at the provincial level, appears to have changed the direction of Canadian climate policy toward more serious emissions reductions. In 2016, the Trudeau administration in support of Canada's NDC under the Paris Agreement announced that all Canadian provinces must establish a carbon pricing scheme (e.g., a carbon tax or a cap-and-trade system) that meets a federal benchmark ("floor price") no later than 2018. If any province fails to do so, the federal government will impose its own carbon price on that province. The 2019 reelection of the Trudeau government affirmed the government's climate and energy policies. The federal government's carbon price started at twenty Canadian dollars (CAD) per metric ton in 2020, and is set to increase to CAD 50 by mid-2022 (or from approximately 14 US dollars to about 35 US dollars). Most Canadian households will receive rebates for some of the taxes paid. If fully implemented, this carbon-pricing scheme will result in a stark difference between Canadian and US federal climate change policy.

Mexico, which early on engaged in modest voluntary initiatives, took several legal and political steps during the 2010s. Legislation enacted in 2012 passed Mexico's Copenhagen Accord commitment into law. This was followed in 2013 by a legal commitment to cut GHGs by 50 percent from 2000 levels by 2050 and the goal of generating 35 percent of energy from renewable sources by 2024.[42] A modest carbon tax supports Mexican GHG reductions. The 2012 and 2013 legislation gave Mexico at the time the distinction of having the most comprehensive federal climate change law in North America, and it established Mexico as a Latin American leader in climate change policymaking and prodded other South and Central American states to move forward.[43] These federal measures alongside actions by Mexican states also form the foundation for Mexico's NDC under the Paris Agreement. In contrast, President Andres Manuel Lopez Obrador, elected in 2018, has prioritized fossil fuel investments and policies over climate change action and renewable energy expansion.

Beyond the federal level, subnational actors have developed an important and diverse set of responses (see Chapter 2). Figures 12.1 and 12.2 show the variance among total and per-capita energy-related CO_2 emissions in US states in 2016.[44] Between 2005 and 2016, energy-related CO_2 emissions fell in 41 US states and increased in 9.[45] The states with the greatest reductions were New Hampshire (−35.3%), Maryland (−29.8%), and Maine (−28.8%). At the other end of the spectrum, CO_2 emissions increased by +11.1% in Nebraska, +12.8% in South Dakota, and +16.3% in Idaho. These differences stem from a host of factors, including climate, substantial variance in the sources of energy used, differential economic and population growth rates and density, diverging transportation infrastructure and needs, and large differences in state and local climate change and energy policies. Some US

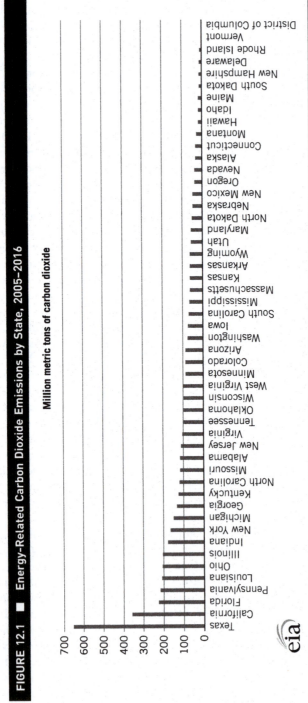

FIGURE 12.1 ■ Energy-Related Carbon Dioxide Emissions by State, 2005–2016

Million metric tons of carbon dioxide

Source: U.S. Energy Information Agency, *Energy Related Carbon Dioxide Emission at the State Level, 2005–2016* (Washington DC: Department of Energy, 2019), available at www.eia.gov/environment/emissions/state/analysis/pdf/stateanalysis.pdf

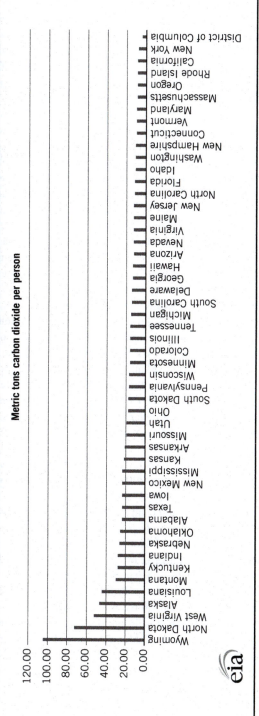

FIGURE 12.2 ■ Per Capita Energy-Related Carbon Dioxide Emissions by State, 2005–2016

Metric tons carbon dioxide per person

Source: U.S. Energy Information Agency, *Energy Related Carbon Dioxide Emission at the State Level, 2005–2016* (Washington DC: Department of Energy, 2019), available at www.eia.gov/environment/emissions/state/analysis/pdf/stateanalysis.pdf

states such as Texas and California emit total CO_2 emissions on par with large industrialized countries, and many other US states emit about as much as smaller industrialized countries. Wyoming was the US state with the highest per capita CO_2 emissions in 2016 at 108 metric tons per person, followed by North Dakota at 72 metric tons. Meanwhile, states at the low end—including New York, California, Massachusetts, and Oregon—each emitted less than 10 metric tons per person.

Many US states have taken policy actions in support of their GHG reduction goals. These include enacting renewable portfolio standards requiring electricity providers to obtain a minimum percentage of their power from renewable sources, formulating ethanol mandates and incentives, pushing to close coal-fired power plants, setting vehicle emissions standards, mandating the sale of more efficient appliances and electronic equipment, and changing land use and development policies to curb emissions as discussed in Chapters 2 and 8. Several states, through their respective attorneys general, were a driving force behind the 2007 Supreme Court decision declaring CO_2 a pollutant. Especially California, building on its tradition of air pollution and energy leadership, has adopted a suite of climate change policies comparable to those of the EU.[46] These policies aim to increase energy efficiency across the state; reduce GHG emissions from power plants, homes, businesses, and transportation; and expand renewable energy generation. Among Canadian provinces, British Columbia distinguished itself (in Canada and globally) by implementing a relatively comprehensive carbon tax in 2008, which climbed to 40 CAD (or about 28 US dollars) by 2019.[47]

Subnational state leaders also enact collaborative standards and policies. Launched in 2009, the Regional Greenhouse Gas Initiative (RGGI) creates a cap-and-trade scheme across 10 states: Maryland, Maine, Vermont, New Hampshire, Massachusetts, Rhode Island, Connecticut, New York, and Delaware (New Jersey, an original member, left for several years and then rejoined). Virginia and Pennsylvania have announced plans to join. RGGI was designed to stabilize CO_2 emissions from power plants between 2009 and 2015 and to achieve a 10 percent reduction by 2019. By 2013, emissions were already far below the cap, prompting states to cut the 2014 emissions cap by 45 percent and plan for it to decline through 2020 (resulting in emissions 68 percent below the 2000–2006 period).[48] By 2019, regional electricity sector emissions were already down by 47 percent (exceeding the 2020 goal), coal use had declined 58 percent between 2005 and 2015, and auctioning emission permits has produced over $2.6 billion for energy efficiency and other public benefits investments, saving consumers an estimated $4.6 billion and creating over 30,000 jobs.[49] RGGI's success helped engender a 2019 initiative by 12 states in the US Northeast to design a new cap-and-trade scheme to help curb transportation emissions.

Under the Western Climate Initiative, British Columbia, California, and Québec work to harmonize GHG mitigation efforts. By 2020, California, Québec, and Nova Scotia were actively coordinating their emissions trading programs (Ontario, previously a member, left the initiative in 2018). California climate change and renewable energy collaboration with Mexico expands subnational efforts into the third NAFTA member. Many Obama administration EPA proposals, if implemented, were designed to further incentivize state action and collaboration. This is not the goal of Trump administration initiatives, which often tried to restrict state-led efforts to address climate change. Canadian Prime Minister Justin Trudeau's climate change and clean energy initiatives are designed to push provinces to take more mitigation and adaptation leadership. Some provinces are also moving beyond Canada's national goal as formulated in its NDC in terms of their own GHG reduction goals. For example, Québec's reduction goals are minus 20 percent by 2020 (from 1990 levels) and minus 37.5 percent by 2030.

In addition, large and small municipalities all over North America are taking action, both individually and through networks such as the United States Climate Alliance, International Council for Local Environmental Initiatives and its Cities for Climate Protection program, the US Conference of Mayors' Climate Protection Agreement, the Federation of Canadian Municipalities' Partners for Climate Protection program, and the C40 group.[50] Although many municipal climate change programs are modest, cities such as New York City, Toronto, and Portland have achieved noteworthy results. North American municipalities are also increasingly developing new GHG reduction and energy efficiency programs that rely in part on innovative private financing. Additionally, a growing number of North American firms are voluntarily reducing GHG emissions and investing in low carbon technology. However, it is important to note that a significant number of firms still take only limited action, while many others continue to fund lobbying efforts to prevent climate policy adoption.[51]

Even as important political and technical precedents for future climate change actions are being set all over the United States and Canada, significant public and private sector resistance to mandatory action and controls continues in North America (more so than in Europe). Polls demonstrate that US voters are deeply split along partisan lines. Also, the fact that many companies were enthusiastic about Trump-era regulatory rollbacks, including auto companies that previously endorsed higher fuel efficiency standards, illustrates continuing corporate opposition to climate change policy. Some states, provinces, and municipalities also delayed or even rolled back enacted climate change policies due to combinations of local political opposition and tougher economic times during the economic downturn starting in the late 2000s. In the US Congress, opponents of climate and clean energy policies (and of climate science) remain determined to limit the EPA's regulatory authority (see Chapter 5)—and even its voluntary programs.

The energy futures of the North American countries are deeply linked. On the one hand, opportunities exist for much more cross-border renewable energy development and trade between Canada and the United States and the United States and Mexico. In all three countries, the cost competitiveness of renewable energy sources such as solar and wind has improved, and renewable energy capacity and generation have increased in recent years. On the other hand, the tar sands extraction in Alberta and the related presidential executive actions to advance approval of the controversial Keystone XL pipeline system connecting Canadian oilfields with US refineries and ports in the Gulf of Mexico—as well as Trump's promise to expand the extraction and use of coal and other fossil fuels—were welcomed and supported by the fossil fuel industry. Other related controversies fester: whether to expand US oil, gas, and coal exports; how many new areas to open for additional fossil fuel extraction; and what, if any, subsidies to offer to renewable energy investments and production. A final, noteworthy development of the Trump era, however, was the continuing decline of coal extraction and coal-fired electricity generation across the United States, despite the frequent pro-coal rhetoric of Trump administration officials.[52] In fact, as of 2020, renewable energy seems likely to permanently surpass coal as the second largest source of electricity in the United States (behind natural gas).[53]

THE DEVELOPING WORLD AND THE BRICS

As the international community attempts to move forward collectively under the Paris Agreement, many developing countries face myriad mitigation and adaptation problems, alongside a multitude of other critical sustainable development issues. The situation of relatively vulnerable countries to climatic changes gives rise to important procedural and distributive social justice issues, from a global equity perspective.[54] Procedural justice refers to the ability of peoples and countries to partake fully in collective decision-making processes focusing on mitigation and adaptation issues, including under the Paris Agreement. Distributive justice concerns how climate change impacts or how mitigation policies affect societies and people differently, in part because countries have greatly varying domestic capacities to deal with these challenges.

For many developing countries, procedural justice issues relate to how international climate change policy is formulated and how these countries' interests are represented and taken into account. Many smaller developing countries face multiple obstacles to engaging actively in multilateral environmental negotiations and assessments.[55] These obstacles include having fewer human, economic, technical, and scientific resources with which to prepare for international negotiations or implement resulting agreements. The significant capacity differences between wealthier industrialized countries and poorer developing countries risk skewing international debates and decision-making in favor of the perspectives and interests of more powerful countries, which are often interested in minimizing their own mitigation costs.

On distributive justice issues, the preambles to both the UNFCCC and the Paris Agreement recognize that many developing countries are "particularly vulnerable" to the adverse effects of climate change. For example, countries in South Asia with vast and densely populated low-lying coastal areas, including Bangladesh and India, will experience many of the first impacts of sea level rise and increased storm intensity, as will the inhabitants of small island states. Changes in seasons and precipitation present a more acute threat to millions of poor, small-scale farmers in Africa and other tropical countries than they do to those in rich countries. Similar issues can be extended to indigenous populations all over the world, who are often among the most vulnerable in any society.[56] The IPCC's fifth assessment report included chapters on "Human Security" and "Sustainable Development and Equity" that catalogue vast and growing risks to citizens and communities in the developing world, as well as the complexity and urgency of equity and social justice issues related to climate change causes, implications, and mitigation and adaptation policymaking.[57]

The developing world has often tried to speak with one voice on climate change through the Group of 77 (G-77).[58] Evident in G-77 pronouncements is the strong interest of member states in preserving the principle of common but differentiated responsibilities as key for assigning GHG reduction requirements and in prioritizing economic development partly based upon the continued use of fossil fuels. The ability of the very diverse group of G-77 members to present a unified front has always been inconsistent, becoming more difficult in recent times. One major change over the last two decades is the rapid economic growth and increasingly assertive positions of major developing countries, often exemplified by the BRICS (Brazil, Russia, India, China, and South Africa). Despite increasing debates about whether rapidly industrializing countries such as China and India are well suited to represent the "Global South"—or even belong in such a category—Indian and Chinese officials remain attached to many of the ideas and principles associated with developing country identity in global politics in the environmental arena and beyond.[59]

Major developing countries, having seen large increases in national emissions over the past three decades, are under growing pressure to reduce GHGs. For example, low-lying islands fear Chinese and Indian emissions as much as those originating from North America and Europe, industrialized countries want a level economic playing field for their firms in international markets, and environmental organizations push for much deeper global emission cuts as a central sustainable development issue. Developing country NDCs include a wide range of national goals and measures, and, like US climate policy, these can change with changes in governments. Brazil, for example, moved from an engaged participant in the Paris Agreement negotiations, to a most hostile and much less cooperative posture under President Jair Bolsonaro's government. In China, as air pollution and public health issues are becoming more salient domestically, the government is stepping up mitigation actions. These include investing heavily in greener technologies, expanding wind and solar power generation capacities, and launching pilot cap-and-trade programs.[60] Among the major goals in China's initial NDC are that its emissions will peak no later than 2030 and then begin to decline. With China's annual GHG emissions exceeding those of all other countries, China and its leaders have become essential and influential participants in all global climate forums—as illustrated by the pre-Paris Agreement US-China bilateral climate change agreement in 2014. India's NDC pledges to focus on renewable energy growth and adaptation. In both India and China, trends in coal use are likely to play a central role in emissions trajectories over the next decades.

Diverse NDC goals and metrics may be politically necessary, but they may also prove difficult to monitor and assess collectively. Furthermore, national GHG emissions under many NDCs are predicted to continue to grow in the short and medium terms. Finally, developing countries pay close attention to contributions to the Green Climate Fund and other major sources of climate finance as an important indication of the level of commitment shown by donor countries and the international community. Many, like Mexico, stipulate that fulfillment of their national GHG reduction targets is at least partially dependent on the delivery of adequate financial and technological assistance from donor countries. While there has been an increase in financial pledges, developing countries have cause to be skeptical about such pledges until funds are actually delivered. In addition, developing countries strongly argue for the principle of "additionality"—that funds going toward climate change mitigation and adaptation should be *in addition* to both development assistance and to funds going to other environmental issue areas.

LOOKING FOR LEADERSHIP

Climate change policy is developing across global, regional, national, and local governance levels, but existing initiatives are not enough to meet the Paris Agreement's temperature goals. Since the 1880s, average global temperatures have increased by roughly 1°C, and past emissions have already built in further warming over the coming decades and centuries. The world's industrialized and developing countries do not face the same challenges, but climate change threatens all. As global GHG emissions continue to rise, the challenge of finding ways to reduce these significantly—by upwards of 80–90 percent by 2050, as is often stated—while simultaneously tackling poverty and promoting sustainable economic and social development cannot be overstated (see Chapter 13). Together with more aggressive actions by industrialized countries, developing countries must first stop their emissions growth and then quickly reduce those emissions if the Paris Agreement's temperature goals are to be reached. At the same time, the vast majority of the world's poorest people live in these countries, facing growing climate-related adaptation problems that hit weaker and poorer communities much more harshly than they do stronger and richer ones. Even after global GHG emissions have been drastically reduced, researchers estimate that related climatic changes will be with us for the next 1,000 years.[61]

The United States re-joining the Paris Agreement under the Biden administration, after Trump's withdrawal, means the return of the world's second largest GHG emitter to global climate change cooperation. However, the United States on-again, off-again relationship with global climate change forums leaves it a fickle and unreliable partner. It also undermines American leadership, as others offer climate change leadership. China and the EU, only 1 day after Trump announced the US withdrawal from the Paris Agreement, signed a declaration to forge a stronger bilateral alliance to address climate change and lead the transition to a low-carbon economy. Parties to the Paris Agreement continue to meet annually and design rules and procedures to implement and review the agreement, with or without meaningful US engagement. Can a Biden administration offer global leadership? Will the EU, China and other parties find a committed long-term partner in the United States? What will be the US climate change posture in other global forums, such as in the ozone layer protection regime, the G7/G20, and the United Nations Security Council? More aggressive GHG reduction goals are required of all major emitters—and massive adaptation challenges remain.

Because of frustration with the UNFCCC process, even after the Paris Agreement, and disappointment in the Trump administration's decision both to leave the Paris Agreement and to reverse federal standards while supporting greater fossil fuel extraction and use, some policy advocates look to the plethora of governance experiences outside of global institutions.[62] Actors at every level of social organization are experimenting with new policies and institutions to address mitigation and adaptation needs in a myriad of ways. Research makes it clear that innovation around climate and energy collaboration is nearly infinite, but questions remain about whether this huge variety of nonstate and private sector governance can meet the climate change challenge without much more significant state action in both the Global North and the Global South. This debate was reignited in the United States in the wake of the Trump administration's many federal and foreign policy actions aimed at weakening previous initiatives supportive of climate change mitigation and adaptation.

Climate finance needs are also daunting. The International Energy Agency estimates that a $53 trillion investment in energy supply and in energy efficiency is needed by 2035 to move the world toward the 2°C goal.[63] This translates into almost $2.5–3 trillion more per year up to 2035. In 2010, the World Bank calculated that the world's developing countries would need to spend between $70 and 100 billion each year between 2010 and 2050 on necessary adaptation projects under a 2°C scenario.[64] A United Nations Environment Programme report suggested that adaptation costs for developing countries under different emission scenarios could be even higher, between $140 and 300 billion in 2030 and rising to between $280 and 500 billion a year by 2050.[65] One thing is clear, however: the longer we delay investments in mitigation and adaptation, the larger the consequences of climate change and the higher the costs of adapting.

The scientific debate about the reality of human-induced climate change is settled, but significant disagreements remain within and among countries, and public and private sector actors, about the allocation of costs and responsibilities for cutting GHG emissions and switching to cleaner technology. International politics and national and local governments are central to addressing major climate change mitigation and adaptation challenges, but large and small firms are also critical players in their roles as investors, polluters, innovators, experts, manufacturers, lobbyists, and employers.[66] Major economic and social changes—such as the drive for low-carbon lifestyles—create both business constraints and opportunities. Collective action is required, but action by firms, consumers, and citizens are needed as well. If the challenges posed by climate change are to be met, we must all take responsibility for our impact on the global climate system, using and expanding our influence over our own behavior and in our local communities, workplaces, and governments.

SUGGESTED WEBSITES

Center for Climate and Energy Solutions (www.c2es.org) This site offers information about international and US climate change policymaking and private sector action.

EcoInternet (www.ecointernet.org) This portal, search engine, and news feed covers climate change issues.

European Commission (http://ec.europa.eu/clima) The EC site on climate action provides information about European perspectives and policy initiatives to address climate change.

Intergovernmental Panel on Climate Change (www.ipcc.ch) This panel of international experts conducts periodical assessments of scientific and socioeconomic information about climate change.

RealClimate (www.realclimate.org) This moderated forum provides commentaries on climate change science news by scientists working in different fields.

UN Climate Change Newsroom (http://newsroom.unfccc.int) This website operated by the UNFCCC Secretariat contains information about meetings and other activities organized under the UNFCCC.

NOTES

1. https://earthobservatory.nasa.gov/images/144510/2018-was-the-fourth-warmest-year-continuing-long-warming-trend
2. Robert O'Neill, "Ban Ki-moon Delivers Call to Action on Global Challenges," *News*, October 22, 2008, available from the Belfer Center for Science and International Affairs, Harvard Kennedy School at http://www.belfercenter.org/publication/ban-ki-moon-delivers-call-action-global-challenges
3. Kerry Emanuel, *What We Know About Climate Change*, 2nd ed., (Cambridge, MA: Boston Review/MIT Press, 2012).
4. Spencer R. Weart, *The Discovery of Global Warming*, 2nd ed., (Cambridge, MA: Harvard University Press, 2008); and Leila McNeill, "This Lady Scientist Defined the Greenhouse Effect But Didn't Get the Credit, Because Sexism," *Smithsonian Magazine*, December 5, 2016, http://www.smithsonianmag.com/science-nature/lady-scientist-helped-revolutionize-climate-science-didnt-get-credit-180961291/
5. https://www.usatoday.com/story/news/world/2019/05/13/climate-change-co-2-levels-hit-415-parts-per-million-human-first/1186417001/
6. For discussions about the latest developments in climate change science, see RealClimate (www.realclimate.org).
7. IPCC, *Climate Change 2013: The Physical Science Basis* (Cambridge: Cambridge University Press, 2013), http://www.ipcc.ch/report/ar5/wg1/
8. For a chart of different global warming potentials, see UNFCCC data at https://unfccc.int/ghg_data/items/3825.php
9. Data taken from: https://ourworldindata.org/co2-and-other-greenhouse-gas-emissions
10. https://ourworldindata.org/co2-and-other-greenhouse-gas-emissions#per-capita-co2-emissions
11. John Cook, Naomi Oreskes, Peter T. Doran, William R. L. Anderegg, Bart Verheggen, Ed W. Maibach, J. Stuart Carlton, Stephan Lewandowsky, Andrew G. Skuce, and Sarah A. Green, "Consensus on Consensus: A Synthesis of Consensus Estimates on Human-Caused Global Warming," *Environmental Research Letters* 11, no. 4 (2016): 1–7.
12. The IPCC reports and other data are available on the IPCC website (www.ipcc.ch).
13. IPCC, 2018: Summary for Policymakers. In: *Global Warming of 1.5°C. An IPCC Special Report on the impacts of global warming of 1.5°C above pre-industrial levels and related global greenhouse gas emission pathways, in the context of strengthening the global response to the threat of climate change, sustainable development, and efforts to eradicate poverty* [Masson-Delmotte V., P. Zhai, H.-O. Pörtner, D. Roberts, J. Skea, P. R. Shukla, A. Pirani, W. Moufouma-Okia, C. Péan, R. Pidcock, S. Connors, J. B. R. Matthews, Y. Chen, X. Zhou, M. I. Gomis, E. Lonnoy, T. Maycock, M. Tignor, and T. Waterfield (eds.)] (https://www.ipcc.ch/sr15/chapter/spm/).
14. UNEP. (2019). Emissions Gap Report 2019: Nairobi: UNEP, (https://www.unenvironment.org/resources/emissions-gap-report-2019).
15. Christian Downie, *The Politics of Climate Change Negotiations* (Northampton, MA: Edward Elgar, 2014); Harro van Asselt, *The Fragmentation of Climate Change Governance* (Northampton, MA: Edward Elgar, 2014); and Thomas Hale, David Held, and Kevin Young, *Gridlock: Why Climate Change Cooperation Is Failing When We Need It Most* (London: Polity, 2013).
16. These were Australia, Austria, Belarus, Belgium, Bulgaria, Canada, Croatia, the Czech Republic, Denmark, Estonia, Finland, France, Germany, Greece, Hungary, Iceland, Ireland, Italy, Japan, Latvia, Lichtenstein, Lithuania, Luxembourg, Monaco, the Netherlands, New Zealand, Norway, Poland, Portugal, Romania, the Russian

Federation, Slovakia, Slovenia, Spain, Sweden, Switzerland, Ukraine, the United Kingdom, and the United States.

17. Miranda A. Schreurs, Henrik Selin, and Stacy D. VanDeveer, eds., *Transatlantic Environment and Energy Politics: Comparative and International Perspectives* (Aldershot, UK: Ashgate, 2009).

18. Raymond Clémonçon, "The Bali Roadmap," *Journal of Environment and Development* 17, no. 1 (2008): 70–94.

19. Detailed reports about all COPs, for the UNFCCC and many other international environmental negotiations, can be found in the *Earth Negotiations Bulletin*, available at www.iisd.ca/enbvol/enb-background.htm

20. Kelly Gallagher and X. Xuan, *Titans of the Climate: Explaining the Policy Process in the US and China* (Cambridge, MA: MIT Press, 2019).

21. Robert Falkner "The Paris Agreement and the New Logic of International Climate Politics," *International Affairs* 92 (5) (2016): 1107–25.

22. Joeri Rogelj, Michel de Eizen, Niklas Höhne, Taryn Fransen, Hanna Fekete, Harald Winkler, Roberto Schaeffer, Fu Sha, Keywan Riahi, and Malte Meinshausen, "Paris Agreement Climate Proposals Need a Boost to Keep Warming Well Below 2°C," *Nature* 534 (June 30, 2016): 631–39.

23. UNEP. (2019). Emissions Gap Report 2019: Nairobi: UNEP, (https://www.unenvironment.org/resources/emissions-gap-report-2019).

24. Paul Bodnar quoted in Graham Norword, "'End of the Beginning': What Was Achieved at COP-21," *NewSecurityBeat*, January 6, 2016; and Joshua Busby, "After Paris: Good Enough Climate Governance," *Current History* 115, no. 777 (January 2016): 3–9.

25. Andrew Jordan, Dave Huitema, Mikael Hildén, Harro van Asselt, Tim J. Rayner, Jonas J. Schoenefeld, Jale Tosun, et al., "Emergence of Policycentric Climate Governance and Its Future Prospects," *Nature Climate Change* 5 (November 2015): 977–82.

26. Tim Boersma and Stacy D. VanDeveer, "Coal after the Paris Agreement," *Foreign Affairs*, June 6, 2016.

27. Henrik Selin and Stacy D. VanDeveer, "Broader, Deeper and Greener: European Union Environmental Politics, Policies and Outcomes," *Annual Review of Environment and Resources* 40 (2015): 309–35.

28. Sebastian Oberthür, "Where to Go from Paris? The European Union in Climate Geopolitics," *Global Affairs* 2, no. 2 (2016): 119–30.

29. Henrik Selin and Stacy D. VanDeveer, *European Union Environmental Governance* (New York: Routledge, 2015).

30. Jon Birger Skjærseth and Jørgen Wettestad, *EU Emissions Trading: Initiating, Decision-Making and Implementation* (Aldershot, UK: Ashgate, 2008).

31. European Commission, *The EU Emissions Trading System (EU ETS)* (Brussels: European Commission, 2013); and Selin and VanDeveer, *European Union Environmental Governance*.

32. European Environment Agency, *Trends and Projections in Europe 2013: Tracking Progress Towards Europe's Climate and Energy Targets until 2020* (Copenhagen: European Environment Agency, 2013).

33. European Environment Agency, *Trends and Projections in Europe 2019: Tracking Progress Towards Europe's Climate and Energy Targets* (Copenhagen: European Environment Agency, 2019).

34. Henrik Selin and Stacy D. VanDeveer, eds., *Changing Climates in North American Politics: Institutions, Policy Making and Multilevel Governance* (Cambridge, MA: MIT Press, 2009).

35. Henrik Selin and Stacy D. VanDeveer, "Climate Change Politics and Policy in the United States: Forward, Reverse and through the Looking Glass," in Rüdiger Wurzel and

Mikael Skou Andersen, eds., *Multilevel and Polycentric Climate Governance: Pioneers, Leaders and Followers in Multilevel and Polycentric Climate Governance* (Routledge, forthcoming).

36. Coral Davenport, "Obama Pursuing Climate Accord in Lieu of Treaty," *New York Times*, August 27, 2014.

37. White House, Office of the Press Secretary, "U.S.-China Joint Announcement on Climate Change," November 12, 2014, http://www.whitehouse.gov/the-press-office/2014/11/11/us-china-joint-announcement-climate-change

38. https://obamawhitehouse.archives.gov/the-press-office/2016/06/29/north-american-climate-clean-energy-and-environment-partnership-action

39. Miranda A. Schreurs, Henrik Selin, and Stacy D. VanDeveer, "Conflict and Cooperation in Transatlantic Climate Politics: Different Stories at Different Levels," in *Transatlantic Environment and Energy Politics: Comparative and International Perspectives*, eds., M. A. Schreurs, H. Selin, and S. D. VanDeveer (Aldershot, UK: Ashgate, 2009), 165–85.

40. Henrik Selin and Stacy D. VanDeveer, "Climate Change Politics and Policy in the United States: Forward, Reverse and through the Looking Glass," in *Climate Governance Across the Globe: Pioneers, Leaders and Followers*, eds., R. K. Wurzel, M. Skou Andersen, and P. Tobin (New York: Routledge, 2020).

41. Nadja Popovich, Kendra Pierre-Lous, and Livia Ripka, "All 98 Environmental Rules the Trump Administration is Revoking or Rolling Back," *New York Times*, May 10, 2020.

42. Rona Fried, "Mexico Unveils National Climate Change Strategy," *SustainableBusiness. com*, June 11, 2013, http://www.sustainablebusiness.com/mexico-unveils-national-climate-change-strategy-51628/

43. Lisa Friedman, "Latin Americans Forge Ahead on CO_2 Reduction Plans," *ClimateWire*, June 9, 2014.

44. https://www.eia.gov/environment/emissions/state/analysis/pdf/stateanalysis.pdf

45. U.S. Energy Information Administration, *Energy-Related Carbon Dioxide Emissions at the State Level, 2000–2014* (Washington, DC: Department of Energy, January 2017), https://www.eia.gov/environment/emissions/state/analysis/pdf/stateanalysis.pdf

46. David Vogel, *California Greenin': How the Golden State Became the Environmental Leader* (Princeton, NJ: Princeton University Press, 2018).

47. https://www2.gov.bc.ca/gov/content/environment/climate-change/planning-and-action/carbon-tax

48. See the RGGI website: www.rggi.org

49. Michael Bradley and Christopher Van Atten, *Power Switch: The Future of the Electric Power System in the Northeast and the Disruptive Power of Innovation* (Concord, MA: MJ Badley & Associates, 2016), http://www.mjbradley.com/sites/default/files/power-switch10-19-2016.pdf; Travis Madsen and Rachel Cross, *Doubling Down on Climate Progress: The Benefits of a Stronger Regional Greenhouse Gas Initiative* (Boston, MA: Environment America, 2017); Acadia Center, "The Regional Greenhouse Gas Initiative: 10 Years in Review (Acadia Center, 2019); and Leigh Raymond, *Reclaiming the Atmospheric Commons* (Cambridge, MA: MIT Press, 2016).

50. Christopher Gore and Pamela Robinson, "Local Government Responses to Climate Change: Our Last, Best Hope?" in *Changing Climates in North American Politics: Institutions, Policy Making and Multilevel Governance*, eds., Henrik Selin and Stacy D. VanDeveer (Cambridge, MA: MIT Press, 2009), 137–58.

51. Charles A. Jones and David L. Levy, "Business Strategies and Climate Change," in *Changing Climates in North American Politics: Institutions, Policy Making and Multilevel Governance*, eds., Henrik Selin and Stacy D. VanDeveer (Cambridge, MA: MIT Press, 2009), 219–40.

52. Tim Boersma and Stacy D. VanDeveer, "World on Fire: Coal Politics and Responsibility among Great Powers," in *Great Powers, Climate Change and Global Environmental*

Responsibilities, eds., Barry Buzan and Robert Falkner (Oxford University Press, forthcoming); and Robinson Meyer, "America's Coal Consumption Entered Free-Fall in 2019," *The Atlantic*, January 2020, at https://www.theatlantic.com/science/archive/2020/01/americas-coal-consumption-entered-free-fall-2019/604543/

53. Brad Plumer "In a First, Renewable Energy is Set to Pass Coal in the U.S.," *New York Times*, May 14, 2020.

54. W. Neil Adger, Jouni Paavola, and Saleemul Huq, "Toward Justice in Adaptation to Climate Change," in *Fairness in Adaptation to Climate Change*, eds., W. Neil Adger, Jouni Paavola, Saleemul Huq, and M. J. Mace (Cambridge, MA: MIT Press, 2006), 1–19.

55. Pamela S. Chasek, "NGOs and State Capacity in International Environmental Negotiations: The Experience of the Earth Negotiations Bulletin," *Review of European Community and International Environmental Law* 10, no. 2 (2001): 168–76; and Ambuj Sagar and Stacy D. VanDeveer, "Capacity Development for the Environment: Broadening the Scope," *Global Environmental Politics* 5, no. 3 (2005): 14–22.

56. Arctic Climate Impact Assessment, *Impacts of a Warming Arctic: Arctic Climate Impact Assessment* (Cambridge: Cambridge University Press, 2004).

57. W. Neil Adger, Juan M. Pulhin, Jon Barnett, Geoffrey D. Dabelko, Grete K. Hovelsrud, Marc Levy, Úrsula Oswald Spring, et al., "Human Security," in *Climate Change 2014: Impacts, Adaptation, and Vulnerability—Contribution of Working Group II to the Fifth Assessment Report of the Intergovernmental Panel on Climate Change* (Cambridge: Cambridge University Press, 2014), 755–91; and Marc Fleurbaey, Sivan Kartha, Simon Bolwig, Yoke Ling Chee, Ying Chen, Esteve Corbera, Franck Lecocq, et al., "Sustainable Development and Equity," in *Climate Change 2014: Mitigation of Climate Change—Contribution of Working Group III to the Fifth Assessment Report of the Intergovernmental Panel on Climate Change* (Cambridge: Cambridge University Press, 2014), 283–350.

58. Adil Najam, "The View from the South: Developing Countries in Global Environmental Politics," in *The Global Environment*, 5th ed., eds., Regina Axelrod and Stacy D. VanDeveer (Washington, DC: CQ Press, 2020) 245–68.

59. Shangrila Joshi, "Understanding India's Representation of North-South Climate Politics," *Global Environmental Politics* 13, no. 2 (2013): 128–47; and Philip Stalley, "Principles Strategy: The Role of Equity Norms in China's Climate Change Diplomacy," *Global Environmental Politics* 13, no. 1 (2013): 1–8.

60. Joanna I. Lewis and Kelly Sims Gallagher, "How China's Domestic Energy and Environmental Challenges Shape Its Global Engagement," in *The Global Environment*, 5th ed., eds., Regina Axelrod and Stacy D. VanDeveer (Washington, DC: CQ Press, 2020): 220–44; and Joanna I. Lewis, *Green Innovation in China* (New York: Columbia University Press, 2013).

61. Susan Solomon, Gian-Kasper Plattner, Reto Knutti, and Pierre Friedlingstein, "Irreversible Climate Change Due to Carbon Dioxide Emissions," *Proceedings of the National Academy of Sciences of the United States* 106, no. 6 (2009): 1704–9, https://www.pnas.org/content/106/6/1704

62. Harriett Bulkely, et al., *Transnational Climate Change Governance* (Cambridge: Cambridge University Press, 2014); Jennifer Green, *Rethinking Private Authority* (Princeton, NJ: Princeton University Press, 2014); and Matthew Hoffmann, *Climate Governance at the Crossroads* (Oxford, UK: Oxford University Press, 2011).

63. International Energy Agency, *Special Report: World Energy Investment Outlook* (Paris: OECD/IEA, 2014), 14.

64. World Bank, *Economics of Adaptation to Climate Change: Synthesis Report* (Washington, DC: World Bank, 2010), xxvii

65. UNEP, *The Adaptation Finance Gap Report* (Nairobi, Kenya: UNEP, 2016), xii.

66. Jones and Levy, "Business Strategies and Climate Change."

ENVIRONMENT, POPULATION, AND THE DEVELOPING WORLD

Richard J. Tobin

Environmental problems occasionally make life in the United States unpleasant, but most Americans tolerate this situation in exchange for the comforts associated with a developed economy. Most Europeans, Canadians, Japanese, and Australians share similar lifestyles, so it is not surprising that they typically also take modern amenities for granted.

When lifestyles are viewed from a global perspective, however, much changes. Consider, for example, what life is like in much of the world. The US gross national income (GNI) per capita was $65,880 per year, or about $1,267 per week, in 2019. In contrast, weekly incomes were less than 5 percent of this amount in twenty-six countries, even when adjusted for differences in prices and purchasing power. In several African countries, real per capita incomes are less than 2 percent of those in the United States. Much of the world's population lives on less than $2 a day.[1]

Low incomes are not the only problem facing many of the world's inhabitants. Before millions of girls reach age fifteen in parts of Africa, including 85 percent of more in Egypt, Guinea, Somalia, and Sudan, they suffer from the indignity and emotional and physical damage of female genital mutilation. In some developing countries, women, often illiterate and with no formal education, marry as young as age thirteen. In several countries, about a quarter of girls are married by the age of fifteen, and in many more countries over 50 percent by age eighteen.[2] Many of these marriages are with much older men who have even less education than their teenage brides. Girls' health and well-being are further placed in jeopardy when they bear children while still in their mid-teens.

During their childbearing years, women in many developing countries will typically deliver as many as five or six babies, most without skilled birth attendants. This absence is not without consequences. The likelihood that a woman will die due to complications associated with pregnancy, childbirth, or an unsafe abortion is many times higher in poor countries than in Western Europe or the United States. In Chad, Sierra Leone, and South Sudan, as an illustration, the rate of maternal mortality is almost one hundred times higher than the rate in many European countries.[3] COVID-19 has added an additional challenge that has stressed and overburdened many health systems that were ill-prepared and understaffed before the virus arrived.

Many of the world's children are also at risk. In recent years about one of thirteen children have died before the age of five in sub-Saharan Africa; in high-income countries that number is about 1 in 185. Half of all deaths under the age of five in 2019 were in sub-Saharan Africa.[4] Every week, tens of thousands of children under age five die in developing countries from diseases that rarely kill Americans. Malaria, diarrhea, and acute respiratory infections cause more than half of these deaths, most of which can be easily and cheaply cured or prevented.

Of the children from these poor countries who do survive their earliest years, millions will suffer brain damage because their pregnant mothers had no iodine in their diets; others will lose their sight and die because they lack vitamin A. Many will face a life of poverty, never to taste clean water, breathe unpolluted air, use a computer, toothbrush, or cell phone, enter a classroom, learn to read or write, visit a doctor or a dentist, have access to even the cheapest medicines, or eat nutritious food regularly. Nearly half of all deaths in children under the age of five are attributable to undernutrition.[5] To the extent that shelter is available, it is typically rudimentary, rarely with electricity or proper sanitary facilities. Hundreds of millions in the developing world will also become victims of floods, droughts, famine, desertification, land degradation, waterborne diseases, infestations of pests and rodents, and noxious levels of pollution because their surroundings have been abused or poorly managed.

As children in developing countries grow older, many will find that their governments cannot provide the resources to ensure them a reasonable standard of living or even a seat in a classroom. Yet all around them are countries with living standards well beyond their comprehension. The United States has about 4 percent of the world's population but consumes about 17 percent of the world's energy and more than three times as much as India, whose population is more than four times larger. An Indian mother might wonder why Americans consume a disproportionate share of the world's resources when she has malnourished children she cannot afford to educate or properly clothe. In short, life in much of the world provides an array of problems different from those encountered in developed nations. Residents of poor countries must cope with widespread poverty, scarce opportunities for employment, and a lack of development. Yet both developed and developing nations often undergo environmental degradation. Those without property, for example, may be tempted to denude tropical forests for land to farm. Concurrently, pressures for development often force people to overexploit their natural and environmental resources.

These issues lead to the key question addressed in this chapter: Can the poorest countries, with the overwhelming majority of the world's population, improve their lot through sustainable development? Sustainable development meets the essential needs of the present generation for food, clothing, shelter, jobs, and health without "compromising the ability of future generations to meet their own needs."[6] Achieving this goal will require increased development without irreparable damage to the environment.

Whose responsibility is it to achieve sustainable development? One view is that richer nations have a moral obligation to assist less fortunate ones. If the former do not meet this obligation, not only will hundreds of millions of people in developing countries suffer, but the consequences will be felt in the developed countries as well. Others argue that poorer nations must accept responsibility for their own fate because outside efforts to help them only worsen the problem and lead to an unhealthy dependence. Advocates of this position insist that it is wrong to provide food to famine-stricken nations because they have exceeded their environment's carrying capacity.[7]

The richer nations, whichever position they take, cannot avoid affecting what happens in the developing world. It is thus useful to consider how events in rich nations influence the quest for sustainable development. At least two related factors affect this quest. The first is a country's population; the second is a country's capacity to support its population.

POPULATION GROWTH: CURE OR CULPRIT?

Population growth is one of the more contentious elements in the journey toward sustainable development. Depending on one's perspective, the world is either vastly overpopulated or capable of supporting as many as thirty times its current population (about 7.8 billion in mid-2020 and increasing at an annual rate of about 77 million per year).[8] Many developing nations are growing faster than the developed nations (Table 13.1), and more than 80 percent of the world's population lives outside the developed regions. If current growth rates continue, the proportion of those in developing countries will increase even more. Between 2020 and 2050, more than 97 percent of the world's population increase, estimated to be about two billion people, will occur in the latter regions, exactly where the environment can least afford such a surge. Many of the new inhabitants will live in countries that are experiencing little, if any, economic development.

Africa is particularly prone to high rates of population growth, with some countries facing increases of 3.5 percent or more per year. This may not seem to be a large percentage until we realize that such rates will double the countries' populations in about twenty years. Fertility rates measure the number of children an average woman has during her lifetime. All countries with fertility rates at 5 or above in 2020 are in sub-Saharan Africa. By comparison, the fertility rate in the United States was 1.7 in that year.

Although many developing countries have altered their attitudes about population growth, many have also realized the immensity of this task. The theory of demographic transition suggests that societies go through three stages. In the first stage, in premodern societies, birth and death rates are high, and populations remain stable or increase at low rates. In the second stage, death rates decline and populations grow rapidly because of vaccines, better health care, and more nutritious foods. As countries begin to reap the benefits of development, they enter the third stage. Infant mortality declines, but so does the desire or need to have large families. Population growth slows considerably.

This model explains events in many developed countries. As standards of living increased, birthrates declined. The model's weakness is that it assumes economic growth; in the absence of such growth, many nations are caught in a demographic trap. They get stuck in the second stage. This is the predicament of many countries today. In some African countries, the situation is even worse. Their populations are growing faster than their economies, and living standards are declining. These declines create a cruel paradox. Larger populations produce increased demands for food, shelter, education, and health care; stagnant economies (and global pandemics) make it impossible to provide them.

The opportunity to lower death rates can also make it difficult to slow population growth. The twenty-five countries with the lowest average life expectancy are all in Africa. If these Africans had access to the medicines, vitamins, clean water, and nutritious foods readily available elsewhere, then death rates would drop substantially. Life expectancies in these countries could be extended by twenty years or more.

TABLE 13.1 ■ Estimated Populations and Projected Growth Rates

Region or Country	Estimated Population (millions)			Rate of Annual Natural Increase in Mid-2020 (%)	Number of Years to Double Population
	Mid-2020	Mid-2035	Mid-2050		
World total	7,773	8,937	9,876	1.1	65
United States	330	362	386	0.3	240
Canada	38	44	49	0.2	360
Mexico	128	142	148	1.2	60
China	1,402	1,424	1,366	0.3	240
India	1,400	1,576	1,663	1.4	51
Bangladesh	170	197	216	1.6	45
Japan	126	124	110	−0.4	—
Angola	33	53	82	3.5	21
Chad	17	26	38	3.2	23
Niger	24	42	66	3.8	19
Uganda	46	70	98	3.3	22

Source: Population Reference Bureau, *2020 World Population Data Sheet Shows Older Populations Growing, Total Fertility Rates Declining* (Washington, DC: Population Reference Bureau, 2020), https://www.prb.org/2020-world-population-data-sheet/.

There are several reasons to expect death rates to decline. Development agencies have attempted to reduce infant mortality by immunizing children against potentially fatal illnesses and by providing inexpensive cures for diarrhea, malaria, and other illnesses. These efforts have met with enormous success, but COVID-19 has affected this trajectory. Reduced mortality rates among children can also reduce fertility rates. Nonetheless, the change will be gradual, and millions of children will be born in the meantime. Most of the first-time mothers of the next twenty years have already been born.

The best-known population programs have been in India and China. India's family planning program started in the early 1950s as a low-key effort that achieved only modest success. The program changed from being voluntary to compulsory in the mid-1970s. The minimum age for marriage was increased, and India's states were encouraged to select their own methods to reduce growth. Through a variety of approaches, India has been able to cut its fertility rate significantly, but cultural and religious resistance may stifle further gains.[9] India is expected to add about 175 million more people by 2035 and surpassed China to become the most populous country by late 2020 or early in 2021.

To reduce the growth rate in China, its government once discouraged early marriages. It also adopted a one-child-per-family policy in 1979. Until the policy was relaxed in January 2016, allowing families to have two children, the government gave one-child families monthly subsidies, educational benefits for their child, preferences for housing and health care, and higher pensions at retirement. Families that had previously agreed to have only one child but then had another were deprived of these benefits and penalized financially. The most controversial elements of China's population policy involved the government's monitoring of women's menstrual cycles, forced sterilizations, and late-term abortions. The one-child policy also increased the likelihood of female infanticide in rural areas.[10]

China's initial efforts lowered annual rates of population growth considerably. Total fertility rates declined from 5.8 in 1970 to 1.5 in 2020, which is well below the replacement rate of approximately 2.1. As a result of its one-child policy, Chinese officials claimed that nearly four hundred million births had been averted.[11] More recently, voluntary decisions not to have multiple children mean that China's population is expected to be lower in 2050 than in 2020.

For many years, the US government viewed rapidly growing populations as a threat to economic development. The United States backed its rhetoric with money; it was the largest donor to international population programs. The US position changed dramatically during the Ronald Reagan administration. Due to its opposition to abortion, the administration said the United States would no longer contribute to the United Nations Population Fund (UNFPA) because it subsidized some of China's population programs. None of the fund's resources are used to provide abortions, but the US ban on contributions nonetheless continued during George H. W. Bush's administration.

Within a day of taking office, President Bill Clinton announced his intention to alter these policies, to provide financial support to the UNFPA, and to finance international population programs that rely on abortions. Just as Clinton had acted quickly, so too did George W. Bush. He reinstated Reagan's policy banning the use of federal funds by international organizations to support or advocate abortions. The cycle continued with President Barack Obama. He reversed the Bush rules and urged

Congress to restore funding for the UNFPA. President Donald Trump reversed Obama's position and halted funding for the UNFPA. In addition, Trump also banned US support for international nongovernmental organizations working on health-related issues, such as the control of malaria, unless they agreed not to perform, discuss, or promote abortions as a method of family planning in their other activities unrelated to malaria.

Concerns about abortion are not the only reason many people have qualms about efforts to limit population increases. Their view is that large populations are a problem only when they are not used productively to enhance development. The solution to the lack of such development is not government intervention, they argue, but rather individual initiatives and the spread of capitalist, free-market economies. Advocates of this position also believe that larger populations can be advantageous because they enhance political power, contribute to economic development and pension systems, encourage technological innovation, and stimulate agricultural production.[12] Other critics of population control programs also ask if it is appropriate for developed countries to impose their preferences on others.

Another much-debated issue involves increased access to abortions, and who chooses to have them. The consequences of efforts to affect population growth are not always gender neutral. In parts of Asia, male children are prized as sources of future financial security, whereas females are viewed as liabilities. In years past, the sex of newborns was known only at birth, and in most countries, newborn males slightly outnumbered newborn females. With the advent of ultrasound, however, the sex of a fetus is easily ascertained months before a child is born. This knowledge can be the basis of a decision to abort female fetuses, notably in parts of China, India, and Eastern Europe.[13]

In sum, the appropriateness of different population sizes is debatable. There is no clear answer about whether growth by itself is good or bad. The important issue is a country's and the world's carrying capacity. Can it ensure a reasonable and sustainable standard of living? Can it do so in the future when the world's population will be substantially larger?

PROVIDING FOOD AND FUEL FOR GROWING POPULATIONS

Sustainable development requires that environmental resources not be overtaxed so that they are available for future generations. When populations exceed sustainable yields of their forests, aquifers, and croplands, however, these resources are gradually destroyed.[14] The eventual result is an irreversible collapse of biological and environmental support systems. Is there any evidence that these systems are now being strained or will be in the near future?

The first place to look is in the area of food production. Nations can grow their own food, import it, or, as most nations do, rely on both options. The Earth is richly endowed with agricultural potential and production. Millions of acres of arable land remain to be cultivated, and farmers now produce enough food to satisfy the daily caloric and protein needs of a world population exceeding twelve billion.[15] These data suggest the ready availability of food as well as a potential for even higher levels of production. This good news must be balanced with the realization that hundreds of millions of people barely have enough food to survive now.

As with economic development, the amount of food available in a country must increase at least as fast as the rate of population growth; otherwise, per capita consumption will decline. If existing levels of caloric intake are already inadequate, then food production (and imports) must increase faster than population growth to meet minimum caloric needs. Assisted by the expanded use of irrigation, improved farming practices, and high-yielding seed varieties, many developing countries, particularly in Asia, have dramatically increased their food production. Asia's three largest countries—China, India, and Indonesia—are no longer heavily dependent on imported food.

Other countries can point to increased agricultural production, but many of these increases do not keep pace with population growth. The consequence is that average caloric consumption declines or imports of food increase, although declining consumption and a growth in imports can happen concurrently. With frequent spikes in food and fuel prices, many countries find themselves without sufficient resources to import enough food to ensure that even minimal levels of nutrition can be maintained. As an illustration, the Food and Agriculture Organization (FAO) estimated that over 55 percent of sub-Saharan Africa's population suffered from moderate or severe food insecurity in 2019. For those facing severe food insecurity, this meant that people were likely to run out of food, to experience hunger, and often have nothing to eat for several days. Throughout sub-Saharan Africa more than 225 million people were believed to face severe food insecurity with another 420 million in Asia.[16] Among the consequences of insufficient food are stunted growth, weakened resistance to illness and disease, and impaired learning abilities and capacity for physical labor.

Agricultural production can be increased, but many countries suffer a shortage of land suitable for cultivation. Other countries have reached or exceeded the sustainable limits of production. Their populations are overexploiting the environment's carrying capacity. Farmers in India, Pakistan, Bangladesh, and West Africa may already be farming virtually all the land suitable for agriculture, and the amount of arable land per capita is declining in many developing countries. Consider the situation in Bangladesh. Its population of about 170 million people in 2020 occupied an area smaller than Iowa, which had a population of just over 3.1 million in that year. Moreover, the FAO estimates that more than half of the land now used for agriculture worldwide is moderately or severely degraded, and an area about twice the size of China is seriously degraded, sometimes irreversibly. This degradation reduces productivity and food security, disrupts ecosystems, adversely affects biodiversity and water resources, and increases carbon emissions and vulnerability to climate change.[17] If these trends continue, millions of acres of barren land will be added to the millions that are already beyond redemption.

Shortages of land suitable for farming are not the only barrier to increased agricultural production. With more people to feed, more water must be devoted to agriculture. To feed the world's population in 2050, the amount of water devoted to agriculture will have to double between 2000 and 2050, one estimate suggests.[18] Doing so will be a challenge. All the water that will ever exist is already in existence, and much of this water is already overused, misused, or wasted in much of the world, including in the United States, one of the world's largest users of water on a per capita basis. Some countries are already desperately short of water, as frequent droughts in Africa unfortunately confirm.

The World Resources Institute has classified seventeen countries, which have about one-quarter of the world's population, as facing "extremely high" levels of baseline water stress, where irrigated agriculture, industries, and municipalities withdraw more than 80 percent of their available supply on average every year. Another forty-four countries, home to one-third of the world, face "high" levels of stress, where on average more than 40 percent of available supply is withdrawn every year. Such withdrawal rates are unsustainable. Once used, this water is often returned to the environment without any treatment. According to the World Bank, more than half the wastewater produced in the Middle East and North Africa is untreated, degrading water quality and availability and increasing the risk of waterborne diseases.[19] Farmers in many developing countries must also contend with climate change that affects the amount, intensity, and variability of rainfall.

Many developing countries rely on fish as a major source of protein. Unfortunately, the condition of many of the world's fisheries is perilous. The FAO estimated in 2018 that 33 percent of the world's marine fish stocks are fished at biologically unsustainable levels compared with only 10 percent in 1974. Over 700 oxygen-starved "dead zones" have been identified in the world's oceans and coastal areas. These zones can barely sustain marine life.[20]

It is important to appreciate, as well, that the nature of diets changes as nations urbanize. Irrespective of differences in prices and incomes, according to the International Food Policy Research Institute, "urban dwellers consume more wheat and less rice and demand more meat, milk products, and fish than their rural counterparts." This preference leads to increased requirements for grain to feed animals, the need for more space for forage, greater demands for water, and increased pollution from animal waste. Changes in the composition of diets can be anticipated in many countries. In fact, in virtually every low-income country, urbanization is increasing faster than overall population growth (in many instances, three to four times faster).

Increased demand for food also has environmental consequences. More grain must be produced to feed livestock and poultry. In a typical year, as much as 35–40 percent of the world's grain production is used for animal feed, but the conversion from feed to meat is not a neat one. As many as ten pounds of grain and about 1,900 gallons of water are required to produce one pound of beef. Production of beef is especially problematic for the environment. In a report completed for the National Academy of Sciences (NAS), researchers concluded that beef production requires twenty-eight, eleven, and five times more land, irrigation water, and greenhouse emissions, respectively, than the average of the other livestock categories.[21]

Ruminant livestock need grazing land, which is already in short supply in many areas. Throughout the world, about twice as much land is devoted to animal grazing as is used for crops. If a land's carrying capacity is breached due to excessive exploitation, then the alternative is to use feedlot production, which requires even higher levels of grain and concentrates waste products in small areas.

Relying on Domestic Production

Imports offer a possible solution to deficiencies in domestic production, but here, too, many developing countries encounter problems. To finance imports, countries need foreign exchange, usually acquired through their own exports or from loans. Few developing countries have industrial products or professional services to export, so they must rely on minerals, natural resources (such as timber or petroleum), or cash crops (such as tea, sugar, coffee, cocoa, and rubber).

Prices for many of these commodities fluctuate widely. To illustrate, prices for sugar, cotton, cocoa beans, nickel, aluminum, natural gas, and crude oil were lower in August 2020 than in August 2010.[22] To cope with oft-declining prices for crops intended for export farmers often intensify production, which implies increased reliance on fertilizers and pesticides, or they expand the area under cultivation to increase production. These seemingly rational reactions can depress prices as supply eventually outpaces demand. As the area used for export crops expands, production for domestic consumption may decline. In contrast, high prices are good for farmers but reduce affordability for cash-strapped consumers.

COVID-19 worsened the situation. The executive director of the UN World Food Program raised the possibility that the virus could lead to an "unprecedented global humanitarian catastrophe" and to "multiple famines of biblical proportions" in more than thirty-five countries. Unless millions of people in developing countries were able to access food and the ability to buy or grow it during the pandemic, as many as 300,000 people could die of starvation every day for several months. In each of ten countries, the director added, more than a million people were already on the brink of starvation.[23] In these and other countries millions of children were not attending school and thus were missing the meals that their schools had provided before the virus had descended on them.

In addition to the effects of COVID-19, farmers in eight East African countries had to contend with a literal plague of desert locusts in 2019 and 2020. As incredible as it might seem, swarms of locusts can be as dense as 380 *million* per square mile and eat as much food in 1 day in that square mile as about 90,000 people.[24]

Opportunities exist to increase exports to developed countries, but economic policies in the developed world can discourage expanded activity in developing countries. Every year, farmers in Japan, Europe, and the United States receive billions of dollars in subsidies and other price-related supports from their governments.

The European Union (EU) provided about $65 billion in 2019 for agricultural support for its farmers, including Queen Elizabeth of the United Kingdom. In some years, nearly 40 percent of the EU's annual budget is devoted to farm subsidies. So large are these supports, the president of the World Bank once noted, that the average European cow received a subsidy of about $2.50 per day, or more than the average daily income of about three billion people.[25] Japanese cows were more privileged. They received a daily subsidy of about $7.50 per day, or more than 1,800 times as much aid as Japan provided to sub-Saharan Africa each day.

Subsidies often lead to overproduction and surpluses, which discourage imports from developing countries, remove incentives to expand production, encourage the use of environmentally fragile land, and can increase prices to consumers in countries that provide the subsidies. Rice, sugar, cotton, wheat, and peanuts are easily and less expensively grown in many developing countries, but the US government subsidizes its farmers to grow these crops or imposes tariffs on their importation. Sugar provides an excellent example. In 2019, the cost of refined sugar in the United States was 129 percent higher than the world refined price. In addition to subsidizing US sugar growers and increasing the cost to American consumers, the government imposes a quota on the amount of sugar that can be imported and imposes a tariff on imports of sugar from all countries except Mexico.

Developing countries are increasingly irritated with trade and agricultural policies that they consider to be discriminatory. In response to a complaint from Brazil, the World Trade Organization (WTO) agreed that European subsidies for sugar exports violated international trade rules. This decision followed another WTO decision in which it ruled that US price supports for cotton resulted in excess production and exports as well as low international prices, thus causing "serious prejudice" to Brazil. African producers of cotton have also called for an end to government support for the production of cotton in developed countries, especially the United States, the world's largest exporter of cotton. Without access to export markets, developing countries are denied their best opportunity for development, which, historically, has provided the best cure for poverty and rapid population growth.

The Debt Conundrum

Developing countries could once depend on loans from private banks or foreign governments to help finance imports, but many low- and middle-income countries are burdened with considerable debts. A common measure of a nation's indebtedness is its debt service, which represents the total payments for interest and principal as a percentage of the country's exports of goods and services. These exports provide the foreign currencies that allow countries to repay their debts, which are denominated in foreign currencies, and to import food, medicines, petroleum, and machinery. When debt service increases, more export earnings are required to repay loans, and less money is available for development. Many developing nations, especially in Africa and Latin America, encounter this problem.

Continued borrowing and COVID-19 have worsened the problem of indebtedness among many developing countries. Loans from multilateral organizations such as the World Bank and the Asian Development Bank have long maturities and low interest rates, but their loans always include requirements that borrowers comply with guidelines on social and environmental safeguards, procurement, and financial management. The funded projects must also be justified economically. These requirements and the banks' oversight of compliance with the guidelines can make the loans seem burdensome to borrowers.

Since 2013, more than sixty-five countries have turned to China for access to capital. Its Belt and Road Initiative (BRI) has lent over $350 billion to build large projects, such as dams, ports, and railroads. In contrast to loans from multilateral organizations, BRI loans are less onerous, have higher interest rates, shorter maturities, and often do not require justification beyond a president's or a prime minister's desire to borrow and build. The Chinese often require borrowers to provide collateral if the loans cannot be repaid, but China has been much less tolerant than the multilateral organizations when countries have trouble with repayments. When Sri Lanka was unable to repay a loan, it granted the Chinese a 99-year lease for a strategic port. More recently, as a result of COVID-19's disastrous economic consequences, some countries have asked the Chinese to restructure the loans, to accept delays in repayments, or to forgive the loans entirely.[26] China has been reluctant to do so.

THE DESTRUCTION OF TROPICAL FORESTS

The rain forests of Africa, Asia, and South America are treasure chests of incomparable biological diversity. These forests provide irreplaceable habitats for as much as 70 percent of the world's species of plants and animals. Viable forests also stabilize soils; reduce the impact and incidence of floods; and regulate local climates, watersheds, and river systems.[27] In addition, increasing concern about climate change underscores the importance of tropical forests. Through photosynthesis, trees and other plants remove carbon dioxide from the atmosphere and convert it into oxygen. More than one-quarter of the prescription drugs used in the United States have their origins in tropical plants.

At the beginning of the 20th century, tropical forests covered approximately 10 percent of the Earth's surface, or about 5.8 million square miles. The deforestation of recent decades has diminished this area by about one-third. If current rates of deforestation continue unabated, only a few areas of forest will remain untouched. Humans will have destroyed a natural palliative for climate change and condemned half or more of all species to extinction.

Causes

Solutions to the problem of tropical deforestation depend on the root causes.[28] One view blames poverty and the pressures associated with growing populations and shifting cultivators. Landless peasants, so the argument goes, invade tropical forests and denude them for fuel wood, for grazing, or to grow crops with which to survive. Tropical soils are typically thin, relatively infertile, and lack sufficient nutrients, so frequent clearing of new areas is necessary. Such areas are ill suited for sustained agricultural production.

Another explanation for deforestation places primary blame on commercial logging intended to satisfy demands for tropical hardwoods in developed countries. Whether strapped for foreign exchange, required to repay loans, or subjected to domestic pressure to develop their economies, governments in the developing world frequently regard tropical forests as sources of ready income. Exports of wood produce billions of dollars in annual revenues for developing countries, and some countries impose few limits in their rush to the bank.

Recognizing the causes and consequences of deforestation is not enough to achieve a solution. Commercial logging can be highly profitable to those who own logging concessions, and few governments in developing countries have the capacity to manage their forests properly. These governments often let logging companies harvest trees in designated areas under prescribed conditions. Too frequently, however, these conditions are inadequate or not well enforced, often due to rampant corruption.

An Alternative View of the Problem

One cause of deforestation is the demand for tropical hardwoods in developed countries, so these countries have been under pressure to reduce that demand. Leaders of developing countries quickly emphasize how ironic it is that developed countries, whose consumption creates the demand for tropical woods, are simultaneously calling for developing countries to reduce logging and shifting cultivation. In addition, developing countries point to Europe's destruction of its forests during the industrial revolution and the widespread cutting in the United States in the 19th century. Why then should developing countries be held to a different standard than the developed ones? Just as Europeans and Americans decided how and when to extract their resources, developing countries insist that they too should be permitted to determine their own patterns of consumption.

Will tropical forests survive? Solutions abound. What is lacking, however, is a consensus about which of these solutions will best meet the essential needs of the poor, the reasonable objectives of timber-exporting and timber-importing nations, and the inflexible imperatives of ecological stability.

There is a growing realization that much can be done to stem the loss of tropical forests, and some progress is being made. The rate of deforestation declined by more than 35 percent between 2015 and 2020 compared with the rate in the 1990s.[29] Many countries have developed national forest programs that describe the status of their forests as well as strategies to preserve them. Unfortunately, implementation of these plans does not always parallel the good intentions associated with them. Likewise, rather than seeing forests solely as a source of wood or additional agricultural land, many countries are now examining the export potential of forest products other than wood. The expectation is that the sale of these products—such as cork, rattan, oils, resins, and medicinal plants—will provide economic incentives to maintain rather than destroy forests.

Other proposed options to maintain tropical forests include efforts to certify that timber exports are from sustainably managed forests. Importers and potential consumers presumably will avoid timber products without such certification. For such initiatives to be successful, however, exporters have to accept the certification process, and there must be widespread agreement about what is meant by sustainable management. Such agreement is still absent. In addition, no country wants to subject itself to the potentially costly process of internationally accepted certification only to learn that its forestry exports do not meet the requirements for certification or that less expensive timber is available from countries that do not participate in a certification program.

Other approaches to sustainable management impose taxes on timber exports (or imports). The highest taxes would be imposed on logging that causes the greatest ecological damage; timber from sustainable operations would face the lowest taxes. Yet another option is to increase reliance on the community-based management of forest resources. Rather than allowing logging companies with no long-term interest in a forest to harvest trees, community-based management places responsibility for decisions about logging (and other uses) with the people who live in or adjacent to forests. These people have the strongest incentives to manage forest resources wisely, particularly when they reap the long-term benefits of their management strategies.

Still another promising initiative is the United Nations Collaborative Programme on Reducing Emissions from Deforestation and Forest Degradation (REDD) in Developing Countries, which supports nationally led REDD+ initiatives, broadening the scope of the original program. An aspect of this international program involves developed countries paying developing countries not to harvest their tropical forests.

CONFLICTING SIGNALS FROM DEVELOPED COUNTRIES

Improvements in the policies of many developing countries are surely necessary if sustainable development is to be achieved. As already noted, however, developed countries sometimes cause or contribute to environmental problems in the developing countries.

Patterns of consumption provide an example. Although the United States and other developed nations can boast about their own comparatively low rates of population growth, developing nations reply that patterns of consumption are the real culprits when it comes to sustaining the world's population. This view suggests that negative impacts on the environment are a function of not only a country's population growth but also its consumption and the technologies, such as automobiles, that enable this consumption.[30]

Applying this formula places major responsibility for environmental problems on rich nations, despite their relatively small numbers of global inhabitants. The inhabitants of these nations own and consume far more of the Earth's resources than their numbers justify. Consider that in mid-2019 the world's richest 10 percent of adults owned over 80 percent of its total wealth whereas the bottom half of adults collectively owned less than 1 percent of this wealth.[31] In addition, residents of developed countries consume a disproportionate share of all meat and fish and much of the world's energy, paper, chemicals, iron, and steel.

During a typical day in the United States, about 330 million Americans consume more petroleum than the 3.30 *billion* people who live in India, Japan, and all the countries in Africa and Central and South America combined. There are more motor vehicles than licensed drivers in the United States. An average American driver uses about five times more gasoline than the typical European. Part of the explanation is that many European cars, often designed by US manufacturers, are more fuel efficient than are US cars. The price of gasoline in much of Western Europe is more than two to three times higher than it is in the United States.

Americans' extravagance with fossil fuels provides part of the explanation for US production of about one-sixth of the emissions that contribute to global warming. The Intergovernmental Panel on Climate Change has assessed that a relatively safe level of carbon dioxide emissions is about 2.25 metric tons per person per year.[32] Each metric ton is about 2,205 pounds. With the exception of Saudi Arabia and a few ministates, no country emits as much carbon dioxide per capita as does the United States. It produced 16.1 metric tons per capita in 2018, seven times higher than what sustainable levels of development would require. In Germany, China, and India, annual per capita emissions in the same year were 9.0, 7.8, and 1.9 metric tons, respectively.[33] Although China's emissions of carbon dioxide increased by over 270 percent between 2000 and

2018, a notable portion of the increase was attributable to the production of goods destined for Europe and the United States.

Americans' patterns of food consumption are also of interest. An average American consumes about 3,750 calories per day, almost the highest level in the world. Among young adults, about 25 percent of these calories are from sweetened beverages. Not surprisingly, almost three-quarters of American adults are either obese or overweight. *The Economist* labeled the United States the "fattest country in the world."[34]

As much as 30–40 percent of all food is wasted in the United States each year, and the country spends nearly $220 billion a year growing, processing, transporting, and disposing of food that is never eaten. Wasting food has several undesirable environmental consequences. It consumes about one-fifth of all fresh water and almost the same proportion of cropland. Energy is required to produce, harvest, transport, market, and prepare food. All this energy is squandered when food is wasted. Food is the largest source of waste in American's trash. When food rots in landfills it contributes to emissions of greenhouse gases. If the developed world reduced the food it wasted by half, the challenge of feeding the world's population in 2050 would vanish.[35]

Due to these kinds of inequalities in consumption, continued population growth in rich countries is a greater threat to the global environment than is such growth in the developing world. As the World Wildlife Fund has explained, "If everyone lived like an average resident of the USA, a total of four Earths would be required to regenerate humanity's annual demand on nature." Other experts have suggested that if Americans want to maintain their present standard of living and levels of energy consumption, their ideal population is about 50 million, far less than the mid-2020 US population of about 330 million.[36]

The data just presented raise questions that merit attention. Are the world's environmental problems due to too many poor people who have few resources and who use them sparingly or to too many rich people who use more resources than their numbers justify and whose significant impact on the environment has global implications? Is it fair and equitable for the relatively small numbers of rich people to consume excessively and waste indiscriminately when hundreds of millions of people in developing countries are undernourished and undereducated, live in slums or their equivalent, are frequent victims of deadly illnesses and diseases without access to their cures, and live amidst environmental harms that they cannot escape but that would be considered scandalous and intolerable in the rich world?

CAUSES FOR OPTIMISM?

There is cause for concern about the prospects for sustainable development among developing countries, but the situation is neither entirely bleak nor beyond hope. The rates of deforestation and population growth are slowing in many developing countries. Smallpox, a killer of millions of people every year in the 1950s, has been eradicated (except in laboratories). Polio may soon be the next scourge to be eliminated, and deaths due to AIDS, diarrhea, measles, and respiratory infections such as tuberculosis have dropped significantly.

The proportion of people living in extreme poverty declined by half between 1990 and 2010. The mortality rate for children under five in sub-Saharan Africa dropped by

50 percent between 2000 and 2018. Fertility rates are declining in most developing countries. Life expectancies increased in all parts of the world over the same time period as have the percentages of children who have been immunized, attend and complete basic education, eat nutritious meals, have access to clean water and proper sanitary facilities, and much more.

To build on these successes the world's nations established seventeen Sustainable Development Goals (SDGs) with 169 targets in late 2015. The goals, to be achieved no later than 2030, focus on economic growth, social inclusion, and environmental protection. Goal 15, as an example, seeks to protect, restore, and promote the sustainable use of terrestrial ecosystems; manage forests sustainably; combat desertification; and halt and reverse land degradation and halt biodiversity loss. Other goals address the need for sustainable consumption and for "urgent action" to combat climate change.

The SDGs demonstrate an increasing awareness of the need to address the problems of the developing world, but much remains to be done. Through the President's Emergency Plan for AIDS Relief (PEPFAR), as an example, the United States is the world's leader in responding to HIV/AIDS in the developing world. PEPFAR, which President George W. Bush initiated in 2003, has helped avert more than eighteen million AIDS-related deaths. Private philanthropic support for development has also grown significantly. The Bill and Melinda Gates Foundation has provided over $50 billion to improve health, education, and the status of women in developing countries. The William J. Clinton Foundation has been similarly active and has successfully negotiated major reductions in the cost of essential drugs in many countries, assisted African farmers to improve their productivity, and worked to reduce deforestation and climate change.

Perhaps the best known target among developed countries is that each country provide at least 0.7 percent (*not* 7 percent, but seven-tenths of 1 percent) of its GNI to official development assistance (ODA) to meet the needs of developing countries. The United Nations General Assembly established this target in 1970, with the expectation that donor countries would meet it by 1975. Sweden and the Netherlands were the first to do so. Despite this admirable target, only Denmark, Luxembourg, Norway, Sweden, the United Arab Emirates, and the United Kingdom met or exceeded this target in 2019. Although the United States typically provides more foreign aid than any other country, its ODA is often less than 0.2 percent of the US GNI, which is among the lowest percentages of the world's major donors. This situation has led some observers to label the United States as a "global Scrooge" based on its seeming unwillingness to share its wealth.

Those who think the United States should continue to scale back its ODA contributions had an ally in President Trump. His initial budget stated that it was time to prioritize the well-being of Americans and "to ask the rest of the world to step up and pay its fair share." Trump proposed large reductions in US foreign assistance and wanted the remaining assistance to be focused in countries of the greatest strategic significance to the United States—rather than in countries with a high level of need. He also proposed the elimination of all US funding of efforts to mitigate the effects of climate change in developing countries, including those implemented by the United Nations.[37] During the height of the COVID-19 pandemic in early 2020, his administration prohibited the use of US foreign assistance to purchase any personal protective equipment for health workers without prior approval from the US Agency for International Development (USAID).

Trump repeatedly proposed the termination of the McGovern–Dole International Food for Education and Child Nutrition Program. Congress rejected each of the proposals. The program supports education, child development, and food security in low-income, food-deficit countries around the globe. This proposal was ironic. Poor nutrition is probably the single most important threat to the world's health, and the program has helped millions of children and pregnant women avoid malnutrition.

Is it important to assist developing countries with food and finance? Americans are often ambivalent in their answers to this question. Surveys of Americans' opinions about foreign aid present an interesting but mixed picture. More than 80 percent of those surveyed in 2008 agreed that developed countries have "a moral responsibility to work to reduce hunger and severe poverty in poor countries." Despite such support, most Americans also believe that the United States is already doing more than its share to help less fortunate countries.

In surveys conducted in 2015 and 2016, respondents estimated that the average amount of the annual federal budget devoted to foreign aid was 31 percent. When told that the actual amount was less than 1 percent, nearly one-third of respondents said this was still too much. Among those not aware how much of the budget is devoted to foreign aid, nearly half of the respondents in a national survey in 2019 said we spend too much.[38]

The United States is among the world's largest donors of emergency food aid. Rather than allowing USAID to purchase less expensive food closer to where it is needed, Congress requires that much of the food be purchased in the United States. In addition, the law requires that at least half of the food be shipped on American-flagged vessels, which are among the most expensive to use. As a result of these requirements, US food aid is more expensive than it would otherwise be, and the food often arrives well after it is most needed.[39] Critics of the program also claim that the US food aid destabilizes local markets and competes with farmers close to affected areas who cannot afford to donate their food.

In contrast to Americans' seeming reluctance to share their wealth, other nations have demonstrated an increased willingness to address globally shared environmental problems. The international community operates a Global Environment Facility, a multibillion-dollar effort to address global warming, the loss of biological diversity, the pollution of international waters, and the depletion of the ozone layer. The Green Climate Fund represents a similar international initiative. Created in 2010 by 194 countries, the fund provides resources for low-emission and climate-resilient projects in developing countries, especially those that are highly vulnerable to the effects of climate change.

Many developing nations recognize their obligations to protect their environments as well as the global commons. India's and China's commitment to reducing their emissions of greenhouse gases provide examples. India produces relatively few of these emissions on a per capita basis, but it has committed to reduce its emissions intensity by as much as 33–35 percent by 2030 from the levels reached in 2005. In addition, although it is facing growing demands for electricity, which could be met by burning more coal, India has committed to increasing its reliance on nonfossil fuels by

40 percent in 2030. China has pledged to reduce its carbon intensity 60–65 percent by 2030 compared with the levels in 2005 and has similarly pledged to increase its reliance on renewable sources of energy. China is among the world's leaders in the use of renewable energy. Forty-five percent of the electric cars and 99 percent of the world's electric buses were in China in 2018.[40] Both India and China are reducing their reliance on fossil fuels more quickly than they had previously projected, leading the *New York Times* to conclude that their progress has been "astonishing."[41]

However desirable improved environmental quality may be as an objective, poor nations cannot afford to address their environmental problems in the absence of cooperation from richer nations. Consumers in rich nations can demonstrate such cooperation by paying higher prices for products that reflect sustainable environmental management. One example of this situation would be Americans' willingness to pay higher prices for forestry products harvested sustainably. In fact, however, Home Depot found that only a third of its customers would be willing to pay a premium of 2 percent for such products.[42]

Another issue of importance is resentment in some countries toward the environmental sermons from developed countries. Climate change provides one of several examples. As leaders of several developing countries have asked, why should they slow or alter their path to development to accommodate high standards of living elsewhere? When India released a policy statement on climate change, its prime minister declared that fairness dictated that everyone deserves equal per capita emissions, regardless of where they live.[43] India is not willing, he noted, to accept a model of development in which some countries maintain high carbon emissions while the options available for developing countries are constrained.

President George W. Bush once identified a growing middle class in India, which is "demanding better food and nutrition," as a cause of higher prices. The reaction from India was understandably negative. "Why do Americans think they deserve to eat more than Indians?" asked one journalist. An Indian public official characterized the US position as "[g]uys with gross obesity telling guys just emerging from emaciation to go on a major diet." This characterization may be indicative of a larger concern. The perception that Americans are global environmental culprits is widespread. When people in twenty-four countries were asked in 2008 which country is "hurting the world's environment the most," majorities or pluralities in thirteen countries cited the United States.[44] President Trump's decision to withdraw from the Paris climate agreement reinforced this perception. President Joe Biden., Jr. is expected to reverse that decision soon after taking office. If he were not to do so, the United States would be the only country in the world not to be a party to the agreement.

Further evidence of America's seeming lack of commitment to international environmental issues can be found in the Center for Global Development's Commitment to Development Index. The index ranks the world's richest countries on their dedication to policies that benefit people living in poorer nations. In 2020, the United States ranked twenty-fourth of these twenty-seven countries in the environmental dimension of the index.[45]

Contentious debates and inflammatory rhetoric about blame and responsibility are not productive. The economic, population, and environmental issues of the developing world dwarf those of the developed nations and are not amenable to quick resolution, especially as COVID-19 has jeopardized much of the progress achieved in recent decades. Now more than ever immediate action is imperative. Hundreds of millions of people are destroying their biological and environmental support systems to meet their daily needs for food, fuel, and fiber. The world will add several billion people in the next few decades, and all of them will have justifiable claims to be fed, clothed, educated, employed, and healthy. To accommodate these expectations, the world may need as much as 50 percent more energy in 2030 than it used in 2010.

Whether the environment can accommodate these unprecedented but predictable increases in population and consumption depends not only on the poor who live in stagnant or slow-growing economies but also on a much smaller number of rich, overconsuming nations in the developed world. Unless developed nations work together to accommodate and support sustainable development everywhere, the future of billions of poor people, many of whom become refugees seeking better lives, will determine the future of Americans as well.

As the authors of the Millennium Ecosystem Assessment concluded, the ability of the planet's ecosystems to sustain future generations is no longer assured.[46] Over the past fifty years the world has experienced unprecedented environmental change in response to ever-increasing demands for food, fuel, fiber, freshwater, and timber. Much of the environmental degradation that has occurred can be reversed, but as these authors warned, "The changes in policy and practice required are substantial and not currently underway."[47]

If these experts are correct, unless the United States acts soon and in collaboration with other nations, Americans will increasingly suffer the adverse consequences of environmental damage caused by the billions of poor people we have chosen to neglect and perhaps even abandon—just as these people will suffer from the environmental damage we inflict on them. In short, there are continuing questions about whether the current economic model that depends on growth and extravagant consumption among a few privileged countries is ecologically sustainable and morally acceptable for everyone.[48]

SUGGESTED WEBSITES

Organisation for Economic Co-operation and Development (OECD) (www.oecd.org/development) This website provides extensive information on aid statistics and effectiveness, environment and development, and gender and development.

UN Development Programme (www.undp.org) The program's website provides links to activities and reports on economic development and the environment, including the Sustainable Development Goals (SDGs).

UN Division for Sustainable Development (http://sustainabledevelopment.un.org) The "Sustainable Development Knowledge Platform" within the United Nations provides useful links to the goals and targets on sustainable development for the post-2015 development agenda.

UN Food and Agriculture Organization (www.fao.org) The FAO focuses on agriculture, forestry, fisheries, and rural development, working to alleviate poverty and hunger worldwide.

World Bank (www.worldbank.org) One of the largest sources of economic assistance to developing nations, the World Bank issues reports on poverty and global economic conditions, including progress toward the SDGs. Its *World Development Indicators* include a wealth of development data.

NOTES

1. World Bank, "GNI per capita, PPP (constant international $)," https://data.worldbank.org/indicator/NY.GNP.PCAP.PP.CD. Due to differences in the costs of goods and services among countries, GNI per capita does not provide comparable measures of economic well-being. To address this problem, economists have developed the concept of purchasing power parity (PPP). PPP equalizes the prices of identical goods and services across countries, with the United States as the base economy. For an amusing explanation of PPP, see the "The Big Mac Index" of *The Economist*, at www.economist.com/news/2020/01/15/the-big-mac-index, which compares the price of a McDonald's Big Mac hamburger in nearly fifty countries.
2. UNICEF, *State of the World's Children 2019* (New York: UNICEF, 2019), Table 11.
3. World Health Organization, *Trends in Maternal Mortality: 2000 to 2017* (Geneva, Switzerland: World Health Organization, 2019).
4. United Nations Inter-Agency Group for Child Mortality Estimation, *Levels and Trends in Child Mortality: Report 2020* (New York: UNICEF, 2020).
5. UNICEF, "Malnutrition," https://data.unicef.org/topic/nutrition/malnutrition/, accessed October 5, 2020.
6. World Commission on Environment and Development, *Our Common Future* (London: Oxford University Press, 1987), 8, 43.
7. John N. Wilford, "A Tough-Minded Ecologist Comes to Defense of Malthus," *New York Times*, June 30, 1987, C3.
8. Population Reference Bureau, *2020 World Population Data Sheet Shows Older Populations Growing, Total Fertility Rates Declining* (Washington, DC: Population Reference Bureau, 2020), www.prb.org/2020-world-population-data-sheet/. For a discussion of the world's carrying capacity, see Jeroen C. J. M. Van Den Bergh and Piet Rietveld, "Reconsidering the Limits to World Population: Meta-analysis and Meta-prediction," *BioScience* 54 (March 2004): 195–204; and Erle C. Ellis, "Overpopulation Is Not the Problem," *New York Times*, September 13, 2013, A19.
9. O. P. Sharma and Carl Haub, "Change Comes Slowly for Religious Diversity in India," Population Reference Bureau, March 2009, www.prb.org/Articles/2009/indiareligions.aspx
10. Joseph Kahn, "Harsh Birth Control Steps Fuel Violence in China," *New York Times*, May 22, 2007, A12; and Jim Yardley, "China Sticking with One-Child Policy," *New York Times*, March 11, 2008, A10.
11. Leo Lewis, "China Looks to a Boom as It Relaxes One-Child Policy," *The Times* (London), November 16, 2013.
12. For example, see Julian Simon, *The Ultimate Resource* (Princeton, NJ: Princeton University Press, 1981).
13. Carl Haub and O. P. Sharma, "India's Population Reality," *Population Bulletin* 61, no. 3 (September 2006), https://www.prb.org/wp-content/uploads/2006/09/61.3Indias PopulationReality_Eng.pdf; and Lisa Jane de Gara, "Asia's Missing Millions: How Policy and Social Pressure Made Millions of Women Disappear," *Education about Asia* 22, no. 3 (Winter 2017), 56–8.

14. Lester R. Brown, "Analyzing the Demographic Trap," in *State of the World 1987*, ed., Lester R. Brown (New York: Norton, 1987), 21.

15. Per Pinstrup-Andersen, former director of the International Food Policy Research Institute, believes the world can easily feed twelve billion people. See "Will the World Starve?" *The Economist*, June 10, 1995, 39; and Janet Ranganathan, Richard Waite, Tim Searchinger, and Craig Hanson, "How to Sustainably Feed 10 Billion People by 2050," World Resources Institute, December 5, 2018, www.wri.org/blog/2018/12/how-sustainably-feed-10-billion-people-2050-21-charts

16. FAO, *The State of Food Security and Nutrition in the World* (Rome, Italy: FAO, 2020), http://www.fao.org/publications/sofi/2020/en/

17. FAO, "Land assessments and impacts," www.fao.org/land-water/land/ldn/en/ (accessed May 20, 2020).

18. Colin Chartres and Samyuktha Varma, *Out of Water: From Abundance to Scarcity and How to Solve the World's Water Problems* (Upper Saddle River, NJ: FT Press, 2011), xvii.

19. Rutger Willem Hofste, Paul Reig, and Leah Schleifer, "17 Countries, Home to One-Quarter of the World's Population, Face Extremely High Water Stress," World Resources Institute (August 6, 2019), www.wri.org/blog/2019/08/17-countries-home-one-quarter-world-population-face-extremely-high-water-stress; and World Bank, *Beyond Scarcity: Water Security in the Middle East and North Africa* (Washington, DC: World Bank, 2018), https://openknowledge.worldbank.org/handle/10986/27659

20. FAO, *The State of World Fisheries and Aquaculture 2020* (Rome, Italy: FAO, 2020), http://www.fao.org/state-of-fisheries-aquaculture; and D. Laffoley and J. M. Baxter, eds., *Ocean deoxygenation: Everyone's problem* (Gland, Switzerland: International Union for Conservation of Nature, 2019), 85.

21. Gidon Eshel, Alon Shepon, Tamar Makov, and Ron Milo, "Land, Irrigation Water, Greenhouse Gas, and Reactive Nitrogen Burdens of Meat, Eggs, and Dairy Production in the United States," *Proceedings of the National Academy of Sciences of the United States* 111, no. 33, (July 2014): 11996–2001, https://doi.org/10.1073/pnas.1402183111

22. IndexMundi, www.indexmundi.com/commodities/

23. David M. Beasley, "Covid-19 could detonate a 'hunger pandemic.' With millions at risk, the world must act." *Washington Post*, April 22, 2020; and Edith M. Lederer, "UN food agency chief: World on brink of 'a hunger pandemic,'" *Washington Post*, April 21, 2020.

24. Neha Wadekar, "Two new generations of locusts are set to descend on East Africa again – 400 times stronger," *Quartz Africa*, April 10, 2020; https://qz.com/africa/1836159/locusts-set-to-hit-kenya-east-africa-again-400-times-stronger/. This website includes an astounding photograph of a locust swarm.

25. Statement of James D. Wolfensohn, cited in David T. Cook, "Excerpts from a Monitor Breakfast on Poverty and Globalization," *Christian Science Monitor*, June 13, 2003.

26. Maria Abi-Habib and Keith Bradsher, "China Lent Billions to Poor Countries: And Now They Can't Pay It Back," *New York Times*, May 19, 2020. There is an extensive and often-critical literature on the BRI. For example, see Elaine K. Dezenski, "Below the Belt and Road: Corruption and Illicit Dealings in China's Global Infrastructure" (Washington, DC: Foundation for Defense of Democracies, 2020), www.fdd.org/analysis/2020/05/04/below-the-belt-and-road/

27. NAS, *Population Growth and Economic Development: Policy Questions* (Washington, DC: NAS, 1986), 31.

28. For discussions of the causes of deforestation, see Helmut J. Geist and Eric Lambin, "Proximate Causes and Underlying Driving Forces of Tropical Deforestation," *BioScience* 52 (February 2002): 143–50; and Antoine Leblois, Olivier Damette, and Julien Wolfersberger, "What has Driven Deforestation in Developing Countries since the 2000s? Evidence from New Remote-Sensing Data," *World Development* 92 (April 2017), 82–102.

29. FAO, *The State of the World's Forests 2020: Forests, Biodiversity and People* (Rome, Italy: FAO, 2020).

30. Paul R. Ehrlich and John P. Holdren, "Impact of Population Growth," *Science* 171 (1971): 1212–17.

31. Credit Suisse Research Institute, *Global Wealth Report 2019* (Zurich: Credit Suisse, 2019), www.credit-suisse.com/about-us/en/reports-research/global-wealth-report.html

32. Commission on Growth and Development, *The Growth Report: Strategies for Sustained Growth and Inclusive Development* (Washington, DC: World Bank, 2008), 85–6, https://openknowledge.worldbank.org/handle/10986/6507

33. PBL Netherlands Environmental Assessment Agency, *Trends in Global CO_2 and Total Greenhouse Gas Emissions: 2019 Report* (The Hague: PBL Publishers, 2020), www.pbl.nl/sites/default/files/downloads/pbl-2020-trends-in-global-co2-and-total-greenhouse-gas-emissions-2019-report_4068.pdf

34. "Obesity: Another Thing It's Too Late to Prevent," *The Economist*, August 15, 2012; and OECD, *Obesity Update 2017* (Paris, France: OECD, 2017).

35. RTS, "Food Waste in America in 2020: Statistics + Facts," www.rts.com/resources/guides/food-waste-america/; and "The 9-billion People Question," *The Economist*, May 26, 2011.

36. World Wildlife Fund, *Living Planet Report 2012: Biodiversity, Biocapacity and Better Choices* (Gland, Switzerland: World Wildlife Fund, 2012), 43; and David and Marcia Pimentel, "Land, Water and Energy Versus the Ideal U.S. Population," *NPG Forum* (January 2005), https://www.npg.org/wp-content/uploads/2013/07/LandWaterAndEnergyVersusTheIdealUSPopulation.pdf

37. For the quotation, see Executive Office of the President of the United States, *America First: A Budget Blueprint to Make America Great Again* (Washington, DC: Office of Management and Budget, 2018), 2.

38. Bianca DiJulio, Mira Norton, and Mollyann Brodie, *Americans' Views on the U.S. Role in Global Health* (Menlo Park, CA: Kaiser Family Foundation, January 2016), http://kff.org/global-health-policy/poll-finding/americans-views-on-the-u-s-role-in-global-health/; and Liz Hamel, Ashley Kirzinger, and Mollyann Brodie, *2016 Survey of Americans on the U.S. Role in Global Health* (Menlo Park, CA: Kaiser Family Foundation, April 2016), http://files.kff.org/attachment/issue-brief-2016-survey-of-americans-on-the-u-s-role-in-global-health/; Kaiser Family Foundation, "Topline: KFF Health Tracking Poll–April 2019," http://files.kff.org/attachment/Topline-KFF-Health-Tracking-Poll-April-2019

39. Ron Nixon, "Typhoon Revives Debate on U.S. Food Aid Methods," *New York Times*, November 22, 2013, A3. See also Randy Schnepf, *U.S. International Food Aid: Background and Issues* (Washington, DC: Congressional Research Service, 2016), https://fas.org/sgp/crs/misc/R41072.pdf

40. David Sandalow, *Guide to Chinese Climate Policy 2019* (New York: Columbia University, 2019).

41. "On Climate, Look to China and India," *New York Times*, May 22, 2017, A24; and Keith Bradsher, "China Turns Economic Engine Toward Clean-Energy Leadership," *New York Times*, June 6, 2017, A1.

42. "The Long Road to Sustainability," *The Economist*, Special Report: Forests, September 25, 2010.

43. Voice of America, "India Rejects Binding Commitment to Cut Greenhouse Gas Emissions," February 7, 2008, https://www.voanews.com/a/a-13-2008-02-07-voa20-66627522/556691.html

44. Heather Timmons, "Indians Find U.S. at Fault in Food Cost," *New York Times*, May 14, 2008, C1; "Melting Asia," *The Economist*, June 7, 2008, 30; and Pew Global Attitudes

Project, "Some Positive Signs for U.S. Image," June 12, 2008, 65, http://www.pewglobal.org/files/pdf/260.pdf

45. For the Commitment to Development Index 2020, see https://www.cgdev.org/publication/commitment-development-index-2020. In previous years the index also ranked the quality of US foreign assistance as below average based on four dimensions of aid quality: maximizing efficiency, fostering institutions, reducing burden, and transparency and learning.

46. Millennium Ecosystem Assessment, *Ecosystems and Human Well-Being: Synthesis* (Washington, DC: Island Press, 2005), https://www.millenniumassessment.org/documents/document.356.aspx.pdf

47. Millennium Ecosystem Assessment, "What Are the Findings of the MA?" available from the Environment and Ecology website, http://environment-ecology.com/millennium-ecosystem-assessment/109-millenium-ecosystem-assesment.html

48. UN Development Programme, *Human Development Report 2007/2008: Fighting Climate Change—Human Solidarity in a Divided World* (New York: UNDP, 2007), http://hdr.undp.org/sites/default/files/reports/268/hdr_20072008_en_complete.pdf. For an informative and provocative discussion on the prevailing emphasis on economic growth, see Freakonomics Radio's podcast 429, "Is Economic Growth the Wrong Goal?" August 12, 2020.

CREATING THE GREEN ECONOMY

Government, Business, and a Sustainable Future

Daniel J. Fiorino

What do we expect from government? A degree of physical security and safety, to be sure. An opportunity to earn a living and meet basic needs for food and shelter, as well as higher-level needs such as education, healthcare, and a good quality of life. And most people would expect that government will create conditions for achieving fairness in society, as well as for providing social safety nets when bad times or misfortune hits. In sum, we count on governments to maintain internal order and protect against external threats, promote national economic well-being, and organize and deliver welfare services. James Meadowcroft expresses these activities respectively as those of the *security* state, *prosperity* state, and *welfare* state.[1] In activities and expectations, these have become the core business of the state, the terms in which political debates occur, policy is defined, institutions are created, and leadership choices are made.

Beginning in the middle of the last century, these expectations and activities expanded to encompass the *ecological* state. A combination of industrial development, urbanization, expanded food production, mobility through fossil fuels, and population growth posed new challenges for the modern state. Air and water pollution, deforestation, conversion of land for commercial uses, chemical production and waste, and energy and water use, among other trends, moved the environment up on policy agendas, and the public called for action. Governments responded with laws, ministries and regulatory agencies, scientific capabilities, and new strategies and policies.[2]

Yet, in nearly all countries, ecological issues have not achieved parity with the other core activities of the state: "the environment remains the most vulnerable of these core domains of state activity because it is the newest and least institutionally embedded."[3] This is unfortunate because ecological issues increasingly are linked to other core activities: climate change, water scarcity, and deforestation create prospects of geopolitical instability, mass migration, and political violence that threaten the security state. Just as compelling, failures in the ecological state increasingly undermine economic prosperity and human well-being. A changing climate and its costs are an obvious global threat. Others are health effects of energy systems based on fossil fuels, ecological degradation from land conversion and deforestation, and overconsumption of resources. Threats to prosperity and security undermine the political stability that enables environmental protection. In sum, a lack of attention to ecological issues threatens other core activities of the modern state.

This chapter focuses on one critical linkage among these activities—that of the prosperity with the ecological state. It reframes the economy–environment relationship from one of inevitable trade-offs (of zero-sums) to one in which complementary, positive-sum relationships are not only possible but increasingly necessary. This reframing is captured in the concept of a *green economy.*

A green economy is one in which economic and political choices account for the effects on the environment—on health, ecosystems, natural resources, and critical ecological services, such as a healthy climate system, vibrant wetlands and estuaries, or the protection of forests and land. It defines both constraints and opportunities. The *constraints* are the limits of the planet in the face of steady, exponential economic and population growth and their consequences. The *opportunities* are for healthier air and water; more vibrant economies and communities; long-term resource preservation (water or materials); family-supporting jobs; and a more equitable world.[4]

This chapter examines the concept of a green economy and its relevance to such issues as climate change, the energy system, water quality and security, natural resources, and ecosystems. The thesis is that human activity is placing more and more stress on planetary capacities, and that the foundations of current and especially future well-being increasingly are at risk. At the same time, economic prosperity is a high priority for all governments, adds to a high quality of life (at least to a point; more on that later), and will continue to occur on a global scale. The solution is to rethink the content of economic growth, how it is achieved, and maybe even its desirability. The next part of the chapter considers how the worlds of growth and planetary limits are colliding. Then the chapter addresses the leading expressions of the green economy, followed by an analysis of the similarities between greening among business firms (a microlevel) and an economy-wide, macrolevel. After that, I turn to a discussion of a policy agenda for a green economy and the need for global, collective action.

TWO WORLDS COLLIDING

The notion that planetary limits exist and are tested by economic and population growth goes back to a 1972 analysis by a team from the Massachusetts Institute of Technology.[5] In an early use of computers, they projected trends in resource use, pollution, and waste accumulation to reach controversial findings. Although specifics varied according to which scenario was used, their message was that "given postulated limits to resource availability, agricultural productivity, and the capacity of the ecosphere to assimilate pollution, some limit was generally hit within 100 years, leading to the collapse of industrial society and its population."[6] This argument challenged a post-World War II growth mindset. Economic growth, captured in the Gross Domestic National Product (GDP, which is the sum of the value of goods and services produced in a year), had become a central goal of modern governments and a sign of the dominance of the prosperity state. The notion that growth had to be limited to accommodate planetary limits was heavily disputed. Although still controversial, assessments conclude that the *Limits to Growth* thesis largely has been confirmed.[7]

A recent version of this thesis reconceives the limits to growth as planetary boundaries and a *safe operating space for humanity.*[8] A team of scientists collaborated in assessing the state of knowledge regarding nine planetary boundaries: atmospheric concentrations of carbon dioxide; biodiversity loss as reflected in extinction rates; stratospheric ozone depletion; ocean acidification; changes in land cover (percentage of land converted to cropland); atmospheric aerosol loading (a measure of air particulate concentrations, a major threat to global health); chemical pollution; and as measures of water pollution, trends in the global nitrogen and phosphorous cycles.

Three of the nine boundaries had already been exceeded: carbon dioxide, biodiversity loss, and the nitrogen cycle. For ocean acidification, changes in phosphorous, freshwater, and land use the boundaries are being stretched, and long-term trends are problematic.

Other evidence of planetary limits is all around us. High coal use in much of the world, along with growing vehicle use and other trends, causes extreme air pollution in major cities. The World Health Organization (WHO) now cites air pollution, especially particulate matter, as a top threat to global health, responsible for some seven million premature deaths annually.[9] The WHO estimates that in the 15 countries emitting the most greenhouse gases, health damages due to air pollution cost more than 4 percent of their annual GDP annually. Much of this harm falls on children, which undermines their well-being and the prospects in those countries for years to come.

Over a decade ago, in its *Millennium Ecosystem Assessment*, the United Nations evaluated the state of ecosystems around the world. Of 24 major ecosystems assessed, fifteen (60 percent) were found to be degraded.[10] Among these are capture fisheries, water supplies, waste treatment and detoxification, water purification, natural hazard protection, regulation of air quality, regulation of regional climate, erosion control, and aesthetic enjoyment. Each defines a category of ecosystem services—critical ecological activities on which well-being depends. The drivers of damages are habitat change from land use, physical modification of rivers, water withdrawal, overexploitation of resources like fish stocks, pollution, and effects of climate change. More recently, a 2019 report from the Intergovernmental Science-Policy Platform on Biodiversity and Ecosystem Services (IPBES) documents the severe impacts land and sea use, climate change, and pollution are having on global ecosystems. These are more than environmental issues. As a statement on the report puts it: "We are eroding the very foundations of our economies, livelihoods, food security, health and quality of life worldwide."[11]

Highlighting the case for a transition to a green economy is climate change, viewed more and more as a climate *crisis*. Climate change is occurring due to an accumulation of greenhouse gases in the atmosphere over the last 170 years. While modern energy systems based on fossil fuels improved human prosperity and quality of life, they also caused instability in the global climate system that is playing out in extreme weather, sea level rise, heat waves, droughts and flooding, new patterns of disease, ecosystem species disruption, and other effects (see Chapter 13). These will only worsen. Indeed, successive reports by the Intergovernmental Panel on Climate Change, the United Nations body that tracks science and policy, predict increasingly dire consequences. An October 2018 update found that avoiding such effects means "transforming the world economy at a speed and scale that has no 'documented historical precedent.'"[12] Emissions globally will have to be reduced by 45 percent from 2010 levels by 2030 and then by 100 percent by 2050.

In sum, there is ample evidence of planetary limits being stretched and often breached. These are not just environmental issues, limited to the activities of the ecological state; they affect and threaten to undermine national and global security, welfare, and especially prosperity. And yet modern governments are dedicated to the pursuit of economic growth, which in its current form will increasingly stress planetary limits. The worlds of limits and growth are on a collision course.

WHAT IS THE GREEN ECONOMY?

Concepts matter in politics and policy. The concept of the *cold war* defined American foreign policy for decades, from the end of World War II to the fall of the Berlin Wall. Proponents of aggressive climate action debate whether we should call it *climate change* or a *climate crisis*. Progressive legislators in the United States linked climate to an ambitious social agenda by invoking President Franklin Roosevelt's policies of the 1930s in the need for a *Green New Deal*.

The concepts of *green economy* and *green growth* also carry symbolic weight and value. Contrary to the standard assertions of opponents of environmental protection, the green economy is focused on the potentially positive relationships among ecological and economic goals. Such positive relationships abound. An obvious one is energy efficiency. Using less energy saves money; it also reduces air pollution, helps with climate mitigation, avoids ecological devastation, and may even support more resilient communities and employment.[13] Renewable energy is another illustration. As costs of wind and solar technologies have fallen, the benefits of clean energy have become more apparent. Conserving water and using it more efficiently offer other examples, as do the opportunities for using land better, protecting critical ecosystems, and smart farming practices (see Chapter 8).

Various definitions of the green economy exist. They were a popular theme in discussions of environmental and economic policy in the wake of the Great Recession a decade ago. The most succinct comes from the Green Economy Coalition, a US-based organization, as an economy "that generates a better quality of life for all within the ecological limits of the planet."[14] This view recognizes the reality of planetary limits, a common feature of green economy formulations, as well as aspirations for a high quality of life, with emphasis on the "for all." The United Nations Environment Program (UNEP), an influential international organization, views the green economy as "one that results in improved human well-being and social equity, while significantly reducing environmental risks and economic scarcities."[15] The Institute for European Environmental Policy envisions "a resource-efficient, low-carbon, equitable economy that stays within a 'safe operating space'—or working within the planet's regenerate capacities and avoiding critical ecological thresholds."[16] Meanwhile, the Organization for Economic Cooperation and Development (OECD) seeks economies for "fostering economic growth and development while ensuring that natural assets continue to provide the resources and environmental services on which our well-being relies."[17]

These definitions accept that at some point in the growth process, planetary limits will be tested, if not breached. Exponential economic growth cannot proceed inexorably without serious consequences for ecosystems, human health, resources, and well-being.

At the same time, they all accept the inevitability of some growth—meaning an increasing scale of economic activity that delivers a better quality of life—as not only unavoidable but desirable. This is most critically important for countries early in the process of economic development or those that simply are poor, where the imperative for economic prosperity is most compelling. These definitions also suggest some elements of this new green economy: it is resource efficient, low-carbon, reduces environmental risk, protects critical ecosystem services, and preserves natural capital. To advocates, this is the path to a far more sustainable world. To critics, it is an unrealistic fantasy.

There are competing storylines to the green economy, to be sure. One seeks to avoid the two worlds colliding scenario by deliberately shrinking economies in rich countries while giving room for poor countries to grow, up to a limit.[18] This school of thought argues that we cannot expand the biophysical capacities of the planet to fit the size of the economy, so we have to adjust the size of the global economies to fit the planet. The term for this line of thinking is *degrowth*. It builds on the work of Herman Daly, a prominent ecological economist, on the need for what he termed a *steady-state* economy. Daly wanted to redesign economies: they should deliver a good quality of life but maintain a constant rate of "throughput," "the lowest feasible rates of matter and energy from the first stage of production to the last stage of consumption."[19] Technology innovation alone would not be enough, he argued; a rethinking of economic growth was needed. At the other extreme is all-out growth, whatever the consequences. Advocates of growth-above-all express faith in human ingenuity, unfettered markets, and technology. Such advocates typically come from sectors having the most to lose in a green economy, such as fossil fuels and industrial agriculture. These advocates for a brown economy minimize threats of climate change, argue against mitigation, and ignore degraded ecosystems. As John Dryzek puts it, this is a world in which "natural resources, ecosystems, and indeed nature itself, do not exist."[20]

The green economy sets out an alternative to these unrealistic and risky (in the case of degrowth) and irresponsible and eventually catastrophic (with continued brown growth) alternatives. It is not a compromise, but an alternative path that seeks and carries out positive-sum relationships.

A major intellectual influence on the green economy concept is ecological economics. This emerged in recent decades as a counterpoint to more traditional economic analysis, which critics say pays too little attention to planetary limits and to the effects of exponential growth. The aim is "to provide an integrated and biophysical perspective on environment-economy interactions" that is missing in more traditional economic analysis.[21] Ecological economics places natural capital—clean air and water, healthy ecosystems and forests—on not just an equal but preferential footing relative to other forms of capital. The field views economies as subsystems of the ecological system and as subject to biophysical, planetary constraints. As one ecological economist puts it: "Continuous exponential growth of any physical sub-system of a finite system is impossible."[22]

Another intellectual influence in the emergence of the green economy is transformative thinking and practices among major business firms. Indeed, it is fair to say that the case for positive relationships among economic and environmental goals first became evident in the greening of business. If firms can find the economy–environment win-wins, then why not entire economies?

THE MICROLEVEL: GREENING THE PRIVATE SECTOR

Well before the concept of the green economy gained wide currency, leading firms were reaching similar conclusions about the interactions among ecological and economic goals. Indeed, starting in the 1980s, claims about a greening of industry started appearing in writing about business and the environment. Many firms realized there were sound financial reasons for not only meeting legal standards but doing more by managing supply chains, reporting voluntarily on environmental performance, setting goals to go beyond compliance (like energy or water efficiency), developing better relationships with communities, redesigning products to reduce their environmental impacts, and even by anticipating more stringent standards in their investments, technologies, and planning.

Trends in the Greening of Business

Why would private business firms do more than what the law requires? One obvious reason is to reduce bottom-line costs. Using less energy, water, and raw materials, or not having to dispose of hazardous waste not generated, saves money. Firms also were becoming more attentive to their reputations and to the effects of bad publicity on shareholders, customers, and communities. These firms sought business advantage in getting ahead of competitors on technology and management practices and improving relationships with state and federal regulators.[23] They also were trying to anticipate future trends, such as limits on carbon emissions, water scarcity, or rising energy costs.

A 1995 essay went further, asserting that the old view of unavoidable trade-offs between financial and environmental success assumes a static world, not the dynamic and competitive one in which firms operate. It argued that "properly designed environmental standards can trigger innovation that may" lower a product's cost or improve its performance.[24] When firms respond to external pressures for improving performance, such as government regulation or rules set by customers and investors, they are pushed to use materials more efficiently, eliminate unneeded activity, and increase the value of their products and services. The authors of this essay distinguish *good* and *bad* regulation. Good regulation sets high standards yet allows flexibility in determining how to meet them. It is reasonably predictable, focused as much as possible on outcomes (results), and aims to support technology and management innovation. In contrast, bad regulation is far less predictable, sets short compliance deadlines, is prescriptive and inflexible, and is poorly designed. It may impede needed levels of technology innovation by locking in old methods and practices.

In sum, it is not hard to find similarities between macro- and microlevel green economy thinking and practice. Both turn the old conception of inevitable economy–environment trade-offs on its head. High levels of environmental performance—improving energy and water efficiency, using cleaner manufacturing processes, reusing materials, protecting critical ecosystems, and constantly finding ways to innovate—all contribute to both economic and environmental success.[25] At both the macro- and microlevel static, short-term thinking is replaced by dynamic, longer-term thinking. Moreover, at both levels, some public policy tools are more effective than others as reflected in distinctions between good and bad public policy, whether regulatory or in other forms.

It is worth examining how the business sector is reframing economy–environment trade-offs into opportunities for business and financial success. These microtrends not only mirror green economy opportunities at the macrolevel but also illuminate the role that business can play in the transition to a global green economy built on a foundation of economy–environment synergies.

A few cases illustrate how firms are doing this. L'Oréal, the cosmetics manufacturer, set out to reduce water consumption "per finished unit of product" by 60 percent from a 2005 baseline by mapping its water consumption systematically, optimizing water use in production processes, reusing industrial wastewater whenever feasible, and recycling water after treatment at various stages of production. Although this required invest-ment, it would pay off not only in reducing water use but using less energy use to heat water, cutting costs for chemicals to treat water, and ensuring continued access to a needed resource in water-scarce areas.[26] The company was promoting a public good—less water use—while also achieving critical business objectives.

A business strategy known as *chemical leasing* also illustrates the alignment of private and public goals in a green economy.[27] Chemical use is pervasive in modern economies. Many of these chemicals pose risks to human health or the environment. From a public perspective, it makes sense to use as few chemicals as possible and substitute *green* chemicals for hazardous ones. The old pattern was that a user (say a company making washing machines or painting cars) would buy needed amounts of chemicals and use them to make products. This rewards inefficient chemical use; the seller is motivated financially to sell as much product as possible, and the user may not know how to optimize it. Chemical leasing defines new relationship: the user pays for the service (such as the area coated), not the amount. This decouples payments from chemical consumption and motivates sellers to optimize use based on an expert knowledge of their products. Among the American companies leading in use of chemical leasing are GM, Lockheed Martin, and Harley-Davidson.[28] A chemical user benefits from lower costs, less liability due to hazardous wastes, and a safer workplace. The seller gains customers with a new business model. Public benefits include less harmful chemicals and safer workers. Benefits are substantial: the WHO estimates that 8.3 percent of deaths each year globally are a result of exposure to hazardous chemicals.[29]

Clean energy is motivating companies to pursue business options that yield societal benefit. Rising consumer interest, stimulated by stringent vehicle standards and gov-ernment subsidies, led Volkswagen, one of the world's largest car manufacturers, to invest $34 billion over 5 years in electric vehicles (EVs), with the aim of being the largest global producer of EVs by the mid-2020s.[30] Utilities are getting into the act. Although known previously for relying heavily on coal, Arizona Public Service joined six other utilities in 2020 in committing to totally carbon-free electricity generation by 2050. It will end coal generation by 2031, 6 years ahead of previous projections.[31]

Business Greening and Public Policy

Should this greening among many private sector firms influence public policy? For many environmental advocates critical of regulation as currently practiced, it should.

Although these reform ideas got lost in the trench warfare of the Donald Trump administration, there was a dynamic environmental innovation agenda in the 1990s and into the 2000s. Proenvironmental critics called for regulation and other policies to recognize changes in business, account for a new generation of problems, and adapt to new technologies and knowledge.[32] The themes in these reforms were to set stringent goals but give flexibility in meeting them, rely more on indicators of performance than just legal compliance, draw upon a diverse range of policy tools, and recognize different levels of performance by firms. Among the policy changes proposed were more use of market incentives, information disclosure, public–private partnerships, and adaptive policy tools.[33]

In many ways, these proposed reforms offered a preview of the green economy policy agenda discussed in the next part of the chapter. They recognized that the old regulatory model of specifying technologies and forcing sources to implement them was a partial solution and one not well suited to new problems and a changing business landscape. Regulatory laws and agencies had adapted over the years, to be sure.[34] In the 1980s, the Environmental Protection Agency (EPA) began to use new policy tools like emissions trading, leading to the acid rain allowance trading program in the 1990 Clean Air Act Amendments. Mandatory information disclosure—forcing firms to track and report on pollution—was adopted in right-to-know laws at state and federal levels (see Chapter 7). Agencies tried new relationships with business and created voluntary programs to cut greenhouse gases, improve water efficiency, and design greener chemicals. These innovations recognized that the behavior of leading firms was changing, and the standard regulatory policy tools were not the most effective or efficient.

Recognizing the limits in the existing regulatory system and ways of getting better results at less cost provided options and experience for policy tools to create a green economy. The array of policy options was much richer as a result. The green economy adds another layer to our thinking about policy innovation. At its core is the idea that we may reframe economy–environment relationships from that of the old lose-lose to a potential win-win. Leaders in the business community had begun this transition. Now it was being scaled up to the macrolevel. The green economy reframes relationships and organizes policy options into a coherent, pragmatic agenda.

How can public policy be designed to create a green economy? A start is by making policies predictable, which has proven challenging in the contentious American system. Uncertainty makes planning difficult; businesses have a hard time deciding what to invest in and how to make business decisions about products, markets, and suppliers. Even an economic tool like a carbon tax loses its effectiveness if the tax changes unpredictably. Policy also should aim for stringent goals, but when necessary be flexible on means. A third design feature is to improve how government, business, and investors collaborate on new technologies. Having government agencies issue technology standards developed in complex rulemaking processes may help diffuse technologies but rarely stimulates invention of new ones. A sixth recommendation is to take advantage of state policy innovation and ways to use federalism better as is discussed in Chapter 2. US leaders in energy and climate innovation recently include states like California, Hawaii, and Virginia; experience in those states may diffuse horizontally to others or become a national model. Indeed, systematic learning across states is essential in creating a green economy.

THE GREEN ECONOMY POLICY AGENDA

The green economy concept is a way of reframing relationships among environmental and economic goals, at both a societal (macro) and a firm (micro) level. It is also a strategic framework for designing and implementing policies in ways that reflect the reframing of these relationships.[35]

This section of the chapter considers the principles and policy tools making up this strategic framework, the green economy policy agenda. The *principles* express the core ideas of and criteria for evaluating tools and their applications. *Policy tools* are mechanisms for changing behavior to achieve goals. These tools should reflect the principles and link them with the desired outcomes.

Principles for a Green Economy Policy Agenda

Public policy experts rely on many generic principles for evaluating policy strategies and tools. These include, as examples, that they should be effective in achieving goals; equitable in not harming or benefitting particular groups unfairly; cost-effective in achieving desired outcomes at the lowest or at minimal costs; scientifically grounded; and feasible in gaining political support. All of these apply to the green economy policy agenda, as well as to most other environmental strategies.

A policy agenda for promoting a greener economy builds upon many current strategies but emphasizes some more than others. This agenda may be viewed in terms of the *principles* behind the policy designs and the *tools* that are used to fulfill these principles. Here are five principles for guiding decision-making toward a greener economy:

1. Seek positive, even synergistic ways of linking environmental and economic goals.
   This is a defining feature as well as a core principle of the green economy. Design policies that seek and implement positive-sum outcomes: energy and water efficiency, green infrastructure, investments in a smart grid or sustainable farming, to name a few. At times, trade-offs are hard to avoid, but by reframing choices and enlarging the scope of decision (how energy is generated or how people and goods are moved more than how pollution is controlled) positive-sums may be found.

2. Grant critical ecosystems and resources a principled priority in decision processes.
   A weakness in the brown economy is giving preference to short-term economic gains that benefit short-term interests over the long-term viability of critical ecosystems. This includes ecosystems and resources that are irreplaceable, provide essential services, have major aesthetic or historical

value, or whose loss is irreversible. These should be granted a *principled priority* in decision-making.[36] This recognizes how essential they are to human well-being. One mechanism for doing this is ecosystem valuation, which aims to determine the actual economic value of ecosystems to human well-being.

3. Create mechanisms for incorporating full cost/social pricing into decision-making.

   Most environmental problems are the result of failures in private markets (see Chapter 10). Traditional market relationships do not force people to account for their air and water pollution (including greenhouse gases) or for damages to common pool resources (such as fishing grounds or wetlands when not privately owned). Government action is required to put a price on these resources and correct the market failure. The most obvious illustration of such a strategy is putting a price on carbon, either with a tax or cap-and trade system.

4. Integrate institutions and policies across policy sectors.

   The reality of modern political economy and governance is that the ecological state usually takes a back seat to the prosperity state. Yet ignoring the protection of ecosystems, health, and natural resources eventually risks the foundations of prosperity. In politics, the short term often dominates the long term, and special interests prevail over general well-being. Only by integrating environmental goals into economic, social, and political decision-making and seeking positive relationships among them may we establish the environment as a core business of modern states. A green economy makes the environment part of infrastructure investments, business investments, development choices, and the design of regulatory programs. In the green economy, nearly every economic decision has environmental implications in the near or long term.

5. Seek outcomes that increase options for family-supporting, career-oriented employment.

   Investments in environmental and energy infrastructure, innovation, and efficiency support what Van Jones calls the *green collar economy*: money spent on environmental progress may generate cobenefits of productive, family-supporting, and career-enhancing jobs.[37] For example, there is evidence that investments in renewable energy and efficiency generate more employment per dollar invested than do investments in traditional fossil fuels. Clean energy is one of the fastest growing sectors for jobs, not only in the United States but globally. Infrastructure investments like building a *smart grid* (using innovative technology, maximizing efficiency, integrating renewable energy), expanding mass transit, and building green water infrastructure would create millions of jobs. Indeed, the Edison Electric Institute, a group representing utilities, estimates that although the costs of a smart grid using advanced technology are formidable ($338–476 billion over 20 years), the benefits to society could amount to two trillion dollars over those two decades.[38]

Policy Tools for a Green Economy Policy Agenda

Policy tools are specific mechanisms for changing behavior to achieve society's goals. In environmental policy, the standard tool is technology or performance-based regulation, but there is growing use of others. Indeed, there has been a transition from an era relying on prescriptive regulation to one using economic tools like emissions (air) and effluent trading (water). Economic tools (trading and eco-taxes) aim to leverage the positive economy–environmental relationships defining the green economy. Still, mandatory performance standards play a role. The policy tools reflect a mix of the old and the new, but overall are designed to promote green economy principles.

Mandatory Performance Standards

Having government define technology or performance standards and ensuring that sources meet them is a mainstay of environmental policy in the United States and most countries. A trend has been to rely less on specific technology requirements and more on performance goals.[39] An example is Renewable Portfolio Standards—legal requirements that utilities derive a specified minimum of their electricity from renewable sources (largely wind and solar) by a given date (see Chapter 8). Some states are aggressive in setting these standards: By the end of 2019, thirteen states or similar entities had committed to 100 percent renewable or clean energy (the latter leaving the option of nuclear energy or hydropower) by the middle of the century.[40] More recently, Virginia became the first southern state to commit to 100 percent renewable electricity generation, with a target of 2050, as well as separate targets for energy storage and offshore wind.[41]

Trading Mechanisms

Policy tools that account for the social costs of pollution—the effects on health, the climate, and ecosystems—deserve a place on the green economy agenda. In economics terms, pollution is an externality: an effect for which unregulated markets do not hold sources accountable. This is why air pollution became a major problem in the 1950s and 1960s, leading to the Clean Air Act of 1970. Mandatory standards are one way to deal with externalities, of course. Government defines the "best available" technology and makes sources install and operate it.

Economic incentive tools are another option. Like mandatory standards, they make sources account for the social costs of pollution, but they do it differently. They put a price on releases into air or water, then allow sources to decide what changes in behavior make economic sense. With trading, government creates artificial markets by setting a cap on air emissions or water effluents and making sources purchase enough permits or allowances to cover their emissions. The US EPA began using emissions trading in the 1970s and 1980s; a cap-and-trade program for acid rain was incorporated into the Clean Air Act Amendments of 1990.[42] Cap-and-trade is used to cut carbon emissions in California, the European Union, and elsewhere.[43] Effluent trading may control water pollution, although this is challenging and faces technical issues.[44]

Eco-Taxes

Like trading, eco-taxes create economic incentives, but in different ways. Instead of setting a cap and then allocating (through auctions or another mechanism) permits to sources, eco-taxes assign a fee to each unit of pollution released.[45] Sources decide how much to reduce emissions or effluents through technology controls, new production processes, or other means to avoid paying the fee. Trading and eco-taxes illustrate two leading economic incentive tools for changing behavior. The difference is that trading caps and allows certainty about the amount of pollution, but prices are uncertain; taxes establish a clear price but resulting pollution levels are uncertain.

Leading proposals for reducing greenhouse gas emissions nationally in the United States focus on a carbon tax.[46] A carbon tax opens up many options for drawing political support. The revenue from the tax could be distributed back to the public (a revenue-neutral tax), used for green investments (energy efficiency, infrastructure, or research), allocated to a just transition for dislocated workers or low-income groups, or used to reduce deficits. The carbon tax thus may be used to meet multiple goals: reducing emissions by increasing the costs of carbon-based fuels; stimulating job creation; realigning public spending; and promoting broader green economy goals, like green infrastructure.

Payments for Ecosystem Services

A fourth policy tool further illustrates the application of green economy principles. One of the premises of ecological economics, discussed earlier, is that ecosystems perform vital services that should be accounted for in economic decision-making. Ecosystems filter water, buffer coastal areas from storms, preserve biodiversity, conserve carbon sinks like tropical forests, detoxify waste in the environment, preserve soils to enable food production, and serve other purposes.[47] The effective price for such services usually is zero because they are not traded in economic markets. Putting economic value on ecosystem services creates incentives for protecting and restoring them.

The policy tool for making this happen is payments for ecosystem services (PES). PES is a system in which beneficiaries of an ecosystem service—a downstream town needing clean water or protection from flooding, a government seeking to protect an estuary—compensate those controlling the natural resources that provide the ecosystem service. The providers of ecosystem service may only have to preserve or not develop the land or resource. Or they may have to do more, so long as "the payment causes the benefit to occur where it would not have otherwise."[48]

PES come in many forms. Often, private or nongovernment organizations purchase land outright to preserve it and the services it provides; examples are the World Wildlife Fund and Nature Conservancy. The Natural Resource Conservation Service of the US Department of Agriculture uses PES through the Agriculture Conservation Easement Program, which pays landowners to preserve land, wetlands, and habitat. Even governments have used the PES; Norway recently committed to a 10-year payment of $150 million to Gabon to preserve rainforests as an ecosystem and carbon sink through the UN's Central African Forest Initiative.[49]

Why This Agenda Promotes a Green Economy

The core idea of the green economy is to identify and carry out positive-sum relationships among economic and environmental goals while promoting social equity. The principles presented here aim to deliver on this positive-sum framing—one with considerable political appeal given the emphasis on the prosperity state—with criteria that shape successful policy strategies and tools. Societies should put the ecological state on a par with the prosperity, security, and welfare states.

Among the green economy principles, the most central are that environmental choices be integrated broadly with economic ones and that critical ecosystems be given principled priority in decision-making. The policy tools are designed to promote these ends. Ecosystem valuation explicitly attaches economic value to sources of critical ecosystem services—the climate system, wetlands, tropical rainforests, and so on—and integrates them with economic decisions. A carbon tax or cap-and-trade assign an economic cost to pollution and account for its social costs. Carbon pricing and other tools link public and private investments and other economic decisions while also promoting environmental goals (less climate impact, healthier air), generating economic benefits, stimulating job creation through infrastructure investments, and promoting social equity.

IS ANYONE ACTUALLY TRYING TO CREATE A GREEN ECONOMY?

The idea of building a green economy recognizes that largely brown growth is unsustainable. To take one illustration, some 80 percent of the global energy system is powered by fossil fuel–based energy. Even setting aside the health and ecological damage of this practice, the accumulation of greenhouse gases from this energy system poses a serious threat. Other issues discussed in this chapter make the same point: continued, exponential growth has consequences. At the same time, political, social, and economic factors driving this growth are unlikely to abate.

Although mentions of the green economy date to the 1990s, the concept gained currency in the wake of the financial crisis and recession in 2008–2010. The crisis called for a reassessment of past policies and major investment to stimulate the economy. The idea of a *Green New Deal* emerged.[50] Extensive public investment was needed to stimulate economies around the world: why not focus on green investments like renewable energy and efficiency, innovative water infrastructure, smart farming, and ecosystem services? A study of 20 countries in 2009 found they had authorized nearly three trillion dollars in stimulus money, of which about 15 percent ($430 million) was aimed at "stabilizing and then cutting global emissions of greenhouse gases" (including water and related investments).[51]

In the United States, the Barack Obama administration used green economy arguments. In framing economy–environment linkages positively, for example, it embraced the *clean energy highway* to create jobs, promote durable growth, and expand business in renewable energy and efficiency. In testimony before Congress in 2014, EPA Administrator Gina McCarthy described climate action as a "spark for business innovation, job creation, clean energy, and broad economic growth."[52] Of course, these arguments were rejected by the Donald Trump administration; it reverted to the old mantra, unsupported by evidence, that economic and environmental goals inevitably will conflict.

The concept of a Green New Deal was invoked in 2019 with proposals from a group of US legislators for ambitious climate action and social reform.[53] This plan called for investment in clean energy, green infrastructure, mass transit, and other green initiatives. It also set out sweeping social goals: universal health care; guaranteed, family-supporting jobs; affordable, safe housing; and higher education available to all, among others. Although reflecting the green economy goal of linking economic and environmental goals and reorienting policies and investment, this framed the green economy within a far bigger social agenda. Doing so raised a strategic issue of how sweeping and aspirational the vision of a green economy could be framed and still be feasible politically. Green New Deal advocates want to link climate change to a broad social agenda (guaranteed jobs, universal health care) as a means of "mobilizing collective action through a utopian vision of the progressive agenda...."[54] Other climate advocates see a greater chance of success in a strategy more specifically tailored to emission cuts, renewable energy, and energy efficiency—one that is less likely to provoke opposition to this much broader social agenda.

Signs of the green growth concept also are evident in the *European Green Deal,* adopted by the European Commission in October 2019.[55] Like the proposal for a Green New Deal in the United States, this frames economy and environment linkages in positive ways while building in equity. Although climate action is the center-piece—making Europe "the first climate neutral continent"—the plan addresses other goals, like eliminating air and water pollution and integrating investment and policy decisions with environmental issues. The European Green Deal aims to raise living standards, guarantee jobs, invest in the future, promote equity, empower communities, support climate justice, and limit temperature increases to 1.5 degrees Celsius. It also reflects a theme among ecological ecologists for alternative measures of well-being beyond GDP growth, including a Genuine Progress Indicator (GPI) that accounts for environmental and social factors.[56]

Governments even are thinking more that well-being includes more than growth rates. The OECD reports on a suite of measures in a *Better Life Initiative,* focused on "understanding what drives well-being of people and nations and what needs to be done to achieve greater progress for all."[57] New Zealand recently decided to form national budgets where spending "is dictated by the 'well-being' of citizens." It relies less on economic measures of growth and productivity and more on "goals like community and cultural connection and equity in well-being across generations."[58] Among the goals is a "transition to a sustainable and low-emission economy" and to outcomes that reflect not only the needs of the present generation but also "long-term impacts for future generations."[59]

Many organizations that promoted the green economy in the early 2010s created a resource to disseminate knowledge and support green transitions. The UNEP, OECD, World Bank, and UN Industrial Development Organization formed the Global Green Growth Institute (GGGI), based in Seoul, South Korea. Recognizing the pressures current growth places on planetary limits, the GGGI created the Green Growth

Knowledge Platform, a "network of experts and organizations dedicated to providing the policy, business, and finance communities with knowledge, guidance, data, and tools to transition to an inclusive green economy."[60] Countries as diverse as China, Kenya, Brazil, and India have used aspects of the Green Growth Knowledge Platform in their planning, although it is likely that countries like Sweden, Denmark, and Norway are closer to having a green economy.

IS THE GREEN ECONOMY A VALID CONCEPT?

The green economy is not without critics. From the left of the political spectrum, critics see it as little more than a justification for continued growth and the dominance of global corporate and financial interests. That it is promoted by an organization like the World Bank, with a history of promoting often environmentally harmful economic growth, is evidence of the concept's bias. At the same time, critics from the right of the political spectrum view it as a means for stunting growth, expanding government, and imposing unneeded constraints on dynamic, productive economies. Some critics see the *economy* part of the concept as a threat; others fear the emphasis on the *green*.

Yet the critics from both sides are missing critical realities. The first is that economies will grow. Expecting countries like China and India to set aside their growth aspirations to make up for the planetary pressures caused by rich countries of North America and Western Europe is a fantasy. Although there is a strong case to be made for rich countries to focus more on the quality than the mere quantity of growth, it is unrealistic to expect them voluntarily to shrink their economies. To deny the fact and likely persistence of economic growth falls short in protecting planetary limits.

Clinging to the idea that countries around the world will at some point recognize the harm from perpetual growth and act decisively to halt and even reverse the historical patterns is a thin reed on which to hang the future of the planet. Political leaders compete on their ability to deliver prosperity. Economic success is a measure of national status and influence. Economic growth has enabled huge gains in the quality of life around the world and expanded access to education, health care, meaningful work, and even more democratic government (see Chapter 13). To put it bluntly, even a 20 percent reduction in the size of the global economy, without changes in the energy, agriculture, transportation, and manufacturing systems, will be insufficient to escape the inevitable breaching of planetary limits. Indeed, the pattern is that when the prosperity state falters, as it did in the Great Recession, support for environmental values falls, governance quality and capacities weaken, and shorter-term thinking that favors the prosperity state over the ecological state takes priority.[61]

This is not to say that a reconceptualization of the role of economic well-being in the quality of life should not be on the table. Research tells us that perceived happiness and well-being level off at some level of income.[62] In economics terms, the marginal benefits of growth begin to decline and even become negative at higher levels of income as the social and environmental costs of exponential growth accumulate: pollution, lost ecosystem services, climate change, economic and social inequality, and pursuit of positional goods (valued for the status they confer, not their actual utility). There is a social as well as an environmental case to be made against too much growth.[63]

Nor is it at all consistent with the evidence to think that continued economic growth along the current path will not, if it has not already, push the scale and scope of human impacts on the environment past planetary limits. The climate is changing; ecosystems and species are being lost at alarming rates; water is less available in many parts of the world; air pollution threatens human health. The facts cannot be denied, no matter how inconvenient they are to advocates of fossil fuels, deforestation, industrial farming, and other harmful activities. These are not just environmental problems; they undermine our well-being. Adapting to climate change will cost lots of money, harm agriculture, threaten human health, and put parts of many cities under water. Indeed, the UNEP estimates the annual costs of adaptation alone may reach $500 billion by 2050.[64] The World Bank calculates that air pollution cost India over 8.5 percent of GDP in 2013.[65] Burying our heads in the sand and assuming we may stay within planetary limits without changing how growth occurs and economies operate is a recipe for disaster. With the 2020 coronavirus pandemic, of course, air quality is improving in countries like India. The issue is whether or not this will lead to a greater appreciation of the need for a green economy or if the old pattern of largely brown economic growth will resume or even accelerate.

The concept of a green economy recognizes that growth will occur, in some form, but that the form and trajectory of growth must change. The solutions are there. Using technology for energy efficiency or modernizing the electricity grid cuts energy costs, improves health, is good for the climate, and spares ecosystems from damages due to coal mining or natural gas fracking. It also creates useful employment; as noted earlier, a dollar invested in clean energy, efficiency, and infrastructure generates more employment than the same investment in fossil fuel resources.[66] An equitable green economy also may respond better to demands for environmental justice by reducing health risks, creating greener infrastructure, and promoting cleaner energy.

Smart farming also illustrates the practical application of green economy ideas. Agriculture is a source of environmental harm; done unsustainably, it leads to water pollution, greenhouse gas emissions, overuse of fertilizers and pesticides, and soil erosion.[67] The goal of smart farming is to think long-term and recognize the value of clean water; the cost of inputs to production like energy, fertilizer, and water; the need to maintain soil quality; and the benefits of clean energy. In addition to reducing costs and maintaining natural capital, farmers can earn income by cutting nonpoint source water pollution and selling water quality credits and by leasing land to wind energy projects.

Economic activity and growth were disrupted dramatically by the coronavirus pandemic that began in 2020. Economic growth rates will be lower than they might have been, although the length and depth of the economic crisis is difficult to predict. As expected, emissions of greenhouse gases and other pollutants fell with less economic activity and growth. Although this reduces short-term pressures on the environment, at some point the drivers of economic growth and pressures on the planet will resume. What matters is not the effects of this short-term respite but the impact of the pandemic on the longer-term transformation of global economies along environmentally sustainable lines.

THE NEED FOR INTERNATIONAL, COLLECTIVE ACTION

The green economy will succeed if it is implemented globally (see Chapter 12). Acting on their own, nations will give in to pressure to gain short-term economic advantage by using fossil fuels, consuming natural resources, and placing short-term growth over longer-term well-being. Global collective action is necessary, or countries will aim to be free riders and derive near-term benefit. As a rich and influential country, the United States plays a role but also is affected by this interdependence.

Rich countries like the United States must be engaged in creating a global green economy. They have a moral obligation, to be sure, by accounting for the most cumulative greenhouse gas emissions and for many other impacts. Beyond this, it is in their self-interest. Deploying clean energy in developing countries, protecting natural capital like tropical forests, and improving air quality not only helps both rich and developing countries environmentally, it promotes political stability, expands access to green technology, enhances national security, and creates paths to cooperation.

US leadership is crucial for global collective action toward a green economy. Its financial, scientific, and diplomatic resources are essential. The contrasting outcomes of the Copenhagen (2009) and Paris (2015) Conferences of the Parties on climate action, the second of which was more substantively successful, are partly due to the large role the United States played in Paris and the Obama administration's commitments. At the same time, the United States may need the rest of the world more, or enough of it to matter.[68] If other countries do not act, domestic opponents of environmental action will argue that the United States is exposing itself to the costs of a transition while others seek short-term advantages. Even moderate critics of clean energy, climate action, and effective regulation worry about production costs, consumer prices, and other bad outcomes if the United States gets too far ahead. Although studies find that environmental policy has minor impacts on the economy, especially given its benefits, short-term impacts on specific sectors are used to oppose action.[69]

What should the United States and other rich countries be doing? They should share technologies and best practices. They should set standards and policies countries at other levels of development may use. They can provide capital for inventing and diffusing technology for energy efficiency, green chemistry, renewable energy, sustainable farming, and other initiatives. They should be helping poor countries achieve a better quality of life—by supporting health, education, and other programs—that focus more on human development than on the sheer quantity of growth (an issue also stressed in Chapter 13 on developing countries).

Rich countries also should be supporting democracy and democratic transitions elsewhere. A range of research suggests that democracies overall are more likely to give priority and respond to environmental issues, including climate.[70] Among the leaders in a green economic transition are countries like Sweden, Denmark, and Norway, which also rank high in measures of democracy. The depth of experience also matters; established democracies tend to perform better than new ones. Indeed, the leaders in global assessments like the Environmental Performance Index (EPI) are established democracies like Switzerland, France, Denmark, Sweden, and the United Kingdom.[71]

Democracies also are more likely to engage in international problem-solving and agreements. There are exceptions: the United States and Australia, for example, have not been impressive recently. Still, as discussed in Chapters 2 and 11, many state and local governments are engaging vigorously in the green economy.

Is the vision of a green economy an illusion, as many critics assert? Is economic growth *in any form* inevitably and irreversibly bad for the planet? In the long run, it is hard to say. The sheer scale of human stress on ecological services, human health, and natural resources may at some point overwhelm human ingenuity and technological prowess, regardless of how green economies become. What is beyond doubt is that the current global brown economy, with its reliance on fossil fuels, inefficient resource use, input-intensive agriculture, destruction of forests and biodiversity, high-volume chemical use, and growing emissions is a road to disaster. The green economy offers an alternative that accepts human aspirations for a better quality of life within the biophysical limits of the planet. As both a policy framing device and as an agenda for policy change, the green economy may be the only politically and economically feasible path to a more sustainable future.

SUGGESTED WEBSITES

Energy Transitions Commission (www.energy-transitions.org) The Energy Transitions Commission is an international organization focused on climate mitigation, access to affordable and sustainable energy, and economic development. It takes a green economy view of energy, with resources on the forms that a global, clean energy transition should take.

Global Green Growth Institute (www.gggi.org) The Global Green Growth Institute is an international, intergovernmental organization dedicated to achieving strong, inclusive, sustainable growth in developing and emerging economies. It includes research, tools, case studies, and other resources on water, clean energy, sustainable landscapes, and green cities.

International Renewable Energy Agency (IRENA) (www.irena.org) IRENA is an international organization dedicated to sustainable energy. Based in Abu Dhabi, it is a rich source of resources on challenges of and opportunities to a global transition to renewable energy.

Niskanen Center (www.niskanencenter.org) The Niskanen Center is a Washington, DC-based think tank that reflects green economy framing and principles, but it is founded on market-based strategies. It offers a right-of-center but responsible approach to green economy issues.

The New Economics Foundation (www.neweconomics.org) The New Economics Foundation is a British think tank advocating transformational change in economic systems toward social and economic equity, political empowerment, and a clean environment. It offers resources for redesigning the relationships among social, economic, and environmental factors.

NOTES

1. "Greening the State?" in *Comparative Environmental Politics*, ed. Paul Steinberg and Stacy VanDeveer (Cambridge, MA: MIT Press, 2012), 63–88.
2. "The Birth of the Environment and the Evolution of Environmental Governance," in *Conceptual Innovation in Environmental Policy*, ed. James Meadowcroft and Daniel J. Fiorino (Cambridge, MA: MIT Press, 2017), 53–76.
3. "Greening the State?" 76.
4. This chapter draws from Daniel J. Fiorino, *A Good Life on a Finite Earth: The Political Economy of Green Growth* (New York: Oxford University Press, 2018).
5. Donella H. Meadows, Dennis L. Meadows, Jorgen Randers, and William H. Behrens III, *The Limits to Growth: A Report for the Club of Rome's Project on the Predicament of Mankind* (New York: Universe Books, 1972).
6. John S. Dryzek, *The Politics of the Earth: Environmental Discourses*, 3rd ed. (New York: Oxford University Press, 2013), 31.
7. Graham M. Turner, "A Comparison of The Limits to Growth with 30 Years of Reality," *Global Environmental Change* 18 (2008), 397–411.
8. Johan Rockstrom et al., "Planetary Boundaries: Exploring the Safe Operating Space for Humanity," *Ecology and Society* 14 (2009), 32.
9. World Health Organization (WHO). *How Air Pollution Is Destroying Our Health*. https://www.who.int/news-room/spotlight/how-air-pollution-is-destroying-our-health#:~:text=The%20health%20effects%20of%20air,of%20eating%20too%20much%20salt
10. Millennium Ecosystem Assessment, *Living Beyond Our Means: Natural Assets and Human Well-Being* (22,005). https://www.millenniumassessment.org/documents/document.356.aspx.pdf
11. Information on the IPBES is at https://ipbes.net/
12. Coral Davenport, "Major Climate Report Describes a Strong Risk of Crisis as Early as 2040," *New York Times* (October 7, 2018).
13. International Energy Agency, *Capturing the Multiple Benefits of Energy Efficiency* (Paris, 2014). https://webstore.iea.org/capturing-the-multiple-benefits-of-energy-efficiency
14. Green Economy Coalition, *The Green Economy Pocketbook: The Case for Acton* (2012). https://www.greeneconomycoalition.org/assets/reports/GEC-Reports/Green-Economy-Pocketbook-the-case-for-action_0.pdf
15. United Nations Environment Program, *Towards a Green Economy: Pathways to Sustainable Development and Poverty Reduction—A Synthesis for Policy Makers* (2011). https://sustainabledevelopment.un.org/content/documents/126GER_synthesis_en.pdf, 9.
16. Patrick ten Brink et al., *Nature and Its Role in the Transition to a Green Economy* (2012). http://img.teebweb.org/wp-content/uploads/2013/04/Nature-Green-Economy-Full-Report.pdf, 34.
17. Organization for Economic Cooperation and Development (OECD), *Towards Green Growth* (Paris: OECD, 2011). https://read.oecd-ilibrary.org/environment/towards-green-growth_9789264111318-en#page1
18. Samuel Alexander, "Planned Economic Contraction: The Emerging Case for Degrowth," *Environmental Politics* 21 (2012), 349–68.
19. Herman E. Daly, *Steady State Economics*, 2nd ed. (Washington, DC: Island Press), 17.
20. Dryzek, *The Politics of the Earth*, 59.
21. Jeroen C. J. M. van den Bergh, "Ecological Economics: Themes, Approaches, and Differences with Environmental Economics," *Regional Environmental Change* 2 (2001), 13.
22. Joshua Farley, "Ecosystem Services: The Economic Debate," *Ecosystem Services* 1 (2012), 43.

23. See Forest Reinhardt, *Down to Earth: Applying Business Principles to Environmental Management* (Cambridge, MA: Harvard University Press, 2000) and Nigel Roome, ed. *Sustainability Strategies for Industry: The Future of Corporate Practice* (Washington, DC: Island Press, 1998).

24. Michael E. Porter and Claas van der Linde, "Toward a New Conception of the Environment-Competitiveness Relationship," *Journal of Economic Perspectives* 9 (1995), 98.

25. Daniel C. Esty and Andrew Winston, *Green to Gold: How Smart Companies Use Environmental Strategies to Innovate, Create Value, and Build Competitive Advantage* (Hoboken, NJ: Wiley and Sons, 2009).

26. A case study on water efficiency reported by the World Business Council for Sustainable Development. https://www.wbcsd.org/Programs/Food-and-Nature/Water/Circular-water-management/Resources/Case-studies/Recycling-and-reuse-of-treated-industrial-wastewater-in-cosmetics-operations

27. On chemical leasing, see the United Nationals Industrial Development Organization report, *Chemical Leasing: A Global Success Story* (Vienna, 2011). https://www.unido.org/sites/default/files/2014-03/ChL_Publication_2011_0.pdf

28. Elizabeth Grossman, "More Companies Clean Up With Chemical Leasing," *Greenbiz* (October 3, 2014). https://www.greenbiz.com/article/more-companies-clean-chemical-leasing

29. A summary of WHO estimates is available at https://www.who.int/gho/phe/chemical_safety/en/

30. Charles Riley, "The Great Electric Car Race Is Just Beginning," *CNN Business* (August 2019). https://www.cnn.com/interactive/2019/08/business/electric-cars-audi-volkswagen-tesla/

31. Robert Walton, "Arizona Public Service Sets 100% Clean Energy Target, But Doesn't Rule Out Carbon Capture for Gas Plants," *Utility Dive* (January 23, 2020). https://www.utilitydive.com/news/arizona-public-service-sets-100-clean-energy-target-but-doesnt-rule-out/570870/. The other six utilities committing to 100% clean energy include Public Service Company of New Mexico (by 2040); Avista and Idaho Power (by 2045); and Duke Energy, Green Mountain Power, and Xcel Energy (by 2050).

32. Daniel J. Fiorino, *The New Environmental Regulation* (Cambridge, MA: MIT Press, 2006).

33. Daniel J. Fiorino, "Streams of Environmental Innovation: Four Decades of EPA Policy Reform," *Environmental Law* 44 (2014), 723–60.

34. Daniel A. Mazmanian and Michael E. Kraft, "The Three Epochs of the Environmental Movement," in *Toward Sustainable Communities: Transition and Transformations in Environmental Policy*, ed. Mazmanian and Kraft (Cambridge, MA: MIT Press), 1–32.

35. Adapted from Chapter 6 of Fiorino, *A Good Life on a Finite Earth*.

36. William M. Lafferty and Eivind Hoven, "Environmental Policy Integration: Towards an Analytical Framework," *Environmental Politics* 12 (2003), 11.

37. Van Jones, *The Green Collar Economy: How One Solution Can Fix Our Two Biggest Problems* (New York: Harper Collins, 2008).

38. Edison Electric Institute, *Estimating the Costs and Benefits of the Smart Grid* (Palo Alto, CA: Electric Power Research Institute, 2011), 1–4.

39. Daniel Fiorino and Manjyot Ahluwalia, "Regulating by Performance, Not Prescription: The Use of Performance Standards in Environmental Policy" in *Handbook of U.S. Environmental Policy*, ed. David M. Konisky (Cheltenham, UK: Edward Elgar, 2020), 217–30.

40. For a status report, see *Progress Toward 100% Clean Energy in Cities and States Across the U.S.* (UCLA Luskin Center for Innovation, November 2019). https://innovation.luskin.ucla.edu/wp-content/uploads/2019/11/100-Clean-Energy-Progress-Report-UCLA-2.pdf

41. Matthew Bandyk, "Virginia Approves 100% Clean Energy Legislation, Pushing State Toward 2.4 GW Storage, RGGI," *Utility Dive* (March 6, 2020). https://www.utility-dive.com/news/virginia-clean-energy-legislation-pushes-state-toward-storage-rggi/572349/

42. Eric M. Patashnik, "The Clean Air Act's Use of Market Mechanisms," in *Lessons from the Clean Air Act: Building Durability and Adaptability into U.S. Climate and Energy Policy*, ed. Ann Carlson and Dallas Burtraw (Cambridge: Cambridge University Press, 2019), 201–24.

43. For a summary of uses around the world, see the World Bank's *State and Trends of Carbon Pricing* (2020). https://openknowledge.worldbank.org/bitstream/handle/10986/33809/9781464815867.pdf?sequence=4&isAllowed=y.

44. Suzie Greenhalgh and Mindy Selman, "Comparing Water Quality Trading Programs: What Lessons Are There to Learn?" *Journal of Regional Analysis and Policy* 42 (2016) 104–25

45. Barry G. Rabe, *Can We Price Carbon?* (Cambridge, MA: MIT Press, 2018).

46. For a recent summary, see Center for Climate and Energy Solutions, "Carbon Pricing Proposals in the 116th Congress (September 2019). https://www.c2es.org/document/carbon-pricing-proposals-in-the-116th-congress/

47. The Economics of Ecosystems and Biodiversity (TEEB), *Mainstreaming the Economics of Nature* (2010). http://www.teebweb.org/publication/mainstreaming-the-economics-of-nature-a-synthesis-of-the-approach-conclusions-and-recommendations-of-teeb/

48. United Nations Environment Program (UNEP), *Payments for Ecosystem Services: Getting Started* (2008). http://wedocs.unep.org/bitstream/handle/20.500.11822/9150/payment_ecosystem.pdf?sequence=1&isAllowed=y

49. Abdi Latif Dahir, "Gabon Will Be Paid by Norway to Preserve Its Forests," *Quartz Africa* (September 23, 2019). https://qz.com/africa/1714104/gabon-to-get-150-million-from-norway-to-protect-its-forests/

50. Edward B. Barbier, *A Global Green Deal: Rethinking the Economic Recovery* (Cambridge: Cambridge University Press, 2010).

51. HSBC Global Research, *A Climate for Recovery: The Colour of Stimulus Goes Green* (2009). https://www.globaldashboard.org/wp-content/uploads/2009/HSBC_Green_New_Deal.pdf

52. Statement before the Senate Committee on Environment and Public Works (January 16, 2014). https://archive.epa.gov/epapages/newsroom_archive/newsreleases/fe40a46647a6007f85257c6200564c15.html

53. Lisa Friedman, "What Is the Green New Deal? A Climate Proposal, Explained," *New York Times* (February 21, 2019). The text of the Green New Deal resolution is at https://www.nytimes.com/2019/02/21/climate/green-new-deal-questions-answers.html.

54. Lachlan Carey, "How to Think about the Green New Deal after Its First Piece of Legislation," Center for Strategic and International Studies (December 18, 2019). https://www.csis.org/analysis/how-think-about-green-new-deal-after-its-first-piece-legislation

55. European Commission, *The European Green Deal.* https://ec.europa.eu/info/strategy/priorities-2019-2024/european-green-deal_en.

56. Ida Kubiszewski et al. "Beyond GDP: Measuring and Achieving Global Genuine Progress, *Ecological Economics* 93 (2013), 57–68.

57. More on the *Better Life Initiative* is at https://www.oecd.org/statistics/better-life-initiative.htm

58. Charlotte Graham-McLay, "New Zealand's Next Liberal Milestone: A Budget Guided by Well-Being," *New York Times* (May 22, 2019).

59. From *The Wellbeing Budget* at https://www.treasury.govt.nz/sites/default/files/2019-05/b19-wellbeing-budget.pdf

60. Information on the Global Green Growth Institute is available at https://gggi.org/. On the Green Growth Knowledge Platform, see https://www.greengrowthknowledge.org/

61. Dennis Chong, "Explaining Public Conflict and Consensus on the Climate," in *Changing Climate Policies: US Policies and Civic Action*, ed. Yael Wolinsky-Nahmias (Washington, DC: CQ Press, 2015), 110–45.

62. Carol Graham, *The Pursuit of Happiness: An Economy of Well-Being* (Washington, DC: Brookings Institution, 2011).

63. For critiques of growth, see Fred Hirsch, *Social Limits to Growth* (Cambridge, MA: Harvard University Press, 1976) and Tim Jackson, *Prosperity without Growth: Economics for a Finite Planet* (Abingdon, UK: Earthscan, 2011).

64. "UNEP Report: Cost of Adapting to Climate Change Could Hit $500B Per Year by 2050" (May 10, 2016). https://www.un.org/sustainabledevelopment/blog/2016/05/unep-report-cost-of-adapting-to-climate-change-could-hit-500b-per-year-by-2050/

65. Bobins Abraham, Air Pollution Cost India 8.5% of Its Economy in 2013, Says World Bank," *India Times* (June 5, 2019). https://www.indiatimes.com/news/india/air-pollution-cost-india-8-5-of-its-economy-in-2013-says-world-bank-261455.html

66. Robert Pollin, James Heintz, and Heidi Garret-Peltier, *The Economic Benefits of Investing in Clean Energy* (Washington, DC: Center for American Progress, 2009). On clean energy globally, see the report from the International Renewable Energy Agency, *Renewable Energy and Jobs: Annual Review 2019.* https://www.irena.org/publications/2019/Jun/Renewable-Energy-and-Jobs-Annual-Review-2019

67. Food and Agriculture Organization of the United Nations, *Climate Smart Agriculture Sourcebook.* http://www.fao.org/climate-smart-agriculture-sourcebook/en/

68. David Victor, "What to Expect from Trump on Energy Policy" (Washington, DC: Brookings Institution, 2016). https://www.brookings.edu/blog/planetpolicy/2016/11/17/what-to-expect-from-trump-on-energy-policy/

69. On the economic effects of environmental regulation see Chapter 3 of Fiorino, *A Good Life on a Finite Earth.*

70. On climate, see Daniel J. Fiorino, *Can Democracy Handle Climate Change?* (Cambridge, UK: Polity Books, 2018). An illustrative study is Kathryn Hochstetler, "Democracy and the Environment in Latin American and Eastern Europe," in *Comparative Environmental Politics*, ed. Steinberg and VanDeveer, 199–230.

71. The EPI is a data-driven ranking of 180 countries, updated biannually, on 24 environmental indicators. The most recent is at https://epi.yale.edu/

CONCLUSION

15

CONCLUSION

Environmental Policy in Crisis

Norman J. Vig, Michael E. Kraft, and Barry G. Rabe

Even our language has to adjust. What once was called "climate change" is now truly a "climate crisis."…We are now seeing unprecedented temperatures, unrelenting storms and undeniable science.

Antonio Guterres, United Nations Secretary General, September 24, 2019

On the day after New Year's in 2020, Nobel Prize–winning economist Paul Krugman published a column in the *New York Times* entitled "Apocalypse Becomes the New Normal." He was not referring to the deadly coronavirus pandemic that would soon engulf the world. Rather, he pointed to the recent string of extreme weather events—including historic flooding in the American Midwest, massive wildfires in Australia, and record heat waves from South Asia to Europe—that had dominated headlines the previous year as evidence that climate change was well underway and already resulting in localized catastrophes. As it turned out, 2020 was indeed a record year for temperatures, wildfires, Atlantic hurricanes, and polar ice melting.

After a half-century of environmental policy development—and decades of dire scientific warnings—no political consensus has emerged in the United States on how to respond to climate change. As noted in Chapter 1 and elsewhere in this volume, ideological and partisan polarization since the mid-1990s has deeply split the country over this issue, which in turn has widened partisan differences over other environmental policies as well (Chapter 3). Support for policies has thus begun to cycle wildly from one presidential administration to the next while Congress remains largely gridlocked (Chapters 4 and 5), threatening the ability of the nation to carry out its environmental commitments from the past and to cooperate with other countries in addressing what is increasingly seen as a global climate crisis.

As during the COVID-19 pandemic, weak national leadership has been partially offset by the initiatives of state and local governments, which have enacted a patchwork of climate policies on their own (Chapter 2). Nevertheless, many states have lagged behind or worked at cross-purposes, and without a major breakthrough to establish a comprehensive national policy it is doubtful that any effective solution will be possible. Later in the chapter we discuss various proposals for next-generation policies that could provide a national climate agenda for the future—but they will require a degree of public support and bipartisan cooperation not seen since the first Earth Day in 1970 and the "environmental decade" which followed.

Despite a rapidly growing body of evidence attributing weather catastrophes to global warming (see below), the Trump administration refused to recognize climate science or to support any measures to slow climate change, as documented throughout this book. Indeed, almost everything the administration did during its first three years in office was calculated to overturn the environmental policies of the previous Obama administration (Chapter 4) and American greenhouse gas (GHG) emissions rose 3.1 percent in 2018 after several years of stabilization or decline.[1] Even after the pandemic struck in early 2020, the Environmental Protection Agency (EPA) and other agencies continued to push to finalize their deregulatory agenda before the upcoming fall elections. Ironically, the EPA attempted to eliminate or greatly reduce the influence of epidemiological studies in assessing risks to public health and used the crisis to relax other regulations that could affect the outcomes of the virus.[2]

The broader impacts of the coronavirus epidemic on environmental policy will play out for years to come. In the short term, the resulting contraction of the US, Chinese, and world economies temporarily reduced carbon emissions and other forms of pollution.[3] The health epidemic could also have longer-lasting benefits as living patterns change and people recognize the need for greater preparedness if we are to reverse the curve of climate change as well as the spread of infectious diseases. Public opinion surveys indicated that people's concern about climate change remained high during the pandemic.[4] And the critical importance of scientific expertise and governmental competence should also be more obvious.

We discuss evidence calling for urgent action on climate change and conflicts over policies to address it in the following section. We then consider recent developments in energy policy and air and water regulation in more detail. Recognizing strong opposition to traditional forms of regulation in some quarters, we also discuss "market" alternatives and other measures that might promote greater political consensus in the future. Finally, we conclude with some preliminary thoughts on lessons that might be learned for future environmental governance from the global health apocalypse.

CLIMATE CHANGE: DIRE WARNINGS

The world's leading authority on climate change, the Intergovernmental Panel on Climate Change (IPCC), has published five comprehensive reports since 1990, each more urgent than the last. The fifth assessment report, released in four installments in late 2013 and 2014, contained the direst warnings yet about current climate trends and future risks throughout the world.[5] The scientific report concluded that it is "extremely likely" (a greater than 95 percent chance) that human activities are "the dominant cause" of the observed global warming since the 1950s. It further documented global changes already occurring, from rapidly melting sea and land ice and rising sea levels to intense heat waves and heavy rains, strains on food production and water supplies, the deterioration of coral reefs, and threats to fish and other wildlife.[6] The final report recognized progress in limiting carbon emissions in some countries, but stated that these gains are being overwhelmed by growing emissions in rapidly developing countries such as China and India.

The report thus called for much greater commitments to both policies to mitigate emissions growth and major investments to allow social and economic adaptation to climate changes that are inevitable in the coming decades. Unless new technologies are put in place by 2030, the report argued, it will probably be too late to avoid catastrophic damage to human life on the planet. "Continued emission of greenhouse gases will cause further warming and long-lasting changes in all components of the climate system, increasing the likelihood of severe, pervasive and irreversible impacts for people and ecosystems," the report stated.[7]

The 2015 Paris Agreement called upon all nations to make voluntary commitments sufficient to hold global warming to "well below" 2 degrees Celsius (3.6 degrees Fahrenheit) by 2100 compared to preindustrial levels, and set an even lower goal of 1.5 degrees C (2.7 degrees F) for the first time (Chapter 12). Ongoing research indicated that warming was occurring more rapidly than scientists had anticipated, and that even a 2-degree C increase could trigger catastrophic climatic effects.[8] The IPCC was thus asked to examine the feasibility of the 1.5-degree goal. Its report, issued in October 2018, proved alarming. It concluded that the world was well on its way to this limit and, without drastic reductions in carbon emissions, could surpass it as early as 2040. It stated further that unless GHG emissions were reduced by 45 percent by 2030, and totally eliminated by 2050 or shortly thereafter, the planet could experience uncontrollable climate changes.[9] The recognition that drastic emission cuts were required *within a decade* created an unprecedented sense of pessimism and crisis by 2019. Other scientific reports on potential threats to the global food chain, on ocean warming and disruption of marine life, and on more rapidly rising sea levels contributed to the new sense of urgency as the decade ended.[10]

A wealth of sectoral data further confirms these trends. The year 2020 was predicted to be the warmest on record, and the 2010 decade as a whole was easily the warmest ever measured.[11] Warming at higher latitudes was especially acute. Arctic sea ice and permafrost were melting at an accelerating rate, as was the Greenland ice cap; and large sheets of Antarctic ice were breaking off in record temperatures there.[12] On land, many areas of the United States have already exceeded a 2-degree C rise, altering seasonal patterns and economic life.[13] The increasing strength and duration of hurricanes in the southeastern United States and the Atlantic have been attributed to warmer ocean waters.[14] Heavy rain events and flooding have also become more frequent, especially in the Midwest.[15] Massive wildfires in California, Europe, and Australia in 2018, 2019, and 2020 were linked in part to extended heat waves and drought conditions.[16]

As serious as these trends are for the United States and other wealthy developed countries, they are potentially far more disastrous for the majority of people living in the developing world. As pointed out by Richard Tobin in Chapter 13, high population growth and rising use of energy, as well as food and freshwater resources threatened by climate change, are already stressing ecosystems to the breaking point.[17] Many of the poorest cities are in low-lying coastal areas that may become uninhabitable by midcentury. According to one recent study, one billion people will have to migrate or adapt to severely changed conditions for every 1 degree C of global average warming.[18] The consequences of climate change will thus affect every aspect of government policy, including foreign policy and national security.

According to a 2019 UN report, global temperatures are on track to rise as much as 3.2 degrees Celsius (5.8 degrees F) by the end of the century, with catastrophic consequences. Another recent study concluded that a doubling of atmospheric carbon dioxide (CO_2) levels could lead to even greater temperature increases.[19] The UN report estimated that to avoid such disastrous warming, GHG emissions would have to begin *falling* by 7.6 percent each year beginning in 2020.[20] Instead, global CO_2 and methane levels continued to rise despite a temporary slowdown in 2020 due to the coronavirus.[21] It will thus take an unprecedented level of both domestic and international policy action in coming years to mitigate these trends.

President Obama made a historic commitment in his negotiations with China and in the 2015 Paris Agreement to reduce US carbon emissions by 26–28 percent below 2005 levels by 2025 (Chapter 12). This national pledge was based heavily on implementation of the Clean Power Plan to reduce carbon emissions from power plants and on raising average fuel economy standards for cars and light trucks to 54.5 miles per gallon by 2025, policies which Trump rejected. It is debatable whether these and other Obama initiatives would have enabled the United States to reach the 26–28 percent target by 2025. US energy-related carbon emissions did, however, decline about 14 percent between 2005 and 2017, primarily due to the closure of coal-fired power plants and substitution of natural gas and renewables in electrical generation and industry. However, global CO_2 emissions rose 21 percent during this same period.[22]

Despite overwhelming evidence, President Trump continued to question climate science and filled his administration with climate change skeptics and deniers, including many lawyers and lobbyists from the oil and gas industry. Moreover, as detailed in Chapters 1, 4, and 5, he repeatedly proposed deep budget cuts to virtually all scientific research programs related to climate change, including those of the EPA, the Department of Energy, National Aeronautics and Space Administration (NASA), and the National Oceanic and Atmospheric Administration (NOAA). Although Congress did not enact most of these cuts, many career scientists, researchers, and other experts left government and were not replaced. Both the EPA and the Interior Department also eliminated numerous scientific advisory committees and replaced academic scientists on many of the remaining panels with members friendlier to business and industry who often lacked scientific qualifications.

When the National Climate Assessment—compiled by thirteen federal agencies as mandated by Congress—came out in 2017, Trump dismissed it summarily and abolished the advisory committee which oversees the assessment process.[23] The president further demonstrated his disdain for science by not appointing a senior White House science advisor for 18 months. By that time, his administration had also disbanded other scientific staffs including the National Security Council's directorate for global health and biodefense, which President Obama had established to prepare for future pandemic outbreaks such as the coronavirus. The damage done to the research infrastructure throughout the federal government could take years to repair.

If allowed to stand, the Trump rollbacks will have major consequences for GHG emissions. For example, the revised fuel economy standards for cars and trucks—now the largest source of carbon emissions—would lower the mileage standards from 54 to 40 miles per gallon by 2026, reducing fleetwide fuel efficiency gains from nearly 5 percent to 1.5 percent annually. The proposed rule would consequently allow vehicles to emit a billion tons more CO_2 over their lifetime than they would have under the Obama standards. According to one analysis, this would be equivalent to emissions from 68 coal power plants during that period.[24] The new rule was challenged in court by more than 20 states on grounds that it lacks both a legal and a scientific basis, but it will encourage automakers to build more large cars, SUVs, and trucks in coming years.[25] In the electricity sector, the deregulatory reversal of the Clean Power Plan will wipe out anticipated reductions for GHGs and many air contaminants. Its replacement by the Affordable Clean Energy rule may actually lead to net increases during the 2020s in emissions from power generation by incentivizing sustained use of coal and natural gas.[26]

President Trump's withdrawal from the Paris Agreement and failure to participate in other international forums on climate change will also have far-reaching effects. Other countries are now less likely to meet their climate pledges (Chapter 12), and the administration's growing hostility toward China threatened previous collaboration on such matters as banning hydrofluorocarbons (HFCs). Trump's attempts to cut off foreign economic assistance to developing countries for birth control and sustainable development projects were also likely to be counterproductive (Chapter 13).

Despite the daunting projections of scientists, Trump's opponents struggled to unify around a common policy alternative. Progressives in the Democratic Party first unveiled their proposal for a "Green New Deal" in December 2018.[27] The plan called for radical changes in line with the recent IPCC warnings, including total reliance on "clean and renewable" sources of energy for electricity generation by 2035, and "zero net emissions" from fossil fuels by 2050. The proposal contained some flexibility in allowing, for example, nuclear power to be considered a "clean" source and in allowing natural gas and other fossil fuels to be used until mid-century with "carbon capture" offsets (carbon capture and storage technology). The plan also proposed net-zero emissions from cars and light duty vehicles by 2030, followed by "a swift phase out of internal combustion engines"; net-zero building standards by 2030; 100 percent fossil-free transportation by 2050; a "green jobs guarantee" to employ millions; and numerous other proposals to protect clean air and water and to promote sustainable economic growth and environmental justice. Many of these principles were embodied in a nonbinding resolution introduced in Congress by Rep Alexandra Ocasio-Cortez (D-NY) and Sen. Edward J. Markey (D-Mass.) in February 2019.[28]

Although the resolution had no chance of passage and was never transformed into full legislative proposals, it did raise climate change and environmental policy to the top tier of issues in the Democratic presidential primary campaign. Almost all the candidates endorsed the concept in principle, while taking different positions on specific issues. For the first time, one candidate, Gov. Jay Inslee of Washington, built his entire campaign around climate change, and most of the other candidates also took strong stances.[29] All candidates pledged to rejoin the Paris Agreement promptly and to reassert American leadership on the world stage. Former Vice President Joseph Biden proposed spending $1.7 trillion to leverage $5 trillion in addition private sector and state and local government spending over 10 years, with a goal of achieving net-zero carbon emissions by 2050. After becoming the presumptive nominee, he adopted a more ambitious platform, including a larger economic infrastructure and recovery package that would also give greater protection to poor and minority communities disproportionately exposed to environmental hazards and affected by the pandemic. He also vowed to use executive powers to implement policies going well beyond the Obama administration in which he served. His running mate, California Senator Kamala Harris, endorsed the general concept of the Green

New Deal and would be expected to be a strong advocate for environmental justice in a Biden administration.[30]

Congressional Democrats have also introduced numerous climate change bills in the House and Senate and have fought to increase funding for environmental agencies (Chapter 5). At least eight separate carbon pricing bills were introduced into the House or Senate (or both) in the 116th Congress (2019–2020).[31] House Speaker Nancy Pelosi (D-Calif.) created the House Select Committee on the Climate Crisis after assuming office in 2019 to channel legislation. The committee recommended, and the House passed, for example, a bill requiring the United States to remain in the Paris Agreement and denying any funding for withdrawal. House Democrats also proposed a large infrastructure plan in early 2020 that included provisions for making all federal buildings carbon neutral and for various "green" transportation initiatives.[32] In the Senate, Sen. Thomas Carper (D-Del.) and 33 cosponsors introduced the "Clean Economy Act of 2020," which would force the EPA to adopt rules to eliminate US contributions to global warming by 2050, and Minority Leader Chuck Schumer proposed phasing out gas-powered cars by 2040.[33] Sen. Brian Schatz (D-Hawaii) and several cosponsors who were then running for president also introduced a bill in late 2019 that would require the Federal Reserve to conduct "climate stress tests" of the largest banks and financial institutions to ensure they gave sufficient assets to handle risks associated with climate change.[34] None of these bills had a chance to pass due to Republican opposition in the Senate. Nonetheless, some Democrats worked with Republicans to pass more limited bills, such as legislation funding climate research and extending tax credits for carbon capture and storage technologies.

Republicans have been much slower to offer alternative policies. Although some polls indicate that a significant share of rank-and-file party members are concerned about climate change and support more vigorous government action, most Republican leaders have accepted President Trump's refusal to recognize the issue and have remained silent in public.[35] Nevertheless, some prominent Republicans have advocated a national carbon tax (discussed later in the chapter), and various proposals have been broached in Congress favoring, among other things, increased energy research and innovation, tax breaks for natural gas exports, revival of the nuclear power industry, and large tree-planting programs to absorb CO_2. Twenty-three Republican representatives and seven senators also belonged to the House and Senate Climate Solutions Caucuses in 2020. These bipartisan caucuses have worked across the aisle on legislation, including carbon tax and rebate proposals, as well as with local organizations such as the nonpartisan Citizens' Climate Lobby.[36]

Many cities and states have continued to oppose the Trump administration's climate change rollbacks (see Chapters 2, 11, and 12). Cities account for up to two-thirds of GHG emissions. Leaders of more than 500 cities, counties, and tribes from every state and 25 state governors joined the America's Pledge movement by 2020, vowing to support the Paris Agreement for emissions within their boundaries. Collectively, these jurisdictions represented 65 percent of the US population and 51 percent of total national GHG emissions in 2019. Full implementation of these policies through 2025 without new federal policies was projected to reduce American emissions 19 percent below the 2005 level, beginning to approach Obama-era national emission pledges.[37]

Nearly thirty states already have climate change policies such as renewable energy portfolio standards and several of them adopted far more ambitious targets of between 50 and 100 percent renewables during 2019 and 2020. Many states have already reduced their carbon emissions from power plants by at least 40 percent compared to 2005 levels, and some states have reduced emissions even more. They include northeastern and mid-Atlantic states that are members of the Regional Greenhouse Gas Initiative (RGGI), which began implementing a cap-and-trade system in 2009 (see Chapters 2 and 10). After reducing CO_2 emissions by 45 percent from 2005 levels by 2020, the RGGI states committed to reduce their power plant emissions by an additional 30 percent by 2030 while expanding their collaborative network to include New Jersey and Virginia.[38] California is also a model for the nation and in 2017 extended its landmark climate change legislation to require a 40 percent reduction in total GHG emissions from 1990 levels. In 2020 it adopted new rules requiring that all heavy trucks sold in the state must have zero emissions by 2045. Governor Gavin Newsom also issued an executive order to develop a plan for phasing out the sale of new gasoline-powered cars and light trucks in the state by 2035.[39]

Numerous private companies, colleges and universities, churches, and other institutions have also adopted their own sustainability plans to achieve carbon neutrality. Many large corporations, such as Apple, Google, Amazon, Walmart, Target, and Anheuser Bush, have pledged to cut their carbon emissions substantially and/or to obtain all or most of their electricity from renewable sources. Apple is now extending these commitments to its supply chains in China and elsewhere. Some companies and financial institutions, such as Microsoft and the world's largest asset manager, Black-Rock, have already made decarbonization a central goal in their investment strategies.[40] Large banks and insurance companies are under increasing pressure to follow suit: for example, JPMorgan Chase, Wells Fargo and Citigroup have all said they will decline funding for Arctic oil and gas drilling projects.[41] According to one business survey, 84 percent of business decision-makers said they were aware of the IPCC reports issued in late 2018, and two-thirds of those said they had reviewed or changed their management strategies in response to the reports.[42]

Despite these pledges, much stronger nationwide policies are likely to be necessary in the future. As the coronavirus pandemic has demonstrated, state and local governments play critical frontline roles, but the country as a whole needs far greater leadership and direction to maximize the effectiveness of our response. Because the climate crisis is becoming more obvious, and because we have more scientific data on which to act, most experts believe there is still time to avert its most devastating consequences if we adopt comprehensive policies in the coming years. Such policies are likely to require a broad range of tools affecting every sector of the economy.[43] We look more closely at energy policies in the following section and other options in later sections.

National Energy Policy

Congress has never been able to agree on a comprehensive national energy policy. It has, however, passed a handful of energy bills since the 1970s to encourage certain types of energy production, as well as numerous other pieces of legislation that impact energy development and use (see Chapter 5 and Appendix 1). The Energy Policy and

Conservation Act of 1975 created a National Petroleum Reserve and a Corporate Average Fuel Economy (CAFE) standard, the latter requiring US auto manufacturers to meet an average mileage requirement for its new cars for the first time. Many of these bills reflected the oil shocks of the 1970s which have led all presidents since Richard Nixon to pursue "energy independence" in one form or another. For example, the Energy Policy Act of 2005 provided massive tax incentives and loan guarantees for energy production, especially oil, natural gas, and nuclear power; while the Energy Independence and Security Act of 2007 raised CAFE standards for the first time in decades and provided subsidies to alternative fuels, especially biofuels such as ethanol. The result was a huge increase in oil and gas production as well as of ethanol for use in gasoline. By 2018, the United States had become the world's largest oil and gas producer and began to export some fuels such as liquid natural gas (LNG) to other countries.[44]

Although the George W. Bush administration was instrumental in this expansion, President Obama also followed an "all-of-the-above" energy strategy that encouraged the development of oil and gas production as well as of energy from renewable sources.[45] Domestic oil and gas production increased rapidly during Obama's tenure, primarily due to a dramatic refinement and expansion of fracking techniques used to unlock massive oil and gas reserves from shale deposits; oil imports dropped from 60 percent of national consumption in 2008 to 17 percent by 2017 and earlier concerns about gas scarcity disappeared amid massive production upswings.

President Obama's American Recovery and Reinvestment Act of 2009 also included some $80 billion in spending, tax incentives, and loan guarantees for renewable energy projects. Energy production from new sources such as wind and solar increased dramatically in subsequent years as costs declined. By 2020 renewable energy accounted for nearly 20 percent of total national electricity production, up from 10 percent in 2010, and was economically competitive with fossil fuels in many energy markets, including Texas and California.[46] Collectively, natural gas and renewables would combine to dramatically reduce coal use for electricity, and wind, solar, and biofuel production employed far more workers than the coal mining industry in 2020.[47]

While presiding over this energy boom, President Obama also placed increasing restrictions on certain kinds of energy production. In addition to the climate change measures noted in the previous section, he tightened rules for offshore oil drilling (in the wake of the Deepwater Horizon oil blowout in the Gulf of Mexico) and placed more areas of the Atlantic and Alaskan coasts off limits to new oil leasing through 2022. The Interior Department declared a moratorium on all new coal leases on public lands; issued new regulations on oil and gas fracking, including restrictions on methane flaring and emissions on public lands; and imposed a new rule to stop coal companies from dumping mining wastes into nearby streams and rivers. After years of delay, Obama also blocked construction of the Keystone XL and Dakota Access pipelines. In one of the last acts of his administration, he invoked the 1953 Outer Continental Shelf Lands Act to ban offshore oil drilling permanently along wide swaths of Alaska's coast and the Atlantic seaboard from Maine to Virginia.[48]

The other thrust of President Obama's energy policy was a strong push for energy conservation and efficiency. In addition to raising mileage standards for cars and light trucks, new regulations established fuel economy standards for large trucks and other heavy vehicles. Standards for building construction and appliance efficiency were also tightened significantly. Consumers became more aware of energy use through programs such as Energy Star ratings for appliances and electronics. Through executive orders and memoranda, Obama also required all federal agencies, including the Department of Defense, to assess their carbon footprints and to reduce energy consumption.

President Trump's "America First" energy strategy, which was largely shaped by executives and lobbyists from the fossil fuel industries, as well as by conservative think tanks such as the Heritage Foundation and the Competitive Enterprise Institute, called for the elimination of these and other restrictions and for opening vast new areas to energy production in order to achieve "energy dominance."[49] Trump made no effort to advance any of these efforts through legislation, even during his first two years when Republicans controlled both chambers of Congress, instead relying exclusively on a range of executive branch tools. His executive order on "Promoting Energy Independence and Economic Growth" of March 28, 2017, directed all executive departments and agencies to "immediately review existing regulations that potentially burden the development or use of domestically produced energy resources and appropriately suspend, revise, or rescind those that unduly burden the development of domestic energy resources beyond the degree necessary to protect the public interest or otherwise comply with the law." The order also directed agencies to revoke Obama's climate change plans, to review fracking and methane regulations, and to end the moratorium on coal leasing on public lands. This executive order was complemented by others, often leading to protracted administrative review processes designed to weaken or eliminate Obama-era provisions.

These efforts included provisions designed to open up large areas of the Atlantic and Arctic Oceans to oil and gas leasing and roll back many of the safety rules for offshore drilling adopted after the Deepwater Horizon disaster in 2010.[50] Millions of acres of public land were leased by the Department of Interior throughout the West as rules governing methane releases and fracking were relaxed, including areas previously protected under the Endangered Species Act or near national parks and wilderness areas.[51] The Bears Ears and Grand Staircase Escalante national monuments in Utah were drastically reduced in size and potentially opened for mining and energy development (Chapter 9). The administration also inserted a provision into its 2017 tax reform bill that would open the remote Arctic National Wildlife Refuge in Alaska to commercial oil and gas drilling for the first time.[52]

The Department of Energy (DOE) also attempted to invoke national security powers to prolong the life of aging coal and nuclear plants in order to bail these industries out, though largely without success.[53] In 2018 DOE officially proclaimed that it was no longer necessary to conserve oil, and the next year even announced plans to roll back a rule dating from the Bush administration that set energy efficiency standards for light bulbs.[54] President Trump also placed a 25 percent tariff on imported solar panels, which cost the US solar industry an estimated 62,000 jobs.[55] Finally, under the 2020 appropriations bill tax credits for solar energy will be phased out by 2022 unless restored by a future Congress, unlike tax benefits for fossil fuel production that retained protection.[56]

As noted in the previous section, and in Chapters 2, 8, 9, and 12, states also play important roles in regulating energy development and the implementation of federal policies. Most states have their own renewable portfolio standards which require utilities to produce a growing share of energy from renewable sources, and many have their own energy efficiency standards and offer incentives to promote sales of electric vehicles and solar systems, as detailed in Chapter 8. Although there has been some backsliding, most states with clean energy requirements have maintained them and many have acted unilaterally or in concert to oppose many of Trump's initiatives. For example, Florida and numerous other coastal states immediately demanded exemptions from proposed offshore drilling expansion, resulting in the renewal of drilling bans off several of them as mentioned at the beginning of Chapter 4. Other states, such as those dependent on tourism or which have already invested heavily in solar and wind energy, also oppose preferential treatment for fossil fuels. "Red" states such as Texas, Oklahoma, Wyoming, and Iowa have been leaders in wind energy production and have generally remained committed to these industries through various tax and regulatory provisions. Other states, such as California and the 14 "bandwagon" states which embraced its rigorous vehicle emission standards, launched an aggressive legal battle led by their attorneys general to retain them, as discussed in Chapter 2. Whether Trump's unprecedented attempt to rescind California's waiver to set higher standards for emissions under the Clean Air Act (CAA) is upheld in the courts will thus have major consequences for the nation.

National and global market forces also play a large, and often dominant, role in energy production. The rapid decline of coal mining and coal power generation in favor of cheaper natural gas was primarily driven by market forces which continued under Trump. In 2020, more electricity was expected to be produced from renewable sources than from coal in the United States for the first time.[57] The world oil glut and collapse of crude oil prices and stock market values during the pandemic crisis discouraged investment in new oil and gas production, at least temporarily.[58] However, low gasoline prices will also encourage consumers to buy less-efficient vehicles and undermine efforts to achieve a low-carbon future. States will be under increased economic pressure to relax their environmental standards and compete for business and resources during the coronavirus pandemic. Additional national intervention, such as a federal carbon tax or renewable energy standard, as well as revised "command and control" regulations, will thus be necessary if the clean energy transition is to accelerate to meet climate change imperatives. We discuss some of these alternatives later in the chapter.

Air and Water Pollution

Our current focus on energy and climate change should not blind us to more traditional environmental problems such as air and water pollution which affect most people directly. These forms of pollution are regulated by the EPA under statutes written largely in the 1970s and, in some cases, revised in the 1980s and 1990s (see Chapters 5 and 7). As pointed out in Chapter 1, air quality has improved greatly over the decades, with aggregate emissions of the six "criteria" pollutants controlled by the CAA falling by 74 percent between 1970 and 2018—despite the fact that our economy increased by more than 241 percent over that period. Emissions of other pollutants such as hazardous chemicals and toxic metals have also declined markedly. Nevertheless, air quality varies greatly in different localities and among different population groups, with many urban areas still suffering high concentrations of ozone, particulates, heavy metals, and chemicals that can cause severe health effects and even death. In 2018, 41 percent of Americans still lived in counties that had at least one major type of air pollution that exceeds EPA standards. Many poor urban neighborhoods are especially polluted, raising environmental justice concerns.

Among the most controversial policies of the George W. Bush administration were efforts to weaken the CAA by exempting power plants from the New Source Review (NSR) provisions of the law which require technological updates when plants increase production or expand capacity. Although the federal courts eventually struck down the rule in 2007, it slowed progress toward controlling the largest sources of conventional air pollutants as well as GHGs during the Bush years. Two other rules required by the CAA—the Clean Air Interstate Rule (CAIR) to limit emissions from power plants that pollute neighboring states in the eastern half of the country and the Clean Air Mercury Rule (CAMR) to regulate mercury and other hazardous air emissions—were proposed by the Bush administration, but both were also found inadequate by the courts.

Under Obama, the EPA issued a new Mercury and Air Toxics Standard (MATS) in 2011, which further tightened emission standards for mercury, lead, asbestos, and other hazardous air pollutants from coal- and oil-fired electric utility plants. The EPA also issued a new Cross-State Air Pollution Rule in 2011 to replace the Bush CAIR. The new rule required power plants in twenty-seven eastern states to limit emissions of pollutants such as sulfur dioxide, nitrogen oxides, and fine particulates that affect downwind states. If implemented, the Obama regulations limiting carbon emissions from new and existing power plants would have cut emissions of all of the conventional air pollutants as well. In proposing the Clean Power Plan in June 2014, for example, the EPA stated that it would have major "co-benefits" by reducing emissions of conventional pollutants by more than 25 percent, which would avoid up to 6,600 premature deaths and 150,000 asthma attacks in children.

The Trump administration chose not to count such co-benefits in its cost–benefit calculations and drastically reduced its estimates of future pollution control benefits generally, as explained in Chapter 7.[59] And although President Trump defined "clean air and water" as the "core mission" of the EPA, his administration attempted to relax many of the Obama-era regulations and celebrated these steps in its annual reports as economic achievements, overcoming environmental extremism.[60] Its proposed Affordable Clean Energy rule for power plants intended to allow them to release more conventional pollutants such as fine particulate matter (PM2.5) and nitrogen oxides as well as more CO_2; by the EPA's own analysis, this could result in up to 1,400 additional premature deaths and 15,000 new cases of upper respiratory disease annually by 2030.[61] The EPA also proposed revising the MATS rule to allow higher limits on mercury and other toxic air emissions from power plants.[62]

In rewriting these rules, the EPA altered its own methods for calculating health benefits in order to justify additional rollbacks, often ignoring its own scientific advisers.[63] In addition, it also attempted to severely limit the kinds of public health research that can be utilized in rulemaking by excluding epidemiological studies that do not publish their raw data due to privacy concerns. If approved, the new "Strengthening Transparency in Regulatory Science" rule could eliminate consideration of the scientific data underpinning most of EPA's air and water regulations, including those controlling chemical releases to the environment.[64] Finally, administration enforcement of New Source Review and other reporting requirements under the CAA declined sharply to allow older, dirtier plants to continue operating with minimal federal oversight.[65] Although full data are not yet available, it appears that ambient air concentrations of at least some air pollutants, such as PM2.5, increased after 2016.[66]

Much the same can be said for water pollution. During the campaign Donald Trump repeatedly promised to rescind the Waters of the United States (WOTUS) rule, which was issued by the Obama administration in 2015 to clarify which streams, smaller tributaries, and wetlands are covered by the Clean Water Act (CWA) in the wake of the Supreme Court's confusing *Rapanos* decision of 2006 (see Chapters 6 and 9). This revised rule was strongly opposed by farmers and ranchers, construction companies, oil and gas drillers, and other business interests as government "overreach" that violated local land use laws, and it was temporarily enjoined in most states by lower federal courts.[67] The EPA announced formal plans to repeal and replace the rule in late 2018, withdrew it and reinstated the old rule in 2019, and issued its replacement rule in January 2020.[68] The new rule, issued by the EPA and the Army Corps of Engineers which jointly administer the law, could exclude millions of miles of smaller seasonal streams and, by some calculations, more than half of the nation's wetlands, from protection under the CWA even if they have periodic surface or subsurface hydrological connections to navigable waters. Individual permits for discharges or draining into these excluded bodies of water will no longer be required, raising public health concerns among many scientists who opposed the change.[69] However, in April 2020 the Supreme Court cast doubt on whether this provision of the rule would stand.[70]

These changes could affect drinking water quality in many areas of the country. As noted in Chapter 1, the Safe Drinking Water Act is largely implemented by state and local governments which often lack resources to detect and control contaminants, especially runoff from sources such as agricultural fields, feedlots, golf courses, and construction sites. Although data on groundwater pollution are sketchy, it appears that millions of wells and other drinking water sources across the country are contaminated with unsafe levels of nitrates and pesticides, especially in poor rural areas.[71] Many other toxic substances are also being found in urban and suburban water supplies, in some cases worsened by increased flooding from major storms.[72] Coal ash from power plants is another potential source of water pollution. Despite the fact that these wastes contain arsenic, lead, mercury and other contaminants that have leaked into both groundwater and watersheds, in 2019 the Trump administration moved to scale back President Obama's 2015 ash disposal regulations and exempt many power plants from them.[73]

High levels of other hazardous industrial toxins, including PFAS "forever" chemicals such as PFOA that have been linked to cancer, birth defects, and other health issues, have also been detected in many areas. After long delays, EPA launched a nationwide "action plan" to deal with these chemicals in 2019, but it is unclear how much priority it will have, how chemical risks will be assessed, and how local water authorities will benefit.[74] In May 2020 the EPA refused to set a safe drinking water standard for perchlorate, a chemical used in rocket fuel that has long been detected in drinking water and which is known to cause brain damage in fetuses and newborns and thyroid problems in adults.[75]

However, in a rare act of bipartisanship Congress passed an omnibus water development act in 2018 (America's Water Infrastructure Act of 2018), which strengthened provisions of the Safe Drinking Water Act for the first time since 1996 and provided funding to assist public water systems in poor and disadvantaged communities (including lead removal).[76] Congress has also continued funding some large watershed programs such as the Great Lakes Restoration Initiative, and voted in 2019 to permanently fund the Land and Water Conservation Fund, which uses money collected from oil and gas drilling to protect public lands and waters.[77]

ALTERNATIVE POLICY APPROACHES

Traditional "command-and-control" regulation has been generally successful in improving environmental quality since the 1960s (Chapter 1). Nevertheless, critics of the EPA have long argued that its "one size fits all" regulations are economically inefficient and often inappropriate for specific local circumstances. Legal mandates under these rules have also resulted in endless litigation and have tended to discourage innovation and voluntary advancements that could improve outcomes (though most laws allow states to set higher standards if they choose to). Many environmental policy scholars have thus advocated new forms of "smart regulation" that utilize more flexible, less intrusive, and more cost-effective methods for reducing pollution and addressing other environmental problems.[78] Economists from across the ideological spectrum have advocated market-based regimes that provide price signals which discourage polluting activities while maximizing economic efficiency, as explained in Chapter 10.[79] These proposals have taken two primary forms.

Cap-and-trade systems set overall limits on the quantity of pollution emissions allowed and lower the number of allowances or permits available for use over time to meet reduction goals. After initial allocations, affected entities can buy and sell emission allowances, with the price being determined by the trading market. As the cap is tightened, the price of allowances should go up, encouraging companies to cut their emissions or incur higher costs for allowances. Theoretically, this form of market pricing and trading gives regulated parties freedom to make their own decisions and should result in the most cost-effective way of lowering pollution, starting with sources with the lowest reduction costs. The first large-scale cap-and-trade program was launched by the G. H. W. Bush administration in 1990, focusing on sulfur dioxide emissions for electric power plants under the CAA, where it received far-reaching recognition as a cost-effective source of environmental protection.[80] The American Clean Energy and Security Act of 2009, which passed in the House of Representatives but died in the Senate in 2010, would have established an economy-wide system of this kind to control GHGs.

Despite the act's failure, two large GHG emissions trading systems are operating today in the United States and a growing number operate around the world.[81] The RGGI market, now including 11 states, began operations in 2009 and the California system launched in 2012 with a foreign partner in the Canadian province of Québec. RGGI focuses exclusively on the electricity sector and has completed several multistate reviews leading to ongoing tightening of its emissions cap. It also pioneered the use of auctions to allocate emission allowances rather than free distribution, thereby producing more than three billion dollars for participating states in its first decade that facilitated energy efficiency and renewable energy development in many of them.[82] The California-Québec system is much broader in scope, cutting across numerous emission sectors and generating a consistently higher auction price during its initial years of operation, while also exploring possible expansion with other states such as Oregon and Washington. As a result of these dual systems, more than 100 million people, or nearly one-third of the American population, are now covered by such systems, which have already contributed to substantial carbon emission reductions. In 2019–2020 New Jersey and Virginia took formal steps to join RGGI, and Pennsylvania Governor Thomas Wolf launched plans to follow their lead. In 2020 a dozen northeastern American states actively explored creation of a transportation-focused sequel to RGGI, through a "cap-and-invest" program intended to implement a pricing system for vehicle transportation in the region and return revenue to participating states for development of more carbon-friendly transit alternatives. The European Union has struggled mightily in its extended attempts to launch a continental cap-and-trade system for carbon but has implemented recent reforms that appear to make it considerably more effective and a centerpiece of its ambitious climate plans for the future.[83] China experimented for nearly a decade with a suite of locally based emission trading programs and transitioned in 2020 toward implementation of a national version. Other Asian nations were exploring possible cap-and-trade development in 2020.[84]

Cap-and-trade has registered some genuine successes but has proven more administratively complex and politically contentious than many of its proponents anticipated. As a result, a number of economists, business leaders, and politicians of different political persuasions have instead embraced *carbon taxes* as an alternative system for pricing carbon emissions.[85] In principle, taxing CO_2 or other forms of pollution via taxes or fees could achieve cost-effective outcomes similar to those of emission trading systems, as explained in Chapter 10. Taxes are also easier to administer and more transparent, thus conveying clearer signals to consumers; and the proceeds can be rebated to taxpayers as a credit or dividend to offset higher fuel prices or invested in energy transition. Five Nordic nations (Denmark, Finland, Iceland, Norway, and Sweden) pioneered this concept in the early 1990s, all adopting carbon taxes with high rates that proved durable and demonstrated an ability to reduce demand for fossil fuels and GHGs. Many other European nations maintain some form of a carbon tax alongside their continental cap-and-trade program and there has been expanding development of such taxes among nations in Asia and South America.[86]

In its landmark climate report of October 2018, the IPCC stated that putting a price on CO_2 emissions must become a central strategy if we are to limit global warming. The report drew heavily on the work of Yale economist William D. Nordhaus, who was awarded a share of the 2018 Nobel Prize in Economics for his decades-long advocacy of carbon taxes as the most efficient method for reducing GHG emissions.[87] In January 2019, a group of 45 prominent economists, including 27 Nobel laureates and almost all former chairs of the Federal Reserve Board and of the Council of Economic Advisers in both Democratic and Republican administrations, signed a letter in the *Wall Street Journal* endorsing a US carbon tax as "the most cost-effective lever to reduce carbon emissions at the scale and speed that is necessary."[88] In late 2019 the International Monetary Fund came out for a global tax of $75 per ton of CO_2 by 2030.[89]

On the other hand, taxes do not guarantee any given level of emissions reduction and could face even greater political opposition. Some of these obstacles might be overcome if these taxes were revenue neutral—that is, part of a larger tax reform in which other taxes such as corporate, income, or payroll taxes were simultaneously reduced to offset the carbon tax—or if the tax revenues were returned to citizens or taxpayers in the form of dividend checks or tax credits. Such a system has been used successfully in British Columbia for more than a decade and is a model for a national system now being implemented across Canada (see Chapter 2).

The latter approach has been gaining support among some prominent Republicans, corporate executives, and conservatives. In 2017 former Treasury Secretaries James A. Baker III, George P. Shultz, and Henry M. Paulson Jr. proposed a tax beginning at $40 per ton of CO_2, which would raise an estimated $200–$300 billion a year. The money would be returned to consumers as a "carbon dividend" estimated at $2,000 annually for an average family of four.[90] A revised version of this proposal was reintroduced in early 2020 on behalf of the Climate Leadership Council, a bipartisan group which includes many economists, corporate executives, and former high government officials. The new plan would impose a "carbon fee" of $43 per ton beginning in 2021 that would increase every year by five percent above inflation, while simultaneously streamlining related regulations to reduce compliance costs. Proponents claimed that this approach would lower carbon emissions more rapidly than Obama's policies and more than meet his pledge for emission reductions under the Paris Agreement.[91] All proceeds would be returned to taxpayers in quarterly dividend checks.

These proposals face an uphill fight, especially following the economic recession and massive deficit spending resulting from the coronavirus pandemic. Many have noted the failure of carbon tax initiatives in Washington State and the "yellow jacket" protests against gas tax increases in France in 2016 and 2018.[92] Critics have long noted the challenges of achieving fairness in imposing costs across various income groups and then determining how to allocate revenues. David Leonhardt has also argued that carbon taxes will remain unpopular because they emphasize the *means* rather than the *ends* of policy and imply a negative need for sacrifice, whereas positive proposals for clean energy performance standards and subsidies for green job creation are far more popular among voters.[93] Actual costs per unit of emission reduction may be considerably higher in these nonpricing options, but it is harder for citizens to see this in their daily lives. Several Democratic candidates for president in 2020 endorsed carbon taxes, although the final two contenders, Joe Biden and Bernie Sanders, were more equivocal about their commitment to this tool than in prior campaigns.[94] Instead they advanced their alternative versions of a Green New Deal that emphasized major new federal spending on green energy and regulation rather than pricing carbon. The idea of a carbon tax or cap-and-trade thus remained one option among many in the early 2020s in the United States but its economic advantages were offset by its political challenges.

Voluntary collaboration and self-policing programs are another alternative to traditional regulation. In this approach, regulated parties such as corporations or industrial sectors voluntarily agree to meet higher performance standards than required by law, in return for greater regulatory flexibility. Most of these programs were created during the Clinton administration as part of its "reinventing government" initiative.[95] Although many of these experiments failed to produce measurably better results than traditional regulation and have been discontinued, they did help some sectors of business and industry to improve their environmental management and reporting, and in some cases their operations. But stronger organization and governmental supervision, or effective oversight by independent bodies, appear necessary if a genuine "greening"—defined as a transformation to sustainable production—is to occur (Chapter 14). One key to stronger environmental performance is public *disclosure of information*. The leading example of such information disclosure is the federal Toxics Release Inventory (TRI), which has operated since the late 1980s. Over twenty thousand industrial facilities a year report on their release of some 650 toxic chemicals to the air, water, or land. The information is made available to the public in a variety of ways, including through the EPA's own TRI website.

The key assumption of such policies is that an informed public may bring some pressure to bear on poorly performing industrial facilities and their parent companies, what some have called regulation by embarrassment. It is just as likely that companies will seek to avoid such public censure by proactively altering their production processes to reduce chemical releases even if they are not subject to any regulatory requirements to do so. Although the disclosed information is often quite technical and difficult to translate into meaningful risks to the public's health, the program has led to significant decreases in the release of toxic chemicals over time, as well as to improved environmental performance on the part of industry.[96]

The success of some information disclosure programs such as the TRI led the federal government to try much the same approach with GHG releases. The EPA maintained a Greenhouse Gas Reporting Program (GHGRP) that collected and released emissions data from facilities in forty-one different source categories and that was easily accessible by the public. Much like the TRI, the GHGRP was created under the assumption that facilities would seek to reduce their emissions precisely because the data are available to the public. However, the Trump administration terminated this program and no longer provides this information. Nonprofit organizations, of course, can play a very similar role, as demonstrated by the Carbon Disclosure Project. Since 2007, it has ranked companies on their carbon emissions. Other GHGs such as methane could also be folded into a climate-focused information disclosure system, particularly as increasingly sophisticated technologies including satellite and drone surveillance add new precision to traditional voluntary reporting of estimated releases. However, no such federal system has ever been attempted, and only a handful of states have taken any steps to explore this option on their own.

Another promising policy approach, *local and regional sustainability planning*, has been used with impressive results in many cities and regions in the United States as well as in other nations. The concept of sustainability or sustainable development came into wide use following the 1987 Brundtland Commission report and later at the Earth Summit of 1992. It also was promoted heavily during the Clinton administration by the President's Council on Sustainable Development (PCSD), which focused on how communities might develop "bold, new approaches to achieve economic, environmental, and equity goals."[97]

As Chapter 11 showed, both large and small cities in the nation have embarked on intriguing programs to pursue economic development in a way that seeks to integrate environmental and equity considerations into the equation. Although some of the cities, such as Seattle, Washington, and Portland, Oregon, are well known and often celebrated for their remarkable sustainability achievements and their highly supportive local citizenry, they are by no means the only examples of successful sustainability planning. New efforts to promote sustainability are found in large cities such as New York, Chicago, and Los Angeles; midsized cities such as San Francisco, Austin, and Boston; and smaller cities such as Boulder, Colorado, and Chattanooga, Tennessee. These often include innovative programs to improve air quality, water quality, building efficiency and energy use, local transportation planning, land use, water conservation, and more. Some cities also are moving ahead in planning for adaptation to climate change. Two other critical needs are a committed investment in *scientific research and development* and the improvement of *public education in science*. As evident throughout the book, little progress is possible on environmental challenges without strong scientific evidence to document the problems the nation faces and to identify potential solutions. The role of science is particularly important today when public trust in science and scientists appears to be in decline and political ideology often distorts scientific findings.[98] The United States historically has been highly supportive of investment in scientific research and technology development, and both Democratic and Republican administrations have given such research a high priority. Yet the Trump administration consistently proposed massive cuts to both basic and applied research programs throughout the federal government. Its fiscal 2021 budget request to Congress continued the pattern. The EPA's science and technology programs were slated for a cut of 37 percent.[99]

ENVIRONMENTAL GOVERNANCE FOR THE FUTURE

The environmental policies adopted following Earth Day in 1970 have served the nation well but have also become matters of deep political disagreement. Over the past quarter-century the regulatory systems established by the CAA, CWA, and other founding statutes have come under increasing attack by political and economic interests determined to undermine or destroy them. The Trump administration represented the culmination of these pressures as the EPA and other environmental agencies were largely captured by industries that have long opposed federal regulation. If the resulting policy changes survive judicial scrutiny and electoral turnovers, they could set our national system of environmental protection back at least a generation.

A post-Trump presidency could quickly reverse many of these changes through executive action, continuing the policy cycling of recent administrations (Chapter 4). But the task for the future will require more fundamental alterations in how policy is made and implemented. Without public and congressional involvement to legitimize policies, they are not likely to prove effective for long. Many of the older laws require updating and revision to eliminate loopholes and reduce bureaucratic rigidities and other inefficiencies. Most importantly, Congress must come to grips with the most challenging problem of climate change by enacting a national strategy and series of policies to sharply limit US carbon emissions in the coming decades. Congress must also provide adequate funding and support for the agencies and personnel in the forefront of this battle, especially the scientific and technical staffs that have the expertise to assess evidence and design responses that will best help to mitigate climate change.

Many alternatives have been discussed in this chapter and throughout this volume. In addition to restoring traditional regulatory standards where necessary, we could draw on the experience of many states within the United States, as well as from numerous other countries mentioned above, in creating new, more economically efficient, regimes for encouraging the transition already underway from reliance on fossil fuels to alternative energy sources. Nearly one-third of Americans already live within cap-and-trade regimes that have proved effective in driving down carbon emissions through market incentives. Congress could once again attempt to enact a national carbon pricing system, perhaps drawing on the new Canadian system established in 2019. This system requires provinces to put a price on carbon emissions either through a cap-and-trade system or via carbon taxes such as British Columbia has had for more than a decade. Many countries throughout the world have different forms of carbon pricing, and it is encouraging that influential figures in both American political parties, as well as many distinguished economists and business leaders, have expressed support for a national carbon tax or "fee" that could be returned to citizens as a carbon "dividend" which could more than offset higher fuel costs. Alternatively, some or all of these tax revenues could be used to fund sustainable energy investments, following the RGGI model in the northeast. Another possibility—especially given the high level of unemployment created by the pandemic—is creation of a new federal jobs program for modernizing our national energy infrastructure, as recommended by former energy secretary Ernest Moniz.[100]

There are many other encouraging signs. Most of the states, cities, corporations, colleges and universities, and other institutions that have made commitments to reduce and eliminate their carbon footprints are making progress toward these goals regardless of the administration in Washington. A strong case can be made for this kind of "bottom up" approach in coming years.[101] The experience of the COVID-19 epidemic has forced state and local governments to take on much greater responsibility for public health and has made people more aware of their own lifestyles. If they can be maintained and strengthened, state and local programs and private actions can have a large impact even when Washington is unable to act. Indeed, alliances among strong states and other institutions with high performance standards may be the most effective drivers of technological innovation in the near term.

In any case, it is increasingly apparent that green economy initiatives are not incompatible with economic growth and prosperity, but rather can be combined in positive-sum strategies, as Daniel Fiorino explains in Chapter 14. Many investments that reduce dependence on fossil fuels and polluting chemicals save companies money, improve their reputation, and create millions of jobs that improve human health and welfare. More and more corporations and financial institutions recognize that future limits to growth are more likely to come from negative ecosystem feedbacks such as climate change and disease vectors than from shortage of capital or opportunities to invest in new technologies.[102] Private, nonprofit, and public alliances began to expand creative efforts to pool financial resources to support these transitions through instruments such as green banks and bonds.[103] Indeed, several states began to consider ways to promote expanded green investment through these types of instruments and cover bond repayment through carbon price revenues, viewing them as environmentally sound initiatives that might also promote economic development.[104]

The United States cannot resolve the climate crisis alone—it must rejoin the other nations of the world in the Paris Agreement and increase support for other international bodies that are collaborating to bring about the global energy transition necessary to avoid the most damaging effects of climate change. The Trump administration's attempt to isolate the United States from global environmental diplomacy was unsustainable for many reasons, not least because whatever happens in the rest of the world will directly affect us. A strong argument can be made on national security as well as economic grounds that the United States must exert leadership in this area if the world is to avoid recurring natural catastrophes, economic disruptions, and political conflicts.[105] The United States has led strong international coalitions in the past, and must do so again. It must reengage with developing countries as well.

Times of national crisis have often created opportunities for new policy agendas in the past, as occurred during the 1960s and 1970s when new environmental policy cycles were launched in the United States and internationally. The COVID-19 catastrophe and associated global economic decline created new conditions that, with sufficient public concern and business leadership, could stimulate new rebuilding programs that include measures to prepare for and mitigate the parallel climate crisis.[106] A number of nations lauded for taking effective steps during the pandemic to minimize public health risk and find ways to sustain economic activity were among those that have been leaders on environmental and climate protection, both historically and in more recent years. Policies for less burdensome environmental regulation, crash research and development programs such as those begun during the coronavirus epidemic, and strong market incentives for technological innovation could be combined and gain greater bipartisan backing in these circumstances. The political obstacles are still formidable, but we believe that the next decade may provide such opportunities, built upon knowledgeable and passionate citizen action at all levels.

POSTSCRIPT

This book was completed before the 2020 presidential and congressional elections of November 3, 2020. However, we received preliminary results of those elections during the final stages of the production process, and thus we can venture some tentative conclusions on the likely implications of the elections for environmental policy in coming years.

During his campaign, Democratic candidate Joseph Biden proposed the most far-reaching climate change plan to date. He pledged to spend some $2 trillion over four years to rapidly cut carbon emissions and build green infrastructure. He espoused national goals of eliminating fossil fuels from electric power generation by 2035 and of achieving "net zero" carbon emissions no later than 2050. In 2020, about 38 percent of U.S. electricity generation came from clean sources, such as wind, solar, hydropower, and nuclear power. Polls continue to tell us that broad swaths of the American public would back such policies.

However, it remains to be seen if Biden's victory over Donald Trump will allow major policy breakthroughs where congressional approval is essential, for example, in the passage of new legislation or appropriation measures. Although the final Senate election results are not available at this writing, it is doubtful that Biden will have a working majority in Congress sufficient to enact major climate legislation in the near term. Nevertheless, with the nation suffering through an historic pandemic crisis and related economic recession, it is probable that further economic recovery measures will be approved. Those measures are likely to include spending and incentives for alternative energy and other "green" infrastructure. Some legislated version of Biden's "Build Back Better" program could thus fulfill some of the critical steps necessary to deal with climate change.

We can also expect a reversion to the "administrative presidency" described in Chapter 4 and elsewhere in this volume. That is, a president can do much on his own without congressional action. Biden will employ the powers of appointment, budgeting, reorganization, and executive orders to reverse numerous Trump policies, most of which also were established in much the same way. Biden indicated that he will rejoin the Paris Agreement on climate change, reversing President Trump's withdrawal from it. Biden also said the nation once again will play an active role in international environmental diplomacy beyond that agreement. He could revoke as many of President Trump's executive orders and directives as possible, and also begin the often-lengthy process of rewriting affected rules and regulations.

In many cases Biden is likely to take unilateral actions that go beyond those of the Obama administration in which he served. He is also expected to restore the original boundaries of national monuments that President Trump had shrunk and limit development in the Arctic National Wildlife Refuge and in other wild areas such as the Tongass National Forest. Biden is also very likely to appoint environmental scientists and other experts to top White House and Cabinet positions and to advisory posts throughout government, and to propose budget and staff increases for the EPA and other agencies to restore their capabilities to what they were prior to the Trump presidency. Early statements by Biden make clear that climate change is likely to regain top priority in his White House and in all relevant agencies, including those of the Departments of Defense, State, Agriculture, Transportation, and Treasury.

President Biden, like other presidents, will face stiff opposition in addressing matters such as fracking and methane releases and on oil and gas leasing on public lands. It will also be difficult to restore major policies such as the Clean Power Plan and the vehicle fuel economy standards that were the centerpieces of Obama's climate change agenda. It will take lengthy administrative procedures to replace rules that were finalized before the election, and any new rules will have to withstand judicial review by a more conservative judiciary than Obama faced. Still, as explained in Chapter 6, courts are more likely to slow down policy changes than to reverse them.

Perhaps more significantly, many states and cities will continue to strengthen and enforce environmental rules and regulations and to tighten energy portfolio requirements and other performance standards, even though the 2020 elections did not produce major shifts in partisan control of state and local governments; these cities and states will no longer face the prospect of presidential counteraction. In addition, California and the enlarged RGGI states are cutting power plant emissions at a significant rate, and California itself plans to eliminate the sale of gas-powered cars in the state by 2035 as well as to require zero-emission trucks by 2045. Large corporations and financial institutions also are moving toward clean energy commitments, and many have set "net zero" emission goals. Both national and international investment markets are likely to further these trends regardless of government policies. The policies of the Trump administration are unlikely to have altered this fundamental trajectory.

Environmental policy appears to have played a significant role in the presidential election for the first time, and we are confident that it will remain near the top of the national political agenda. The interrelationships of climate change, public health, racial justice, and economic growth also are likely to become more apparent in the 2020s. We thus believe that continuing progress is likely at all levels of government and in the private sector in coming years.

NOTES

1. U.S. Environmental Protection Agency, *EPA Inventory of U.S. Greenhouse Gas Emissions and Sinks: 1990–2018* (Washington, DC: EPA, 2020).
2. Lisa Friedman, "Coronavirus Doesn't Slow Trump's Deregulatory Rollbacks," *New York Times*, March 25, 2020; Friedman, "E.P.A., Citing Coronavirus, Drastically Relaxes Rules for Polluters," *New York Times*, March 26, 2020; Michael R. Bloomberg and Gina McCarthy, "How Trump's EPA Is Making Covid-19 More Deadly," May 4, 2020, https://www.bloomberg.com/opinion/articles/2020-05-04/how-trump-s-epa-is-making-covid-19-more-deadly.
3. Brad Plumer, "Emission Declines Will Set Records This Year. But It's Not Good News," *New York Times*, April 30, 2020. "Global carbon dioxide emissions declined 17 percent in April 2020 and were expected to fall 4–7 percent for the year." Chris Mooney, Brady Dennis and John Muyskens, "Global Emissions Plunged an Unprecedented 17 Percent During the Coronavirus Pandemic," *Washington Post*, May 19, 2020.
4. John Schwartz, "Americans See Climate as Concern, Even Amid Coronavirus Crisis," *New York Times*, May 19, 2020.
5. This and other IPCC reports are available at www.ipcc.ch.
6. Justin Gillis, "Climate Panel Cites Near Certainty on Warming," *New York Times*, August 20, 2013; Gillis, "U.N. Climate Panel Seeks Ceiling on Global Carbon Emissions," *New York Times*, September 28, 2013; Gillis, "Panel's Warning on Climate Risk: Worst Is Yet to Come," *New York Times*, March 30, 2014.
7. Quoted in Justin Gillis, "U.N. Panel Warns of Dire Effects from Lack of Action Over Global Warming," *New York Times*, November 2, 2014.
8. See Yagyang Xu, Veerabhadran Ramanathan, and David G. Victor, "Global Warming Will Happen Faster Than We Think," *Science* 564 (2018): 30–2; Naomi Oreskes, Michael Oppenheimer, and Dale Jamieson, "Scientists Have Been Underestimating the Pace of Climate Change," *Scientific American*, August 19, 2019; Henry Fountain, "Climate Change Is Accelerating: 'Things Are Getting Worse,'" *New York Times*, December 4, 2019.
9. Intergovernmental Panel on Climate Change, *Global Warming of 1.5°C*, October 6, 2018; Coral Davenport, "Major Climate Report Describes a Strong Risk of Crisis as Early as 2040," *New York Times*, October 7, 2018. For an excellent summary and perspective, see Nathan Hultman, "We're Almost Out of Time: The Alarming IPCC Climate Report and What to Do Next," *Brookings*, October 16, 2018.

10. Chris Mooney and Brady Dennis, "New U.N. Climate Report: Monumental Change Already Here for World's Oceans and Frozen Regions," *Washington Post*, September 25, 2019; Dennis, Steven Mufson, and Scott Clement, "Americans Increasingly See Climate Change as a Crisis, Poll Shows," *Washington Post*, September 13, 2019.

11. Andrew Freedman, "This Year Is on Track to Be Earth's Warmest on Record, Beating 2016, NOAA Says," *Washington Post*, April 21, 2020; Brady Dennis, Freedman, and John Muyskens, "2019 Capped Off the World's Hottest Decade in Recorded History," *Washington Post*, January 15, 2020.

12. Andrew Freedman, "The Arctic May Have Crossed Key Threshold, Emitting Billions of Tons of Carbon into the Air, in a Long-Dreaded Climate Feedback," *Washington Post*, December 10, 2019; Jason Samenow and Freedman, "The Greenland Ice Sheet Is in the Throes of One of Its Greatest Melting Events Ever Recorded," *Washington Post*, July 31, 2019; Freedman, "Siberian Heat Streak and Arctic Temperature Record Virtually 'Impossible' Without Global Warming, Study Says," *Washington Post*, July 15, 2020; Carolyn Kormann, "A Disastrous Summer in the Arctic," *New Yorker*, June 27, 2020; Henry Fountain, "Loss of Greenland Ice Sheet Reached a Record Last Year," *New York Times*, August 20, 2020.

13. Steven Mufson, Chris Mooney, Juliet Eilperin, and John Muyskens, "Extreme Climate Change Has Arrived in America," *Washington Post*, August 13, 2019; Eilperin, Carolyn Van Houten and Muyskens, "This Giant Climate Hot Spot Is Robbing the West of Its Water," *Washington Post*, August 7, 2020.

14. Chris Mooney and Brady Dennis, "Hurricanes Are Strengthening Faster in the Atlantic, and Climate Change Is a Big Reason Why, Scientists Say," *Washington* Post, February 7, 2019; Andrew Freedman and Jason Samenow, "The Strongest, Most Dangerous Hurricanes Are Now Far More Likely Because of Climate Change, Study Shows," *Washington Post*, May 18, 2020; Henry Fountain and John Schwartz, "Hurricane Sally's Fierce Rain Shows How Climate Change Raises Storm Risks," *New York Times*, September 16, 2020; and Tristram Korten, "Hurricanes Near U.S. Coast Forecast to Worsen and Multiply Due to Global Warming," *Washington Post*, September 27, 2020.

15. Mitch Smith and John Schwartz, "'Breaches Everywhere': Flooding Bursts Midwest Levees, and Tough Questions Follow," *New York Times*, March 31, 2019; Sarah Almukhtar, Blacki Migliozzi, John Schwartz, and Josh Williams, "The Great Flood of 2019: A Complete Picture of a Slow-Motion Disaster," *New York Times*, September 11, 2019.

16. Carolyn Korman, "How Climate Change Contributed to This Summer's Wildfires," *The New Yorker*, August 1, 2018; Kendra Pierre-Louis, "The Amazon, Siberia, Indonesia: A World of Fire," *New York Times*, August 28, 2019; Darryl Fears, "On Land, Australia's Rising Heat Is 'Apocalyptic.' In the Ocean, It's Worse," *Washington Post*, December 27, 2019; and Fears, Faiz Siddiqui, Sarah Kaplan and Juliet Eilperin, "Heat Is Turbocharging Fires, Drought and Tropical Storms This Summer," *Washington Post*, August 21, 2020.

17. Somini Sengupta and Weiyi Cai, "A Quarter of Humanity Faces Looming Water Crises," *New York Times*, August 6, 2019; Christopher Flavelle, "Climate Change Threatens the World's Food Supply, United Nations Warns," *New York Times*, August 8, 2019; Chris Mooney, "Scientists Triple Their Estimates of the Number of People Threatened by Rising Seas," *Washington Post*, October 29, 2019.

18. Andrew Freedman, "Global Warming to Push Billions Outside Climate Range That Has Sustained Society for 6,000 Years, Study Finds," *Washington Post*, May 4, 2020.

19. Andrew Friedman and Chris Mooney, "Major New Climate Study Rules Out Less Severe Global Warming Scenarios," *Washington Post*, July 22, 2020; John Schwartz, "How Much Will the Planet Warm if Carbon Dioxide Levels Double?," *New York Times*, July 22, 2020.

20. Brady Dennis, "In Bleak Report, U.N. Says Drastic Action Is Only Way to Avoid Worst Effects of Climate Change," *Washington Post*, November 26, 2019.

21. Andrew Friedman and Chris Mooney, "Earth's Carbon Dioxide Levels Hit Record High, Despite Coronavirus-Related Emissions Drop," *Washington Post*, June 4, 2020; Brad Plumer and Nadja Popovich, "Emissions Are Surging Back as Countries and States Reopen," *New York Times*, June 17, 2020; and Hiroko Tabuchi, "Global Methane Emissions Reach a Record High," *New York Times*, July 14, 2020.

22. Chris Mooney and Brady Dennis, "The World Has Just Over a Decade to Get Climate Change Under Control, U.N. Scientists Say," *Washington Post*, October 7, 2018.

23. Juliet Eilperin and Brady Dennis, "EPA Dismisses Half of Its Scientific Advisers on Key Board, Interior Suspends More Than 200 Advisory Panels in Sweeping Review," *Washington Post*, May 8, 2017; Brady Dennis and Chris Mooney, "Major Trump Administration Climate Report Says Damages Are 'Intensifying Across the Country'," *Washington Post*, November 23, 2017.

24. Juliet Eilperin and Brady Dennis, "Trump Administration to Finalize Weaker Mileage Standards, Dealing a Blow to Obama-Era Climate Policy," *Washington Post*, March 30, 2020; Coral Davenport, "Trump Administration, in Biggest Environmental Rollback, to Announce Auto Pollution Rules," *New York Times*, March 30, 2020; and Nadja Popovich and Brad Plumer, "What Trump's Environmental Rollbacks Mean for Global Warming," *New York Times*, September 17, 2020.

25. See Juliet Eilperin and Brady Dennis, "EPA Staff Warned that Mileage Rollbacks Had Flaws. Trump Officials Ignored Them," *Washington Post*, May 19, 2020.

26. Amelia T. Keyes, Kathleen F. Lambert, Dallas Burtraw, Jonathan J. Buonocore, Jonathan I. Levy, and Charles T. Driscoll, "The Affordable Clean Energy Rule and the Impact of Emissions Rebound on Carbon Dioxide and Criteria Air Pollutant Emissions," *Environmental Research Letters* 14 (April 2019): 1–14.

27. The text of the full Green New Deal proposal can be found at www.dataforprogress.org/green-new-deal. See also Elise Viebeck and David Weigel, "Green 'New Deal' Divides Democrats Intent on Addressing Climate Change," *Washington Post*, December 28, 2018.

28. Lisa Friedman and Glenn Thrush, "Liberal Democrats Formally Call for a 'Green New Deal,' Giving Substance to a Rallying Cry," *New York Times*, February 7, 2019; Friedman and Trip Gabriel, "A Green New Deal is Technologically Possible. Its Political Prospects Are Another Question." *New York Times*, February 21, 2019.

29. For a more nuanced look at different issues, see John Muyskens and Kevin Urhmacher, "Where 2020 Democrats Stand on Climate Change," *Washington Post*, January 13, 2020.

30. Matt Viser, "Biden's Vision Comes Into View, and It's Much More Liberal Than It Was," *Washington Post*, July 12, 2020; Brady Dennis and Dino Grandoni, "How Joe Biden's Surprisingly Ambitious Climate Plan Came Together," *Washington Post*, July 31, 2020; Christopher Flavelle, "Democrats Detail a Climate Agenda Tying Environment to Racial Justice," *New York Times*, June 29, 2020; Dino Grandoni, "The Energy 202: This Is How Biden's Climate Plan Stacks Up to the Green New Deal," *Washington Post*, October 9, 2020; and Grandoni, "The Energy 202: Kamala Harris Brings Record on Climate Change and Environmental Justice to Biden Ticket," *Washington Post*, August 12, 2020.

31. Jason Ye, *Carbon Pricing Proposals in the 116th Congress* (Arlington, VA: Center for Climate and Energy Solutions, 2019).

32. Kathryn A. Wolfe, "House Democrats Unveil $760B Infrastructure Plan with an Ambitious Climate Agenda," *Politico*, January 29, 2020.

33. Chuck Schumer, "Chuck Schumer: A Bold Plan for Clean Cars," *New York Times*, October 29, 2019.

34. Dino Grandoni, "2020 Candidates Back Bill to Test Whether Big Banks Can Withstand Climate Risk," *Washington Post*, November 20, 2019.

35. See Arlie Hochschild and David Hochschild, "Republicans Care About the Climate," *New York Times*, December 30, 2018; Justin Gillis, "The Republican Climate Closet," *New York Times*, August 12, 2019.

36. See https://citizensclimatelobby.org/climate-solutions-caucus/ for the membership of the caucuses.

37. Nathan Hultman et al. *Accelerating America's Pledge: Going All-In to Build a Prosperous, Low-Carbon Economy for the United States* (New York, NY: Bloomberg Philanthropies, 2020).

38. Lucas Bifera, *Regional Greenhouse Gas Initiative (RGGI)* (Arlington, VA: Center for Climate and Energy Solutions, 2013), www.c2es.org/us-states-regions/regional-climate-initiatives/rggi; "States Dare to Think Big on Climate Change," editorial, *New York Times*, August 28, 2017.

39. Hiroko Tabuchi, "California Set to Require Zero-Emission Trucks," *New York Times*, June 25, 2020; Brad Plumer and Jill Cowan, "California Plans to Ban Sales of New Gas-Powered Cars in 15 Years," *New York Times*, September 23, 2020.

40. Dino Grandoni, "Here's How Apple Is Trying to Curb Greenhouse Gas Emissions," *Washington Post*, April 11, 2019; David McCabe and Karen Weise, "Amazon Pledges to Be Carbon Neutral by 2040," *New York Times*, September 18, 2019; Steven Mufson, "More U.S. Businesses Making Changes in Response to Climate Concerns," *Washington Post*, June 11, 2019; Kara Swisher, "When Will Companies Finally Step Up to Fight Climate Change?" *New York Times*, January 23, 2020; and Rachel Siegel, "BlackRock Makes Climate Change Central to Its Investment Strategy," *Washington Post*, January 14, 2020.

41. Zack Colman, "Climate Groups Turn Up the Heat on Big Banks, Insurance Companies," *Politico*, January 13, 2020; Bill McKibben, "Money Is the Oxygen on Which the Fire of Global Warming Burns," *New Yorker*, September 17, 2019.

42. Steven Mufson, "More U.S. Businesses Making Changes in Response to Climate Concerns," *Washington Post*, June 11, 2019.

43. For examples of some large-scale plans from nongovernmental organizations, see Elliot Diringer et al., "Getting to Zero: A U.S. Climate Agenda," Center for Climate and Energy Solutions, November 2019, https://indd.adobe.com/view/6a1d2b29-5251-4c23-8aba-2b02901bd863; Foundation for Climate Restoration, "Climate Restoration: Solutions to the Greatest Threats Facing Humanity and Nature Today," September 2019, https://foundationforclimaterestoration.org/wp-content/uploads/2019/09/20190916b_f4cr4_white-paper.pdf.

44. Department of Energy, "U.S. Becomes World's Largest Crude Oil Producer and Department of Energy Authorizes Short Term Natural Gas Exports," September 13, 2018, www.energy.gov/articles/us-becomes-world-s-largest-crude-oil-producer-and-department-energy-authorizes-short-term/.

45. White House, *Blueprint for a Secure Energy Future* (Washington, DC: White House, March 30, 2011).

46. Ivan Penn, "Oil Companies Are Collapsing, but Wind and Solar Energy Keep Growing," *New York Times*, April 7, 2020.

47. The U.S. Bureau of Labor Statistics predicted that solar voltaic installers and wind turbine service technicians would be the fastest growing occupations in the U.S. between 2018 and 2028, https://www.bls.gov/ooh/fastest-growing.htm.

48. Darryl Fears and Juliet Eilperin, "President Obama Bans Oil Drilling in Large Areas of Atlantic and Arctic Oceans," *Washington Post*, December 20, 2016.

49. Rick Perry, Ryan Zinke, and Scott Pruitt, "Paving the Way to U.S. Energy Dominance," *Washington Times*, June 27, 2017.

50. Daniel Jacobs, *BP Blowout: Inside the Gulf Oil Disaster* (Washington, DC: Brookings Institution Press, 2016); Kellie Lunney, Timothy Cana, and Nick Sobcyzk, "Congressional Acton Still Lags a Decade After Gulf Spill," *E&E News*, April 20, 2020.

51. Eric Lipton and Hiroko Tabuchi, "Driven by Trump Policy Changes, Fracking Booms on Public Lands," *New York Times*, October 28, 2018; Juliet Eilperin, "More Methane: Interior Eases Rules Curbing Leaks From Oil and Gas Leases on Federal Land," *Washington Post*, September 18, 2018; Lisa Friedman, "E.P.A. to Roll Back Regulations on Methane, a Potent Greenhouse Gas," *New York Times*, August 29, 2019.

52. Steven Mufson and Juliet Eilperin, "Trump Administration Opens Huge Reserve in Alaska to Drilling," *Washington Post*, September 13, 2019.

53. Steven Mufson, "Trump-Appointed Regulators Reject Plan to Rescue Coal and Nuclear Plants," *Washington Post*, January 8, 2018; Scott DiSavino, "President Trump Can't Stop U.S. Coal Plants From Retiring," *Reuters*, January 13, 2019.

54. Ellen Knickmeyer, "US Says Conserving Oil Is No Longer an Economic Imperative," *Associated Press*, August 18, 2018; John Schwartz, "Trump Administration Blocks Energy Efficiency Rule for Light Bulbs," *New York Times*, December 20, 2019.

55. Miranda Green, "Analysis: Trump Solar Tariffs Cost 62k US Jobs," *The Hill*, December 3, 2019.

56. Dino Grandoni, "Clean Energy Loses Big in Year-End Spending Bill," *Washington Post*, December 18, 2019; Bernard Avishai, "A Lost Chance to Bring the Green New Deal Home," *The New Yorker*, February 18, 2020.

57. Brad Plumer, "In a First, Renewable Energy Is Poised to Eclipse Coal in U.S.," *New York Times*, May 13, 2020. In 2019 almost 8,000 jobs were lost in coal-generated electricity, whereas over 9,100 were added in natural gas generation and 10,900 jobs were created in renewable technologies. See National Association of State Energy Offices and Energy Futures Initiative, *2020 U.S. Energy & Employment Report*, www.usenergyjobs.org.

58. Dino Grandoni, "Oil Glut on Coronavirus May Sink Trump's 2020 Message on 'Energy Dominance,'" *Washington Post*, March 10, 2020; Steven Mufson, "Trump Faces Big Decisions on Energy Industry Rescue as U.S. Runs Out of Places to Store Abundance of Oil," *Washington Post*, April 27, 2020. Crude oil prices briefly dropped to negative territory in April 2020 and remained at historically low levels.

59. Brad Plumer, "Trump Put a Low Cost on Carbon Emissions. Here's Why It Matters," *New York Times*, August 23, 2018. Trump's EPA set the social cost of carbon dioxide emissions at $1–7 per ton, compared to about $40 per ton under the Obama administration.

60. U.S. Environmental Protection Agency, *Working Together: FY 2018-2022 U.S. E.P.A. Strategic Plan* (September 2019); U.S. Environmental Protection Agency, *EPA Year in Review 2017–2018*.

61. Lisa Friedman, "Cost of New E.P.A. Rules: Up to 1,400 More Deaths a Year," *New York Times*, August 21, 2018; Juliet Eilperin, "EPA's Scientific Advisers Warn Its Regulatory Rollbacks Clash With Established Science," *Washington Post*, December 31, 2019.

62. Lisa Friedman, "E.P.A. Proposes Rule Change That Would Let Power Plants Release More Toxic Pollution," *New York Times*, December 28, 2018; Friedman and Coral Davenport, "E.P.A., Tweaking Its Math, Weakens Control on Mercury," *New York Times*, April 16, 2020.

63. Lisa Friedman, "E.P.A. Plans to Get Thousands of Deaths Off the Books by Changing Its Math," *New York Times*, May 20, 2019; Jennifer A. Dlouhy, "EPA Plans to Rewrite Costs and Benefits of Anti-Pollution Rules," May 21, 2019, https://www.bloomberg.com/news/articles/2019-05-21/epa-plans-to-rewrite-costs-and-benefits-of-anti-pollution-rules.

64. Lisa Friedman, "E.P.A. to Limit Science Used to Write Public Health Rules," *New York Times*, November 11, 2019.

65. Eric Lipton, "E.P.A. Rule Change Could Let Dirtiest Coal Plants Keep Running (and Stay Dirty)," *New York Times*, August 24, 2018; Juliet Eilperin and Brady Dennis, "Civil Penalties for Polluters Dropped Dramatically in Trump's First Two Years, Analysis Shows," *Washington Post*, January 24, 2019; Friedman, "E.P.A., Citing Coronavirus, Drastically Relaxes Rules for Polluters" (note 2).

66. Nadia Popovich, "America's Air Quality Worsens, Ending Years of Gains, Study Says," *New York Times*, October 24, 2019. Wildfires may also have contributed to this increase.

67. See Sharon Zhang, "Rolling Back Water Rules Doesn't Help Most Farmers—It's for Big Polluters," *Truthout*, January 26, 2020, for numerous references to interests supporting the repeal.

68. Coral Davenport, "Trump Prepares to Unveil a Vast Reworking of Clean Water Protections," December 10, 2018; Juliet Eilperin and Brady Dennis, "Administration Finalizes Repeal of 2015 Water Rule Called 'Destructive and Horrible,'" *Washington Post*, September 11, 2019; Davenport, "Trump Removes Pollution Controls on Streams and Wetlands," *New York Times*, January 22, 2020.

69. See, for example, S. Mazeika, P. Sullivan, Mark C. Rains, and Amanda D. Rodewald, "Opinion: The Proposed Change to the Definition of 'Waters of the United States' Flouts Sound Science," *PNAS*, July 5, 2019, https://www.pnas.org/content/116/24/11558.

70. Robert Barnes, "Supreme Court Rejects Trump Administration's View on Key Aspect of Clean Water Act," *Washington Post*, April 23, 2020.

71. Brad Plumer and Nadja Popovich, "Poor Americans Exposed to Unsafe Water, Study Shows," *New York Times*, February 13, 2018; Maura Allaire, Haowei Wu, and Upmanu Lall, "National Trends in Drinking Water Quality Violations," *PNAS*, February 27, 2018, https://www.pnas.org/content/115/9/2078; Jack Healey, "Rural America's Own Private Flint: Well Water Too Polluted to Drink," *New York Times*, November 4, 2018.

72. Hiroko Tabuchi, Nadja Popovich, Blacki Migliozzi, and Andrew W. Lehren, "Mixing Water and Poison: Thousands of Chemical Sites at Risk in Floods' Path," *New York Times*, February 7, 2018.

73. Steven Mufson and Brady Dennis, "Report Finds Wide-Spread Contamination at Nation's Coal Ash Sites," *Washington Post*, March 4, 2019; Juliet Eilperin and Dennis, "EPA to Scale Back Federal Rules Restricting Waste from Coal-Fired Power Plants," *Washington Post*, November 3, 2019; and Dennis and Eilperin, "Trump Administration Rolls Back Obama-Era Rule Aimed at Limiting Toxic Wastewater From Coal Plants," *Washington Post*, August 31, 2020.

74. Brady Dennis, "EPA Vows National Action on Toxic 'Forever' Chemicals," *Washington Post*, February 4, 2019; Coral Davenport, "E.P.A. Will Study Limits on Cancer-Linked Chemicals. Critics Say the Plan Delays Action," *New York Times*, February 14, 2019. In April 2020, the White House intervened to weaken the rule under consideration; see Ellen Knickmeyer, "White House Moves to Weaken EPA Rule on Toxic Compounds," *Associated Press*, April 18, 2020.

75. Rachel Van Dongen, "EPA Backs Away From Regulating Controversial Chemical in Drinking Water," *Washington Post*, May 15, 2020.

76. John Barrasso and Tom Carper, "Water Infrastructure Act Is a Bipartisan Win for All Americans, From Farms to Cities," *USA Today*, October 23, 2018.

77. See EPA, "Great Lakes Restoration Initiative," Action Plan III," https://www.epa.gov/sites/production/files/2019-10/documents/glri-action-plan-3-201910-30pp.pdf. However, EPA was threatened with law suits for failure to carry out its Chesapeake Bay plan;

see Darryl Fears and Brady Dennis, "Two States, D.C. Plan to Sue EPA for Failing to Enforce Chesapeake Bay Cleanup Plan," *Washington Post*, May 18, 2020.

78. See, for example, Robert F. Durant, Daniel J. Fiorino, and Rosemary O'Leary, eds., *Environmental Governance Reconsidered*, 2nd ed. (Cambridge, MA: MIT Press, 2017).

79. Gilbert E. Metcalf, *Paying for Pollution: Why a Carbon Tax Is Good for America* (New York, NY: Oxford University Press, 2019); Lawrence Goulder and Mark Hafstead, *Confronting the Climate Challenge: U.S. Policy Options* (New York, NY: Columbia University Press, 2018).

80. Eric M. Patashnik, "The Clean Air Acts' Use of Market Mechanisms," in *Lessons from the Clean Air Act*, eds. Ann Carlson and Dallas Burtraw (New York, NY: Cambridge University Press, 2019).

81. Barry G. Rabe, *Can We Price Carbon?* (Cambridge, MA: MIT Press, 2018).

82. Leigh Raymond, "Policy Perspective: Building Political Support for Carbon Pricing—Lessons from Cap-and-Trade Policies," *Energy Policy* 134 (2019): 1–7.

83. Jörgen Wettestad and Torbjörg Jevnaker, *Rescuing EU Emissions Trading: The Climate Policy Flagship* (London: Palgrave Macmillan, 2016).

84. Lawrence Goulder, Xianling Long, Jieyi Lu, Richard D. Morgenstern, *China's Unconventional Nationwide CO2 Emissions Trading System*, Resources for the Future Working Paper 20-02 (January 2020); Harvard Project on Climate Agreements, *Subnational Climate Change Policy in China* (2020).

85. Heather Long, "'This Is Not Controversial': Bipartisan Group of Economists Calls for Carbon Tax," *Washington Post*, January 16, 2019.

86. Rabe, *Can We Price Carbon?*, Chapter 8.

87. Brad Plumer, "New U.N. Climate Report Says Put a High Price on Carbon," *New York Times*, October 18, 2018; Coral Davenport, "After Nobel in Economics, William Nordhaus Talks About Who's Getting His Pollution-Tax Ideas Right," *New York Times*, October 13, 2018.

88. "Economists' Statement on Carbon Dividends," *Wall Street Journal*, January 17, 2019.

89. Chris Mooney and Andrew Freedman, "The World Needs a Massive Carbon Tax in Just 10 Years to Limit Climate Change, IMF Says," *Washington Post*, October 10, 2019.

90. John Schwartz, "'A Conservative Climate Solution': Republican Group Calls for a Carbon Tax," *New York Times*, February 7, 2017; George P. Shultz and Lawrence H. Summers, "This Is One Climate Solution That's Best for the Environment—and for Business," *Washington Post*, June 19, 2017; Schwartz, "Exxon Mobil Lends Its Support to a Carbon Tax Proposal," *New York Times*, June 20, 2017.

91. George P. Shultz and Ted Halstead, "The Winning Conservative Climate Solution," *Washington Post*, January 16, 2020; and, for a more detailed version, James A. Baker, Shultz and Halstead, "The Strategic Case for U.S. Climate Leadership: How Americans Can Win With a Pro-Market Solution," *Foreign Affairs*, May/June 2020; See also Climate Leadership Council, "The Pricing Advantage," January 2020, https://clcouncil.org/pricing-advantage.pdf.

92. See, for example, Justin Gillis, "Forget the Carbon Tax for Now," *New York Times*, December 27, 2018.

93. David Leonhardt, "The Problem with Putting a Price on the End of the World," *New York Times*, April 9, 2019.

94. Emma Newburger, "'Hit Them Where It Hurts': Several 2020 Democrats Want a Carbon Tax on Corporations," *CNBC*, September 20, 2019, www.cnbc.com/2019/09/20/buttigieg-warren-harris-yang-and-others-embrace-carbon-tax.html.

95. Daniel J. Fiorino, *The New Environmental Regulation* (Cambridge, MA: MIT Press, 2006), Chapter 5; Fiorino, "Regulatory Innovation and Change," in *Environmental Governance Reconsidered*, eds. Durant, Fiorino, and O'Leary, 307–36.

96. See Michael E. Kraft, Mark Stephan, and Troy D. Abel, *Coming Clean: Information Disclosure and Environmental Performance* (Cambridge, MA: MIT Press, 2011); Robert F. Durant, "Regulation-by-Revelation," in *Environmental Governance Reconsidered*, eds. Durant, Fiorino, and O'Leary, 337–69.

97. The quotation is taken from the archives of the PCSD, at https://clinton whitehouse2.archives.gov/PCSD/Overview/index.html. For a fuller history of sustainability concepts and actions, see Daniel A. Mazmanian and Michael E. Kraft, eds., *Toward Sustainable Communities: Transition and Transformations in Environmental Policy*, 2nd ed. (Cambridge, MA: MIT Press, 2009); Kraft, "Sustainability and Environmental Policy," in *Environmental Governance Reconsidered*, eds. Durant, Fiorino, and O'Leary, 75–100; Kent E. Portney, *Taking Sustainable Cities Seriously: Economic Development, the Environment, and Quality of Life in American Cities*, 2nd ed. (Cambridge, MA: MIT Press, 2013).

98. See, for example, Aaron M. McCright, Katherine Dentzman, Meghan Charters, and Thomas Dietz, "The Influence of Political Ideology on Trust in Science," *Environmental Research Letters* 8 (2013), http://iopscience.iop.org/1748-9326/8/4/044029; Riley E. Dunlap, ed., "Climate Change Skepticism and Denial," special issue, *American Behavioral Scientist* 57, no. 6 (June 2013): 691–837.

99. Jeffrey Mervis, "Trump's New Budget Cuts All but a Favored Few Science Programs," *Science* 367 (February 14, 2020): 723–724. According to the EPA website, this represented a 32 percent cut from their estimated 2020 expenditure.

100. Ernest J. Moniz, "How an Energy Jobs Coalition Could Help the US Economy Bounce Back," *The Hill*, April 3, 2020.

101. See Nathan Hultman, "Building an Ambitious US Climate Policy From the Bottom Up," *Brookings*, December 9, 2019.

102. For a list of major companies in the Science Based Targets Initiative that are committed to the 1.5 degree C UN campaign, see https://ungc-communications-assets.s3.amazonaws.com/docs/publications/recover-better-statement.pdf.

103. Coalition for Green Capital, *Green Banks in the United States* (Washington, DC: Coalition for Green Capital, 2019).

104. Dirk Heine, Willi Semmler, Mariana Mazzucato, João Paulo Braga, Michael Flaherty, Arkady Gevorkyan, Erin Hayde, and Siavash Radpour, *Financing Low-Carbon Transitions through Carbon Pricing and Green Bonds*, Policy Research Working Paper 8991 (Washington, DC: World Bank Group, 2019); Cameron Hepburn, Brian O'Callaghan, Nicholas Stern, Joseph Stiglitz, and Dimitri Zenghelis, "Will COVID-19 Fiscal Recovery Packages Accelerate or Retard Progress on Climate Change?" *Oxford Review of Economic Policy* 36 (2020).

105. Baker, Shultz and Halstead, "The Strategic Case for U.S. Climate Leadership." See also "National Defense Strategy: Climate Change in the Age of Great Power Competition," https://www.americansecurityproject.org/climate-change-in-the-age-of-great-power-competition; Michael T. Klare, *All Hell Breaking Loose: The Pentagon's Perspective on Climate Change* (New York, NY: Henry Holt and Company, 2019).

106. Implications of COVID-19 for the climate crisis are discussed in John F. Kerry, "The Parallels Between the Coronavirus and the Climate Crisis," *Boston Globe*, April 21, 2020; Ishaan Tharoor, "The Pandemic Could Be a Call to Action on Climate Change," *Washington Post*, April 23, 2020; Marina Pitofsky, "Dozens of Corporations Call for Climate Change Considerations in Coronavirus Recovery Effort," *The Hill*, May 19, 2020.

MAJOR FEDERAL LAWS ON THE ENVIRONMENT, 1969–2020

Legislation	Implementing Agency	Key Provisions
Nixon Administration		
National Environmental Policy Act of 1969, PL 91-190	All federal agencies	Declared a national policy to "encourage productive and enjoyable harmony between man and his environment"; required environmental impact statements; created Council on Environmental Quality.
Resources Recovery Act of 1970, PL 91-512	Health, Education, and Welfare Department (later Environmental Protection Agency)	Set up a program of demonstration and construction grants for innovative solid waste management systems; provided state and local agencies with technical and financial assistance in developing resource recovery and waste disposal systems.
Clean Air Act Amendments of 1970, PL 91-604	Environmental Protection Agency (EPA)	Required administrator to set national primary and secondary air quality standards and certain emissions limits; required states to develop implementation plans by specific dates; required reductions in automobile emissions.
Federal Water Pollution Control Act (Clean Water Act) Amendments of 1972, PL 92-500	EPA	Set national water quality goals; established pollutant discharge permit system; increased federal grants to states to construct waste treatment plants.
Federal Environmental Pesticide Control Act of 1972 (amended the Federal Insecticide, Fungicide, and Rodenticide Act [FIFRA] of 1947), PL 92-516	EPA	Required registration of all pesticides in US commerce; allowed administrator to cancel or suspend registration under specified circumstances.
Marine Mammal Protection Act of 1972, PL 92-532	EPA	Regulated dumping of waste materials into the oceans and coastal waters.

Legislation	Implementing Agency	Key Provisions
Coastal Zone Management Act of 1972, PL 92-583	Office of Coastal Zone Management, Commerce Department	Authorized federal grants to the states to develop coastal zone management plans under federal guidelines.
Endangered Species Act of 1973, PL 93-205	Fish and Wildlife Service, Interior Department	Broadened federal authority to protect all "threatened" as well as "endangered" species; authorized grant program to assist state programs; required coordination among all federal agencies.
Ford Administration		
Safe Drinking Water Act of 1974, PL 93-523	EPA	Authorized federal government to set standards to safeguard the quality of public drinking water supplies and to regulate state programs for protecting underground water sources.
Toxic Substances Control Act of 1976, PL 94-469	EPA	Authorized premarket testing of chemical substances; allowed the EPA to ban or regulate the manufacture, sale, or use of any chemical presenting an "unreasonable risk of injury to health or environment"; prohibited most uses of polychlorinated biphenyls (PCBs).
Federal Land Policy and Management Act of 1976, PL 94-579	Bureau of Land Management, Interior Department	Gave Bureau of Land Management authority to manage public lands for long-term benefits; officially ended policy of conveying public lands into private ownership.
Resource Conservation and Recovery Act of 1976, PL 94-580	EPA	Required the EPA to set regulations for hazardous waste treatment, storage, transportation, and disposal; provided assistance for state hazardous waste programs under federal guidelines.
National Forest Management Act of 1976, PL 94-588	US Forest Service, Agriculture Department	Gave statutory permanence to national forestlands and set new standards for their management; restricted timber harvesting to protect soil and watersheds; limited clear-cutting.

Legislation	Implementing Agency	Key Provisions
Carter Administration		
Surface Mining Control and Reclamation Act of 1977, PL 95-87	Interior Department	Established environmental controls over strip mining; limited mining on farmland, alluvial valleys, and slopes; required restoration of land to original contours.
Clean Air Act Amendments of 1977, PL 95-95	EPA	Amended and extended Clean Air Act; postponed deadlines for compliance with auto emissions and air quality standards; set new standards for "prevention of significant deterioration" in clean air areas.
Clean Water Act Amendments of 1977, PL 95-217	EPA	Extended deadlines for industry and cities to meet treatment standards; set national standards for industrial pretreatment of wastes; increased funding for sewage treatment construction grants and gave states flexibility in determining spending priorities.
Public Utility Regulatory Policies Act of 1978, PL 95-617	Energy Department, states	Provided for Energy Department and Federal Energy Regulatory Commission regulation of electric and natural gas utilities and crude oil transportation systems in order to promote energy conservation and efficiency; allowed small cogeneration and renewable energy projects to sell power to utilities.
Alaska National Interest Lands Conservation Act of 1980, PL 96-487	Interior Department, Agriculture Department	Protected 102 million acres of Alaskan land as national wilderness, wildlife refuges, and parks.
Comprehensive Environmental Response, Compensation, and Liability Act of 1980 (CERCLA), PL 96-510	EPA	Authorized federal government to respond to hazardous waste emergencies and to clean up chemical dump sites; created $1.6 billion "Superfund"; established liability for cleanup costs.

Legislation	Implementing Agency	Key Provisions
Reagan Administration		
Nuclear Waste Policy Act of 1982, PL 97-425; Nuclear Waste Policy Amendments Act of 1987, PL 100-203	Energy Department	Established a national plan for the permanent disposal of high-level nuclear waste; authorized the Energy Department to site, obtain a license for, construct, and operate geologic repositories for spent fuel from commercial nuclear power plants. Amendments in 1987 specified Yucca Mountain, Nevada, as the sole national site to be studied.
Resource Conservation and Recovery Act Amendments of 1984, PL 98-616	EPA	Revised and strengthened EPA procedures for regulating hazardous waste facilities; authorized grants to states for solid and hazardous waste management; prohibited land disposal of certain hazardous liquid wastes; required states to consider recycling in comprehensive solid waste plans.
Food Security Act of 1985 (also called the farm bill), PL 99-198; renewed in 1990, 1996, 2002, 2008, and 2014	Agriculture Department	Limited federal program benefits for producers of commodities on highly erodible land or converted wetlands; established a conservation reserve program; authorized Agriculture Department technical assistance for subsurface water quality preservation; revised and extended the Soil and Water Conservation Act (1977) programs through the year 2008. The 1996 renewal of the farm bill authorized $56 billion over seven years for a variety of farm and forestry programs. These include an Environmental Quality Incentives Program to provide assistance and incentive payments to farmers, especially those facing serious threats to soil, water, grazing lands, wetlands, and wildlife habitat. Spending was increased substantially in 2002.
Safe Drinking Water Act of 1986, PL 99-339	EPA	Reauthorized the Safe Drinking Water Act of 1974 and revised EPA safe drinking water programs, including grants to states for drinking water standards enforcement and groundwater protection programs; accelerated EPA schedule for setting standards for maximum contaminant levels of eighty-three toxic pollutants.

Legislation	Implementing Agency	Key Provisions
Superfund Amendments and Reauthorization Act of 1986 (SARA), PL 99-499	EPA	Provided $8.5 billion through 1991 to clean up the nation's most dangerous abandoned chemical waste dumps; set strict standards and timetables for cleaning up such sites; required that industry provide local communities with information on hazardous chemicals used or emitted.
Clean Water Act Amendments of 1987, PL 100-4	EPA	Amended the Federal Water Pollution Control Act of 1972; extended and revised EPA water pollution control programs, including grants to states for construction of wastewater treatment facilities and implementation of mandated nonpoint-source pollution management plans; expanded EPA enforcement authority; established a national estuary program.
Global Climate Protection Act of 1987, PL 100-204	State Department	Authorized the State Department to develop an approach to the problems of global climate change; created an intergovernmental task force to develop US strategy for dealing with the threat posed by global warming.
Ocean Dumping Ban Act of 1988, PL 100-688	EPA	Amended the Marine Protection, Research, and Sanctuaries Act of 1972 to end all ocean disposal of sewage sludge and industrial waste by December 31, 1991; revised EPA regulation of ocean dumping by establishing dumping fees, permit requirements, and civil penalties for violations.
George H. W. Bush Administration		
Oil Pollution Act of 1990, PL 101-380	Transportation Department, Commerce Department	Sharply increased liability limits for oil spill cleanup costs and damages; required double hulls on oil tankers and barges by 2015; required federal government to direct cleanups of major spills; required increased contingency planning and preparedness for spills; preserved states' rights to adopt more stringent liability laws and to create state oil spill compensation funds.

Legislation	Implementing Agency	Key Provisions
Pollution Prevention Act of 1990, PL 101-508	EPA	Established Office of Pollution Prevention in the EPA to coordinate agency efforts at source reduction; created voluntary program to improve lighting efficiency; stated waste minimization was to be the primary means of hazardous waste management; mandated source reduction and recycling report to accompany annual toxics release inventory under SARA in order to encourage industries to reduce hazardous waste voluntarily.
Clean Air Act Amendments of 1990, PL 101-549	EPA	Amended the Clean Air Act of 1970 by setting new requirements and deadlines of three to twenty years for major urban areas to meet federal clean air standards; imposed new, stricter emissions standards for motor vehicles and mandated cleaner fuels; required reduction in emission of sulfur dioxide and nitrogen oxides by power plants to limit acid deposition and created a market system of emissions allowances; required regulation to set emissions limits for all major sources of toxic or hazardous air pollutants and listed 189 chemicals to be regulated; prohibited the use of chlorofluorocarbons (CFCs) by the year 2000 and set phaseout of other ozone-depleting chemicals.
Intermodal Surface Transportation Efficiency Act of 1991 (ISTEA, also called the highway bill), PL 102-240	Transportation Department	Authorized $151 billion over six years for transportation, including $31 billion for mass transit; required statewide and metropolitan long-term transportation planning; authorized states and communities to use transportation funds for public transit that reduces air pollution and energy use consistent with Clean Air Act of 1990; required community planners to analyze land use and energy implications of transportation projects they review.

Legislation	Implementing Agency	Key Provisions
Energy Policy Act of 1992, PL 102-486	Energy Department	Comprehensive energy act designed to reduce US dependency on imported oil. Mandated restructuring of the electric utility industry to promote competition; encouraged energy conservation and efficiency; promoted renewable energy and alternative fuels for cars; eased licensing requirements for nuclear power plants; authorized extensive energy research and development.
The Omnibus Water Act of 1992, PL 102-575	Interior Department	Authorized completion of major water projects in the West; revised the Central Valley Project in California to allow transfer of water rights to urban areas and to encourage conservation through a tiered pricing system that allocates water more flexibly and efficiently; mandated extensive wildlife and environmental protection, mitigation, and restoration programs.
Clinton Administration		
Food Quality Protection Act of 1996, PL 104-170	EPA	A major revision of FIFRA that adopted a new approach to regulating pesticides used on food, fiber, and other crops by requiring EPA to consider the diversity of ways in which people are exposed to such chemicals. Created a uniform "reasonable risk" health standard for both raw and processed foods that replaced the requirements of the 1958 Delaney Clause of the Food, Drug, and Cosmetic Act that barred the sale of processed food containing even trace amounts of chemicals found to cause cancer; required the EPA to take extra steps to protect children by establishing an additional tenfold margin of safety in setting acceptable risk standards.

Legislation	Implementing Agency	Key Provisions
Safe Drinking Water Act Amendments of 1996, PL 104-182	EPA	Granted local water systems greater flexibility to focus on the most serious public health risks; authorized $7.6 billion through 2003 for state-administered loan and grant funds to help localities with the cost of compliance; created a "right-to-know" provision requiring large water systems to provide their customers with annual reports on the safety of local water supplies, including information on contaminants found in drinking water and their health effects. Small water systems are eligible for waivers from costly regulations.
Transportation Equity Act for the Twenty-First Century (also called ISTEA II or TEA 21), PL 105-178	Transportation Department	Authorized a six-year, $218 billion program that increased spending by 40 percent to improve the nation's highways and mass transit systems; provided $41 billion for mass transit programs, with over $29 billion coming from the Highway Trust Fund; provided $592 million for research and development on new highway technologies, including transportation-related environmental issues; provided $148 million for a scenic byways program and $270 million for building and maintaining trails; continued support for improvement of bicycle paths.

George W. Bush Administration

Legislation	Implementing Agency	Key Provisions
The Small Business Liability Relief and Brownfields Revitalization Act of 2002, PL 107-118	EPA	Amended CERCLA (Superfund) to provide liability protection for prospective purchasers of brownfields and small business owners who contributed to waste sites; authorized increased funding for state and local programs that assess and clean up such abandoned or underused industrial or commercial sites.

Legislation	Implementing Agency	Key Provisions
The Healthy Forests Restoration Act of 2003, PL 108-148	Agriculture Department, Interior Department	Intended to reduce the risks of forest fires on federal lands by authorizing the cutting of timber in selected areas managed by the Forest Service and the Bureau of Land Management; sought to protect communities, watersheds, and certain other lands from the effects of catastrophic wildfires; directed the Secretary of Agriculture and the Secretary of the Interior to plan and conduct hazardous fuel reduction programs on federal lands within their jurisdictions.
The Energy Policy Act of 2005, PL 109-58	Energy Department	Intended to increase the supply of energy resources and improve the efficiency of energy use through provision of tax incentives and loan guarantees for various kinds of energy production, particularly oil, natural gas, and nuclear power. Also called for expanded energy research and development, expedited building for new energy facilities, improved energy efficiency standards for federal office buildings, and modernization of the nation's electricity grid.
Energy Independence and Security Act of 2007, PL 110-140	Energy Department, Transportation Department	Set a national automobile fuel-economy standard of 35 miles per gallon by 2020, the first significant change in the Corporate Average Fuel Economy (CAFE) standards since 1975. Also sought to increase the supply of alternative fuel sources by setting a renewable fuel standard that requires fuel producers to use at least 36 billion gallons of biofuels by 2022; 21 billion gallons of that amount are to come from sources other than corn-based ethanol. Included provisions to improve energy efficiency in lighting and appliances and for federal agency efficiency and renewable energy use.

Legislation	Implementing Agency	Key Provisions
Obama Administration		
The American Recovery and Reinvestment Act of 2009, PL 111-5	Energy Department, Transportation Department, Treasury Department	Although not a stand-alone environmental or energy policy, the economic stimulus bill contained about $80 billion in spending, tax incentives, and loan guarantees to promote energy efficiency, renewable energy sources, fuel-efficient cars, mass transit, and clean coal, including $3.4 billion for research on capturing and storing carbon dioxide from coal-fired power plants, $2 billion for research on advanced car batteries, $17 billion in grants and loans to modernize the nation's electric grid and increase its capacity to transmit power from renewable sources, and nearly $18 billion for mass transit, Amtrak, and high-speed rail.
Omnibus Public Lands Management Act of 2009, PL 111-11	Interior Department, Agriculture Department	Consolidated 164 separate public lands measures that protect two million acres of wilderness in nine states; establish new national trails, national parks, and a new national monument; provide legal status for the 26-million-acre National Landscape Conservation System that contains areas of archaeological and cultural significance; and protect 1,100 miles of 86 new wild and scenic rivers in eight states. Together the measures constitute the most significant expansion of federal land conservation programs in fifteen years.
The Frank R. Lautenberg Chemical Safety for the 21st Century Act of 2016, PL 114-182	EPA	Modified the Toxic Substances Control Act of 1976 to require the EPA to evaluate existing chemicals and set deadlines for doing so; established a new risk-based safety standard; increased public transparency for chemical information.
Trump Administration		
The John D. Dingell, Jr. Conservation, Management, and Recreation Act of 2019, PL 116-9	Interior Department	Set aside more than 1.3 million acres of wilderness area, expanded several national parks, created four national monuments, and made permanent the Land and Water Conservation Fund.

Legislation	Implementing Agency	Key Provisions
Great American Outdoors Act of 2020, PL 116-152	Interior Department, Agriculture Department	Authorizes a National Park and Public Lands Legacy Restoration Fund to address deferred maintenance needs on federal lands and creates permanent funding for the Land and Water Conservation Fund.

Note: As of 2020, no other major laws had been approved by Congress and signed by President Trump. As always, for an update on legislative developments, consult *CQ Magazine* or other professional news sources, or Congressional Quarterly's annual *Almanac*, which summarizes key legislation enacted by Congress and describes the major issues and leading policy actors.

BUDGETS OF SELECTED ENVIRONMENTAL AND NATURAL RESOURCE AGENCIES, 1980–2020 (IN BILLIONS OF NOMINAL AND CONSTANT DOLLARS)

Agency	1980	1990	2000	2010	2020 (Est.)
Environmental Protection Agency (EPA) Operating Budget[a]	1.269	1.901	2.465	3.889	4.703
(Constant 2020 dollars)	3.635	3.833	3.193	4.613	4.703
Interior Department Total Budget	4.674	6.669	8.363	12.843	15.272
(Constant 2020 dollars)	13.393	13.446	10.833	15.235	15.272
Selected Natural Resource Agencies					
Bureau of Land Management	0.919	1.226	1.616	1.074	1.682
(Constant 2020 dollars)	2.632	2.472	2.093	1.274	1.682
Fish and Wildlife Service	0.435	1.133	1.498	1.588	2.932
(Constant 2020 dollars)	1.247	2.284	1.940	1.884	2.932
National Park Service	0.531	1.275	2.071	2.289	4.115
(Constant 2020 dollars)	1.522	2.570	2.683	2.715	4.115
Forest Service	2.250	3.473	3.728	5.297	7.586
(Constant 2020 dollars)	6.446	7.003	4.829	6.284	7.586

Source: Office of Management and Budget, *Budget of the United States Government,* fiscal years 1982, 1992, 2002, 2012, 2021 (Washington, DC: Government Publishing Office, 1981, 1991, 2001, 2011, and 2021), and agency websites.
Note: The upper figure for each agency represents budget authority in nominal dollars, that is, the real amount for the year in which the budget was authorized. The lower figure represents budget authority in constant 2020 dollars to permit comparisons over time. These adjustments use the implicit price deflator for nondefense federal expenditures as calculated by the Bureau of Economic Analysis, Department of Commerce.
For consistency, all figures in the table are taken from the president's proposed budget for the respective years, and all represent final budget authority.
[a]The EPA operating budget, which supplies funds for most of the agency's research, regulation, and enforcement programs, is the most meaningful figure. The other two major elements of the total EPA budget historically have been Superfund allocations and sewage treatment construction or water infrastructure grants (now called state and tribal assistance grants). We subtract both items from the total EPA budget to calculate the agency's operating budget. The EPA and the White House define the agency's operating budget differently. They do not exclude all of these amounts and arrive at a slightly different figure. President Trump's proposed change in the fiscal 2021 budget called for a 26 percent cut in the EPA's total budget.

EMPLOYEES IN SELECTED FEDERAL AGENCIES AND DEPARTMENTS, 1980, 1990, 2000, 2010, AND 2020

Agency/Department	Personnel[a]				
	1980	1990	2000	2010	2020 (Est.)
Environmental Protection Agency	12,891	16,513	17,416	17,417	14,172
Bureau of Land Management	9,655	8,753	9,328	12,741	9,847
Fish and Wildlife Service	7,672	7,124	7,011	9,252	8,269
National Park Service	13,934	17,781	18,418	22,211	17,615
Office of Surface Mining Reclamation and Enforcement	1,014	1,145	622	521	398
Forest Service	40,606	40,991	33,426	35,639	30,372
Army Corps of Engineers (civil functions)	32,757	28,272	22,624	23,608	32,000
US Geological Survey	14,416	10,451	9,417	8,600	6,779
Natural Resources Conservation Service (formerly Soil Conservation Service)	15,856	15,482	9,628	11,446	12,000

Source: U.S. Senate Committee on Governmental Affairs, "Organization of Federal Executive Departments and Agencies," January 1, 1980, and January 1, 1990; and Office of Management and Budget, *Budget of the United States Government,* fiscal years 1982, 1992, 2002, and 2012 (Washington, DC: Government Publishing Office, 1981, 1991, 2001, 2011, and 2021), and agency websites.

[a]Personnel totals represent full-time equivalent employment, reflecting both permanent and temporary employees. Data for 2000 are based on the fiscal 2002 proposed budget submitted to Congress by the Bush administration in early 2001, and data for 2010 are taken from agency sources as well as the administration's proposed fiscal 2012 budget submitted to Congress in early 2011, and the fiscal 2021 budget submitted to Congress in early 2020. Because of organizational changes within departments and agencies, the data presented here are not necessarily an accurate record of agency personnel growth or decline over time. The information is presented chiefly to provide an indicator of approximate agency size during different periods.

FEDERAL SPENDING ON NATURAL RESOURCES AND THE ENVIRONMENT, SELECTED FISCAL YEARS, 1980–2020 (IN BILLIONS OF NOMINAL AND CONSTANT DOLLARS)

Budget Item	1980	1990	2000	2010	2020 (Est.)
Water resources	4.085	4.332	4.800	6.813	9.761
(Constant 2020 dollars)	11.705	8.734	6.218	8.082	9.761
Conservation and land management	1.572	4.362	6.604	11.933	16.881
(Constant 2020 dollars)	4.504	8.794	8.554	14.155	16.881
Recreational resources	1.373	1.804	2.719	3.809	4.665
(Constant 2020 dollars)	3.934	3.637	3.522	4.518	4.665
Pollution control and abatement	4.672	5.545	7.483	10.473	9.062
(Constant 2020 dollars)	13.387	11.179	9.693	12.423	9.062
Other natural resources	1.395	2.077	3.397	6.629	7.326
(Constant 2020 dollars)	3.997	4.188	4.400	7.864	7.326
Total	13.097	18.121	25.003	39.657	47.695
(Constant 2020 dollars)	37.527	36.534	32.387	47.043	47.695

Source: Office of Management and Budget, *Historical Tables, Budget of the United States Government Fiscal Year 2020* (Washington, DC: Government Publishing Office, 2020).

Note: The upper figure for each budget category represents budget authority in nominal dollars, that is, the real budget for the given year. Figures for 1980 are provided to indicate pre–Reagan administration spending bases. The lower figure for each category represents budget authority in constant 2020 dollars. These adjustments are made using the implicit price deflator for the federal nondefense expenditures as calculated by the Bureau of Economic Analysis, Department of Commerce. The natural resources and environment function in the federal budget reported in this table does not include environmental cleanup programs within the Departments of Defense and Energy, which are substantial.

INDEX

Pages followed by b, f, n, or t indicate box, figure, note, or table respectively.